DESIGN OF GEOSYNCHRONOUS SPACECRAFT

DESIGN OF GEOSYNCHRONOUS SPACECRAFT

Brij N. Agrawal, Ph.D.

International Telecommunications Satellite Organization (INTELSAT)
Washington, D.C.

PRENTICE-HALL, INC., *Englewood Cliffs, NJ 07632*

Library of Congress Cataloging-in-Publication Data

Agrawal, Brij N. (date)
   Design of geosynchronous spacecraft.

   Includes bibliographies and index.
   1. Geostationary satellites—Design and construction.
2. Artificial satellites in telecommunication.
I. Title.
TK5104.A27   1986   629.44   85-30097
ISBN 0-13-200114-4

Editorial/production supervision: Sophie Papanikolaou
Interior design: Sophie Papanikolaou
Cover design: 20/20 Services, Inc.
Manufacturing buyer: Gordon Osbourne

Printed in the United States of America

10 9 8 7 6 5 4 3 2

ISBN   0-13-200114-4   025

Prentice-Hall International (UK) Limited, *London*
Prentice-Hall of Australia Pty. Limited, *Sydney*
Prentice-Hall Canada Inc., *Toronto*
Prentice-Hall Hispanoamericana, S.A., *Mexico*
Prentice-Hall of India Private Limited, *New Delhi*
Prentice-Hall of Japan, Inc., *Tokyo*
Prentice-Hall of Southeast Asia Pte. Ltd., *Singapore*
Editora Prentice-Hall do Brasil, Ltda., *Rio de Janeiro*
Whitehall Books Limited, *Wellington, New Zealand*

To the memory of my parents

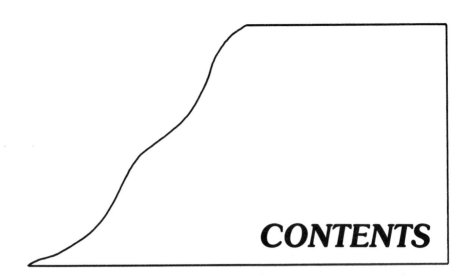

CONTENTS

4   SPACECRAFT STRUCTURES 179

5   THERMAL CONTROL 265

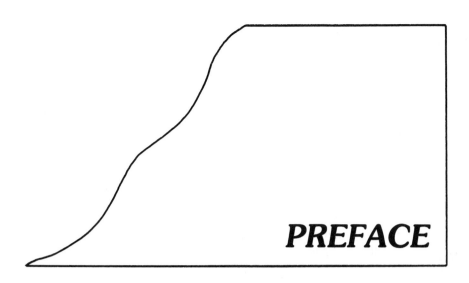

PREFACE

The concept of geosynchronous spacecraft, proposed by Arthur Clark in an article in 1945, became a reality in 1965 with the successful launch and operation of Early Bird for international communications. Since then, there has been an impressive growth in the application of geosynchronous spacecraft in communications, which includes international, domestic, maritime, direct TV broadcast, and military; in photography of cloud motion to improve weather prediction; and in the navigation of aircraft and ships. At the end of 1985, there were more than 80 communications satellites in geosynchronous orbit. There have also been revolutionary changes in the design of geosynchronous spacecraft in terms of complexity, size, and capability. As an example, INTELSAT satellites have increased in size and communications capacity by more than an order of magnitude. In view of this trend, the design of geosynchronous spacecraft is likely to occupy engineers and scientists for many years to come.

This book is the outgrowth of my personal notes on spacecraft design that I started making after joining COMSAT in 1969. As the notes expanded and matured, it became evident that they could be incorporated into a textbook to meet the need of the engineers entering into this challenging and multidisciplinary field. The book brings together information on the basic design principles of geosynchronous spacecraft from technical journals and reports, and my personal knowledge and experience of over sixteen years in this area.

This book is intended to serve two objectives. First, it can serve as a textbook for final year undergraduate and postgraduate students, providing them with a basic understanding of spacecraft design. Second, it can serve as a reference book for spacecraft system designers. Because spacecraft design involves the interaction of a variety of disciplines, this book is also intended for specialists who may need a basic understanding of related fields.

The emphasis in this book has been on the basic aspects of current design practices. Derivations of design equations are included to the extent that they provide clarity and understanding of the assumptions used and of the final formulas. Some of the material is presented in the form of tables and graphs for the solution of typical design problems.

The material selected reflects current design practices for geosynchronous spacecraft and it also has general applicability to other satellite systems. It is arranged as follows.

Chapter 1 provides a review of spacecraft configurations for geosynchronous spacecraft, functional aspects of subsystems, and the impact of stabilization and launch vehicle constraints on spacecraft configurations. The trade-offs of payload mass versus available electric power are also discussed.

Chapter 2 presents basic orbital dynamics, with emphasis on the geosynchronous orbit. Orbit perturbations due to the sun, moon, and earth, and the effects of orbit imperfections are discussed. North-south and east-west station keeping, station repositioning, launch windows, and typical launch sequences are reviewed. Examples are given to calculate velocity requirements for perigee and apogee motors.

Chapter 3 provides an introduction to attitude dynamics and control. Stability conditions for single-spin and dual-spin stabilized spacecraft are derived. Attitude dynamics and controls for momentum-biased and three-axes reaction wheel systems are presented. Sensors for attitude determinations, transfer orbit, and synchronous orbit are discussed. Basic equations for the calculation of propellant mass and a review of hydrazine, electrothermal, and electric thrusters are given. Locations of thrusters for spin-stabilized and three-axis-stabilized spacecraft are reviewed.

Chapter 4 provides a brief discussion of structural loads, stress analysis, dynamic analysis, and materials. Critical loads for instability of columns, flat panels, and shells are given. Vibration responses due to sinusoidal, random, and acoustic excitations are derived. Characteristics of lightweight materials, such as beryllium, and advanced composite materials are discussed. Structural design verification tests are also reviewed.

Chapter 5 discusses spacecraft thermal control. Basic equations for heat transfer by conduction and radiation are discussed, with emphasis on radiation, which is a primary mode of heat transfer in a spacecraft. Calculations of external heat flux from the sun, albedo, and thermal radiation of the earth are given. Heat balance equations, thermal modelling techniques and steady-state and transient responses are presented. Both passive (insulation and coating) and active (heat pipe) thermal control elements are discussed. The thermal control design for typical spin-stabilized and three-axis-stabilized spacecraft is reviewed. Thermal design verification tests are also discussed.

Chapter 6 covers electric power subsystems. Basic characteristics of solar cells, different types of solar cells (conventional, violet, and black), radiation degradation, and solar cell parameters are discussed. The discussion on spacecraft batteries includes consideration of their electrical, charging and discharging characteristics, life expectancy, and reconditioning. The discussion on power control units covers series dissipative, shunt dissipative, pulse width modulated, and partial shunt regulators. Typical electric power subsystem configurations are also discussed in this chapter.

Chapter 7 provides an introduction to satellite payload designs through a discussion of satellite communications. It covers basic definitions, ITU radio regulations, link considerations, antennas, modulation techniques, interference, and satellite capacity calculations.

I wish to acknowledge my indebtedness to all individuals who have contributed both to the contents and quality of this work. In particular, I would like to thank Dr. J. E. Allnutt, Mr. W. J. Billerbeck, Mr. A. F. Dunnet, Dr. G. D. Gordon, Mr. N. L. Hyman, Dr. M. H. Kaplan, Dr. W. D. Kinney, Prof. L. Meirovitch, Mr. A. Ramos, Mr. J. H. Reisenweber, and Mr. P. R. Schrantz for their help in reviewing the drafts in their areas of specialization and for providing valuable suggestions. The cooperation of Dr. G. J. Lo, co-author of Chapter 7, is very much appreciated. Special thanks goes to Mrs. M. Gupta for her support and for editing the manuscript.

My sincere thanks to INTELSAT for granting me permission to publish this book. The support of Mr. J. R. Owens, Dr. J. N. Pelton, Dr. G. Porcelli, and Mr. D. K. Sachdev is sincerely appreciated. I also wish to express my appreciation to all the organizations that granted me permission to use their material in this book and in particular to Mr. R. E. Berry, of FACC; Dr. R. Stoolman, of HAC; Dr. G. Hyde, of COMSAT, and Mr. R. Ashiya, of ISRO. The cooperation of Mr. B. M. Goodwin, Acquisitions Editor, and Ms. S. Papanikolaou, Production Editor, both of Prentice-Hall in coordinating this work is warmly acknowledged.

Finally, my sincere appreciation to my wife, Shail, and children, Vivek, Raka, and Sachin for their support and patience.

Brij N. Agrawal

DESIGN OF GEOSYNCHRONOUS SPACECRAFT

1

SPACECRAFT CONFIGURATION

1.1 INTRODUCTION

Earth satellites offer a significant capability improvement compared to terrestrial techniques in many fields. The fundamental advantage of a satellite is its ability to obtain a global look at a large portion of the earth's surface. As shown in Fig. 1.1, a system of three satellites in geosynchronous orbit can cover almost all of the earth's surface. This has led to the application of satellites in several areas, such as communications by high-frequency line-of-sight radio links, photography for meteorology and earth resources, and navigation for aircraft and ships.

As shown in Fig. 1.2, a spacecraft consists of two basic parts: the mission payload and the spacecraft bus. The payload performs the missions of the spacecraft, such as communications by radio links, earth images for weather forecasting, and high-resolution photography for assessment of visible earth resources. The spacecraft bus supports the payload by providing the required orbit and attitude control, electric power, thermal control, mechanical support, and a two-way command and data link to the ground.

A communications satellite is basically an active repeater that receives a signal from a point, A, on the earth, amplifies it, translates its frequency, and sends it back to point B on the earth. The payload for a communications satellite consists of antennas to receive and transmit and a transponder to amplify and translate the frequency of the signal.

The primary payload of a meteorological satellite is a very high resolution radiometer (VHRR). VHRR takes visual and infrared images of the earth and of cloud motion which assist in improved weather forecasting. In addition, the satellite is normally used to relay the weather information from dispersed measuring instruments to central headquarters.

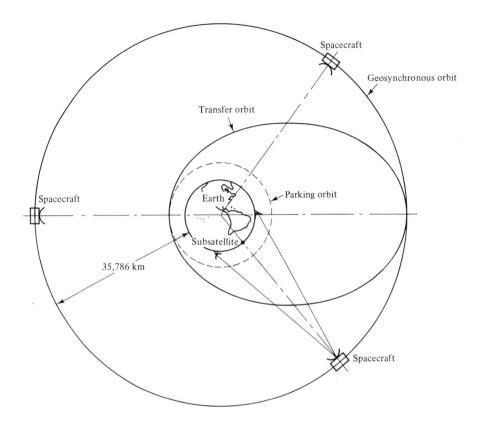

Figure 1.1 Global International Communications Satellite System.

The primary payload for the earth resource satellite, Landsat, is a multispectral scanner (MSS) which creates simultaneous images in four spectral bands: two in the visible frequency band (green and red) and two in the infrared band.

A spacecraft bus can be divided into six subsystems as shown in Fig. 1.2. A brief description of these subsystems is provided here. The attitude control subsystem maintains the spacecraft's attitude, or orientation in space, within the limits allowed. It consists of sensors for attitude determination and actuators, such as thrusters and/or angular momentum storage devices, for providing corrective torques. The attitude disturbance torques come from many sources, such as solar pressure, the gravity gradient, and any misalignment of the thrusters.

The propulsion subsystem injects the spacecraft into the desired orbit and keeps the orbital parameters within the limits allowed. The disturbance forces that will alter the spacecraft's orbit arise from the gravitational forces of the sun and moon, and the earth's ellipticity.

The electrical power subsystem provides electric power to the spacecraft during all mission phases. The primary power is provided through the

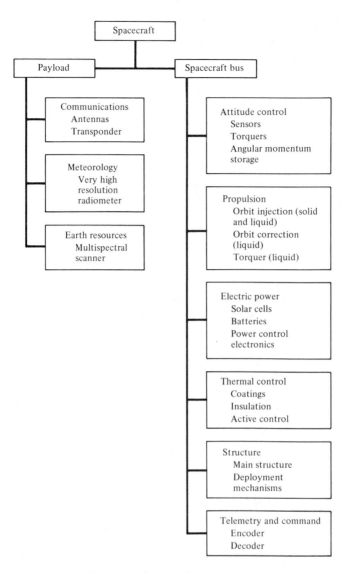

Figure 1.2 Spacecraft block diagram.

conversion of light energy from the sun into electrical energy by the use of solar cells. During the eclipse period when there is no sunlight, electrical power is provided by rechargeable batteries. The batteries are charged by the solar array during the sunlit period. Power control electronics control the bus voltage.

The thermal control subsystem keeps the temperatures of spacecraft equipment within specified ranges by using the correct blend of surface coatings, insulation, and active thermal control devices. In a communications satellite, the major tasks for the thermal control subsystem are to maintain the temperatures of the batteries within narrow limits, to prevent the hydrazine fuel from freezing, and to radiate into space the large amount of heat generated by the TWTs.

The structural subsystem provides the mechanical interface with the launch vehicle, gives mechanical support to all spacecraft subsystems, sustains launch loads, and provides precise alignment where needed, such as for antennas and thruster jets.

The telemetry and command subsystem maintains the two-way command and data link to the ground. The command portion receives the command data, decodes them, and carries out the stored command sequences. The data-handling portion accepts data from the subsystem sensors and, after encoding them, telemeters them to the ground.

A spacecraft design is an iterative process. It can be broadly divided into three phases: preliminary design, detailed design, and final design. At first, a feasibility study is made to determine whether the mission performance requirements can be met within the mass and size constraints of the launch vehicle. The first step is to select a spacecraft configuration which provides a general arrangement of the subsystems. The mass and power requirements of the subsystems are estimated, based on preliminary analysis and extrapolation of the existing designs. After the feasibility of the mission is confirmed, a detailed design of the subsystems starts with detailed analyses and tests carried out at the unit and subsystem levels. The spacecraft design is qualified at the subsystem and system levels by conducting performance, thermal, and vibration tests. Units that do not meet the performance requirements during the tests are redesigned and retested. After successful completion of the qualification tests, the spacecraft design is finalized and the required number of flight spacecraft are fabricated. The flight spacecraft are subjected to acceptance tests to detect manufacturing and assembly defects.

A spacecraft configuration is highly influenced by the performance requirements of the mission payload, the launch vehicle, and the attitude stabilization system selected. This chapter provides a review of these three major elements: satellite missions, potential launch vehicles, and spin stabilization versus three-axis stabilization. The section on satellite missions covers the functions of meteorological, earth resource, and communications satellites. The section on launch vehicles gives the major launch vehicle constraints and payload load capabilities of the U.S. Space Shuttle and the European Ariane. The next section covers the various design considerations, including spin stabilization versus three-axis stabilization and their influence on the design of spacecraft subsystems. As an example, the spacecraft configurations of Intelsat V, three-axis stabilized, and Intelsat VI, dual-spin stabilized, are described. A method for the preliminary estimation of mass and power for spacecraft subsystems is described in the final section. At the end, the current trends in communications satellites are discussed.

1.2 SATELLITE MISSIONS

Spacecraft missions can be divided into two broad categories: scientific exploration and earth application. The scientific satellites uncover new scientific knowledge of the earth, its environment, the solar system, and the stars that cannot be obtained by ground-based instruments. The earth application satellites are used basically for meteorology, earth resources, and communication. The scientific satellites have a wider variation in performance of mission payloads, orbits, and spacecraft configurations. Therefore, a review of these satellites is beyond the scope of this book. The majority of earth application satellites are communications satellites in geosynchronous orbit (24-hour period). This section provides a brief review of the evolution of application satellites, with the main emphasis on communications satellites.

Meteorological Satellites

Conventionally used ground-based or airborne weather observation systems include radiosonde stations, meteorological radars, weather ships, and so on. However, these do not provide global coverage and suffer from serious limitations in making certain types of observations. In contrast, satellite observation of meteorological phenomena and parameters can provide nearly continuous global coverage and it overcomes the problems of the installation and maintenance of, and the transmission of data from, meteorological sensing instruments in remote land areas, oceans, or high altitudes in the atmosphere.

Satellites obtain a full map of the coverage region using a radiation wavelength in either the visible or infrared range. Radiation in the visible range is absorbed or reflected differently by atmospheric gases, masses of water vapor, cloud, snow, land, or water. The visible images, however, can be obtained only when the underlying region of the earth is illuminated by the sun. This limitation is surmounted by using infrared (IR) images. By taking visual and IR images of the entire earth, the motion of a cloud formation over a large portion of the globe can be measured.

There are two types of meteorological satellites: low-altitude sun-synchronous and high-altitude geosynchronous orbit satellites. A typical low-altitude sun-synchronous satellite uses a polar orbit with a 1.92-hour period and cycles back to cover the same earth surface every 12 hours. On the other hand, the geosynchronous orbit satellite has a continuous view of one area of the earth's surface. The optical and spacecraft attitude control requirements for viewing the thin (19 km) tropospheric/stratospheric shell of the earth are quite formidable for the geosynchronous-altitude spacecraft but are within the present state of the art. For example, to achieve a ground resolution of 1 km in the visible light range for an imaging system corresponds to a field of view of 0.2 mrad and a maximum allowable jitter of the optical control and thus an attitude control of no more than 2.5 μrad during the picture-taking sequence. Silicone photodiodes (solid state) are good for the visible light range and require no cooling. For infrared detectors to work efficiently, cryogenic temperatures are required. Hence infrared detectors require special coolers.

The first meteorological satellite was Tiros 1, launched on April 1, 1960. Tiros, and the later Nimbus and ESSA satellites, are all low-altitude meteorological satellites. ATS I, a spin-stabilized satellite, launched on December 7, 1966, was geosynchronous and provided for the first time a full-disk photograph of the earth and its cloud cover.

The sixth World Meteorological Congress (1971) urged the member countries of the World Meteorological Organization (WMO) to conduct comprehensive atmospheric and meteorological observations in their geographical regions. The primary function of WMO is to plan, produce, and distribute worldwide weather data. Five satellites form the geosynchronous satellite portion of this network: West Europe's Meteosat is positioned at 0°, the Soviet Union's GOMS at 70 to 80°E, Japan's GMS at 140°E, and the U.S. GOES at 75° and 135°W longitudes.

The Geostationary Meteorological Satellite (GMS), shown in Fig. 1.3, is a dual-spin-stabilized spacecraft and was launched by a Delta 2914 launch vehicle on July 14, 1977. The primary payload was the visible and infrared

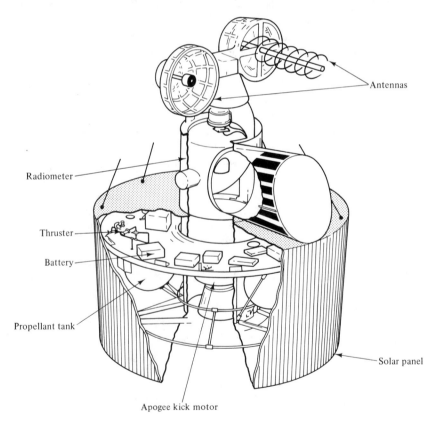

Figure 1.3 Geostationary meteorological satellite. (Courtesy of Hughes Aircraft Co.)

spin scan radiometer (VISSR). It also has communications capabilities in the S band (1.7 GHz) and UHF (400 MHz). The VISSR produced an image of the earth in both the visible and infrared spectra. Extremely sensitive infrared detectors operated at temperatures of about 95 K ($-178°C$). A radiation cooler cooled the detectors by radiating heat into cold space. The VISSR was on the spun section of the satellite, thus providing a west-to-east scan of the earth for each revolution of the satellite. North-to-south scan is provided by using motor-actuated mirror stepping. The mirror step was 140 μrad, thus requiring 2500 steps to scan a region 20° in elevation. At a spin rate of 100 rpm, it took 25 minutes to generate a complete picture. An additional 5 minutes was required for the scan mirror to retrace its steps and for any attitude error corrections. Thus one picture cycle was completed in about 30 minutes. The IR detectors were mercury–cadmium telluride with a field of view (FOV) of 140 microrad, and thus provided 5-km resolution on the earth's surface. The visible detectors were photomultiplier tubes with a FOV of 35 microrad and provided 1.25 km resolution. One IR and four visible detectors were used in the normal scanning mode. Data were also collected from surface data collection platforms by GMS.

Earth Resource Satellites

The work on the study of the earth's resources by satellites began in 1960 with the initiation of the Earth Resources Technology Satellite (ERTS) Program, later called Landsat. The first satellite under this program, Landsat 1, was launched in July 1972. It was equipped with a multispectral scanner (MSS) which transmitted data on water, soil, vegetation, and minerals to earth stations for interpretation and analysis by scientists in the United States and 30 other countries. Landsat 2 and 3, launched in 1975 and 1978, respectively, have sent data to more than 100 nations participating in the program. The new satellite, Landsat-D, shown in Fig. 1.4, provides further improvements in the scanning capability with its thematic mapper (TM) sensor. The data collected by Landsat are used in a wide range of disciplines, including agricultural crop classifications, forest inventories, land-use assessments, map productions, mineral resource explorations, geological structure identifications, water resource determinations, oceanographic resource locations, and environmental impact studies.

The Landsat 1, 2, and 3 satellites orbit in the sun-synchronous orbit, which has an 81° inclination to the equator and at an altitude of 919 km. The orbit period is 103 minutes. The near-polar orbit is sun synchronous, always crossing the equator from north to south at 9:30 A.M. suntime. Each successive orbit is shifted westward by 2875 km at the equator. The 14 orbits each day are shifted 159 km (at the equator) westward relative to the 14 orbits on the preceding day. After 18 days, the orbital cycles repeat. Thus it is possible to map the earth completely in 18 days and to generate fairly high resolution images of 185 km × 185 km areas in a variety of spectral bands. Landsat-D operates at an altitude of 705 km.

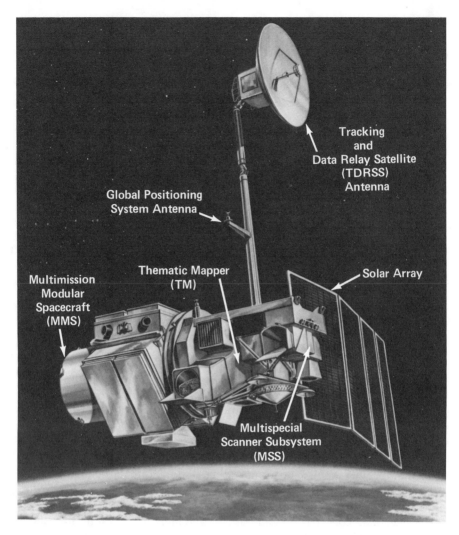

Figure 1.4 Earth resource satellite, LANDSAT-D. (Courtesy of NASA)

The multispectral scanner consists of a vibrating mirror, a telescope, and a bank of 24 detectors to give four colored images in four different spectral bands: green, red, and two infrared bands. A single image is approximately square, 185 × 185 km, and made up of 7.5 million picture elements, each being 80 m square. Each picture element is quantized into one of 64 intensity levels and the intensity value is transmitted to the ground. One full frame is sent every second.

The first three satellites, Landsat 1, 2, and 3, used a Nimbus-type bus, developed originally for meteorological satellites. Landsat-D uses the Multi-Mission Spacecraft bus. It has two sensors, an MSS and a thematic mapper (TM). The thematic mapper is a new sensor which provides images of 30-m resolution.

The thematic mapper is essentially bigger and better than MSS. It images in seven spectral bands versus four for the scanner. The mapper has 100 imaging detectors compared with 24 for the multispectral scanner. Landsat-D has the capacity to generate 800 frames a day (550 MSS and 250 TM), compared to 190 MSS frames per day for Landsat 1, 2, and 3.

Communications Satellites

The microwave frequency range (1 to 30 GHz) is best suited for carrying the large volume of communications traffic existing today. The signals of this frequency band are not appreciably deflected by the earth's ionosphere, limiting the line of sight to about 100 km. In addition, these signals will not travel through the earth. Over land and over short water links, microwave systems may be constructed using repeater stations approximately every 50 km with small dish-type antennas that receive the beamed signal and using a second antenna retransmit it toward the next station in the chain. The placing of such a chain across one of the world's major oceans, such as the Atlantic Ocean, presents insurmountable engineering difficulties. Another approach to transmitting a microwave signal across the Atlantic Ocean would be by erecting a repeater tower nearly 800 km high in the center of the ocean. This project would pose even greater engineering difficulties.

In the space age, it is possible to have a repeater in the sky without building the tower. A satellite launched into a circular orbit of altitude 35,786 km above the equator, called the geosynchronous orbit, would appear to be stationary from a point on the earth. A satellite in this position can view about one-third of the earth's surface. Any earth station in that coverage area directed toward that satellite can potentially communicate with any other earth station. The first well-known article on communications satellites was written in 1945 by Arthur C. Clark. He showed that three geosynchronous satellites, powered by solar energy converted into electricity using silicon cells, could provide worldwide communications. It took major advances in rocket technology for launching satellites and communications equipment to translate Clark's concept into reality.

The first human-made active communications satellite was SCORE, Signal Communications by Orbiting Receiving Equipment, and was launched in December 1958. It was designed to operate for 21 days of orbital life. The operating life was, however, limited to 12 days due to the failure of the batteries. During the late 1950s and early 1960s, the relative merits of passive and active communications satellites were often discussed. Passive satellites merely reflect incident signal, whereas active satellites have equipment that receives, processes, and retransmits incident signal. Project ECHO used two large spherical passive satellites, one 30.48 m (100 ft) in diameter and the second 41.5 m (135 ft) in diameter. They were launched in 1960 and 1964, respectively. The initial orbits were 1520 × 1688 km with 48.6° inclination and 1032 × 1316 km with 85.5° inclination, respectively. The satellite that paved the way for commercial communications satellites was Telstar. Telstar was a low-earth-orbit active repeater satellite. It was soon followed by a geostationary active repeater satellite, Syncom. The Syncom satellites were spin stabilized. Syncom 1 was launched in February 1963. The intended orbit was at the geosynchronous altitude with a 32° inclination. The satellite, however, failed during apogee motor fire. Syncom 2 was successfully launched in July 1963 by a Delta launch vehicle.

A pictorial view of the evolution of communications satellites to 1980 is shown in Fig. 1.5. Reference 6 provides further details on these satellites. Communications satellites can be divided into three categories on the basis of their applications: fixed service, broadcast, and mobile. In the fixed satellite services (FSS), the earth station locations are fixed geographically and provide a fixed-path, two-way communications link (receive/transmit) to the satellite. These satellites are further divided into international, regional, and domestic service satellites, depending on the area of use. Intelsat provides the oldest, largest, and most complex satellite communications network in the world for international communications, as well as substantial domestic service leases.

A traditional international communication space link is shown in Fig. 1.6. Individual telephone calls are processed at local telephone exchanges and relayed to the satellite earth station via a microwave link. At the earth station, these signals are stacked in frequency to form a composite baseband signal that frequency modulates the radio carrier wave sent up to the spacecraft. This is called frequency-division multiplexing. The spacecraft amplifies the received signal, translates its frequency, and sends it back to the receiving earth station, where it is passed into the local terrestrial telephone system for transmission to users after demultiplexing. The current trend in space communications is to eliminate a major portion of the terrestrial system in the link. This has been made possible by the increase in the power of satellite signals, to the level where smaller antennas on customer premises can receive and transmit signals to the satellite.

In broadcast satellites, the satellite receives the TV/radio signal from a central station and transmits it back, after very high signal power amplification, to small, receive-only antennas, usually of less than 1 m diameter, on the roofs of private houses. In mobile satellites, the mobile terminals on the sea, air, and ground, such as ships, airplanes, and automobile/trucks, communicate with a fixed or mobile terminal through the mobile satellite, as shown in Fig. 1.7. In the following section the evolution of the Intelsat system, Inmarsat system, and broadcast satellite system is discussed in more detail.

The Intelsat system. The creation of Intelsat was preceded by the formation of Comsat with the U.S. Communications Satellite Act of 1962, enacted in August 1962 during President Kennedy's term, which authorized the formation of the Communications Satellite Corporation (Comsat) to carry out the objectives set forth in this act. The act directed that (1) an

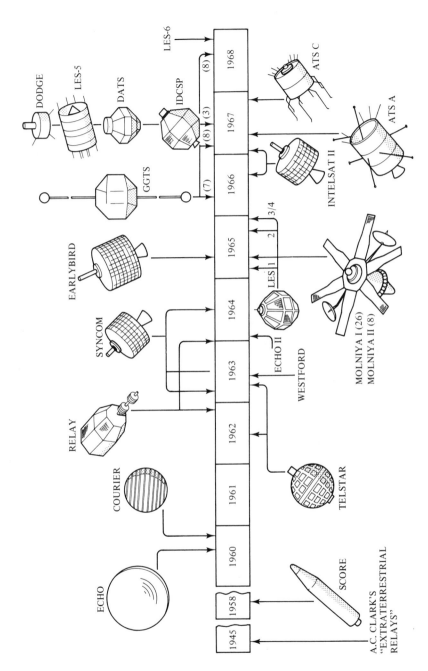

Figure 1.5 Evolution of communications satellites (Ref. 5).

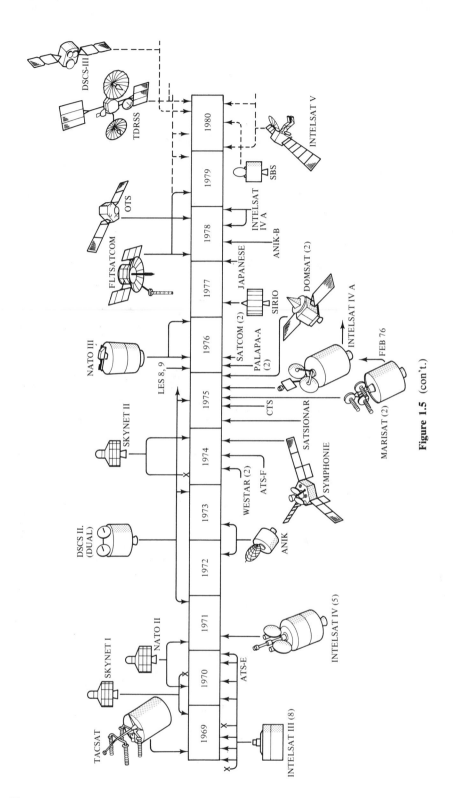

Figure 1.5 (con't.)

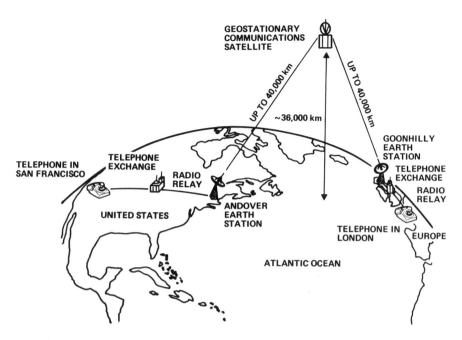

Figure 1.6 Network for international communication (Ref. 9).

international commercial communications satellite system be established as quickly and expeditiously as practicable, in conjunction and cooperation with other countries of the world; (2) the services of the system be made available to all countries of the world without discrimination; and (3) the benefits of the new technology be reflected in the quality and charges for such services. Comsat was incorporated on February 1, 1963. Intelsat (International Tele-communications Satellite Organization) was born in July 1964 and has subsequently grown to over 100 member countries. Comsat represents the interests of the United States in Intelsat.

The major decision to be made in 1963 was whether the first Intelsat satellite system should be based on satellites in a medium-altitude orbit of 10,000 to 13,000 km above the earth or in a geosynchronous orbit at 35,786 km (22,236 miles). The available experience at that time favored the medium-altitude system since Telstar had been so successful and Syncom 1 had failed during launch on February 1963. However, it was decided that whereas a successful launch of one or two medium-altitude satellites would at best allow only experimental use of the system (since 18 to 24 of such satellites would be needed for a fully operable commercial system), the successful launch of a single geosynchronous satellite would permit immediate operational use on one ocean region. Accordingly, a contract was awarded in early 1964 to the Hughes Aircraft Co. for the construction of a geostationary communication satellite (Early Bird). When Early Bird successfully attained its geosynchronous orbit, it was renamed Intelsat I.

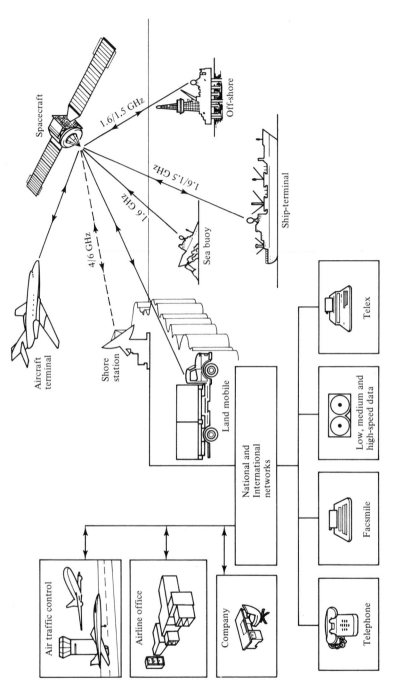

Figure 1.7 Network for mobile services.

Space Segment. The first commercial satellite, Intelsat I, built by Hughes, initially known as the Early Bird, was launched in April 1965. It weighed only 38 kg in orbit and was spin stabilized. It had a fixed, omnidirectional antenna which transmitted in a full 360° arc. It carried 240 two-way telephone circuits or one transatlantic TV channel. It did not, however, allow multiple access; that is, only one pair of earth stations could operate with the satellite at any one time. It provided service across the North Atlantic, between Europe and North America.

The Intelsat II series of satellites, also built by Hughes, were spin stabilized and weighed 86 kg in orbit. They carried 240 two-way telephone circuits or one TV channel. An important innovation in the Intelsat II satellite series was the multiple-access capability, enabling many pairs of earth stations to be connected simultaneously by the same satellite. The first successful Intelsat II launch took place in October 1966 and with these satellites, Intelsat coverage was eventually provided over both the Atlantic and Pacific Ocean areas.

The Intelsat III series satellites, built by TRW, provided a dramatic increase in capacity, from 240 circuits for Intelsat I and II to 1,500 circuits plus a TV channel. They incorporated a mechanically despun antenna which always pointed toward the earth, so the transmission power could be focused in the direction of the earth instead of radiating some of the power into space. They weighed 152 kg in orbit. Intelsat III satellites were placed over all three major ocean areas: the Atlantic, the Pacific, and the Indian. So in 1969, the Intelsat system achieved full global coverage for the first time; it was a translation of Clark's concept into reality.

The Intelsat IV series satellites were built by Hughes and launched by the Atlas-Centaur vehicle. They weighed 832 kg in orbit and were dual-spin-stabilized spacecraft. Intelsat IV provided 3,750 two-way telephone circuits and two TV channels. The major advance of Intelsat IV over its predecessors was the use of additional "zone-beam" antennas, which covered selected small portions of the visible earth. The resulting concentration of radiated power in these zone beams contributed to the increased capacity. Intelsat IV satellites also had four global coverage horns. The first Intelsat IV was launched in January 1971. The Intelsat IV satellite was bandwidth limited rather than power limited, the first time that had ever happened to a communications satellite.

Intelsat IV-A was a derivative of Intelsat IV. These satellites employed a frequency reuse technique for the first time by using spatially separated beams. This permits simultaneous use of the same portion of the spectrum in two separate areas: for example, in the Atlantic region, one shaped beam covered Europe and Africa while the other covered North and South America. With approximately the same weight and power as Intelsat IV and using the same 6/4-GHz frequency band, the Intelsat IV-A satellite has a 50% greater capacity (i.e., about 6,000 circuits) plus two TV channels.

The Intelsat V satellite series, built by Ford Aerospace and Communications Corporation, are three-axis-stabilized spacecraft. Intelsat V uses fourfold frequency reuse, employing both spatially separated beams and dual polarization within the 500-MHz available bandwidth at 6/4 GHz. Additionally, Intelsat V uses the 14/11-GHz band with twofold frequency reuse of the available 500-MHz bandwidth by beam separation. These developments have enabled the IV-A satellite's capacity of 6,000 two-way telephone circuits to be doubled to 12,000 two-way circuits plus two TV channels.

Intelsat V-A is a derivative of Intelsat V. One of the key features of the V-A spacecraft is the addition of two steerable 4-GHz zone beams which can be employed for domestic leases. The average communications capacity is 15,000 two-way telephone circuits plus two television channels.

Intelsat VI is a dual-spin-stabilized spacecraft. In comparison to Intelsat V, it will have twice the payload power, useful mass and number of transponders, and nearly three times the communications capacity. It is compatible for launch by either the Space Shuttle or by Ariane IV. The major technological advancements of Intelsat VI include a sixfold reuse of the 6/4-GHz bands, the dynamic interconnection of six of the satellite's antenna beams for use with satellite-switched time-division multiple access (SS/TDMA), and a 10-year design life (up from 7 years). The average communications capacity is 40,000 two-way telephone circuits plus two TV channels.

The evolution of Intelsat satellites is shown in Fig. 1.8. The communications capability from Early Bird through Intelsat VI represents an increase in capacity by a factor of more than 150 during a period of some 21 years. The Intelsat system has maintained an amazing reliability factor of greater than 99.9%. Furthermore, it has achieved significant reductions in utilization charges. When adjustment is made for global inflationary trends, the cost of a full-time Intelsat circuit ($9,360 per year for 1983) is in effect 18 times less than the cost of service some 18 years ago. The number of earth stations has grown from a handful to some 600. As small earth terminals up to a 3.5-m antenna diameter are adopted for digital business communications services within the next few years, one can expect the number of earth stations to take a quantum leap.

Ground Segment. The ground segment that operates the Intelsat satellites consists of hundreds of receiving and transmitting earth stations located in countries around the world. There are several types of standard earth stations, the three original types used for trunk services being the Standard A, B, and C earth stations. The Standard A earth station uses 30-m or larger-diameter parabolic antennas. It is the oldest and presently the most widely used in the system. The large antenna enables it to use the satellite's capacity most efficiently. It operates in the 6/4 GHz frequency bands. The Standard B earth stations, with their 11-m antennas, are a lower-cost alternative to the Standard A earth stations and are particularly suited for areas with small traffic demands. Because Standard B stations are less efficient users of satellite capacity, charges for satellite circuits through them are usually higher than for Standard A's. The Standard C earth stations,

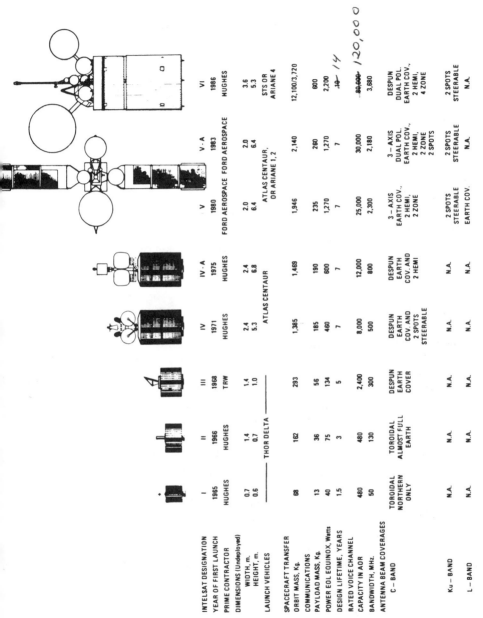

INTELSAT DESIGNATION	I	II	III	IV	IV-A	V	V-A	VI
YEAR OF FIRST LAUNCH	1965	1966	1968	1971	1975	1980	1983	1986
PRIME CONTRACTOR	HUGHES	HUGHES	TRW	HUGHES	HUGHES	FORD AEROSPACE	FORD AEROSPACE	HUGHES
DIMENSIONS (Undeployed) WIDTH, m.	0.7	1.4	1.4	2.4	2.4	2.0	2.0	3.6
HEIGHT, m.	0.6	0.7	1.0	5.3	6.8	6.4	6.4	5.3
LAUNCH VEHICLES	THOR DELTA			ATLAS CENTAUR		ATLAS CENTAUR OR ARIANE 1, 2		STS OR ARIANE 4
SPACECRAFT TRANSFER ORBIT MASS, Kg.	68	162	293	1,385	1,469	1,946	2,140	12,100/3,720
COMMUNICATIONS PAYLOAD MASS, Kg.	13	36	56	185	190	235	260	600
POWER EOL EQUINOX, Watts	40	75	134	460	600	1,270	1,270	2,200
DESIGN LIFETIME, YEARS	1.5	3	5	7	7	7	7	~~10~~ 14
RATED VOICE CHANNEL CAPACITY IN AOR	480	480	2,400	8,000	12,000	25,000	30,000	~~80,000~~ 120,000
BANDWIDTH, MHz.	50	130	300	500	800	2,300	2,180	3,680
ANTENNA BEAM COVERAGES C–BAND	TOROIDAL NORTHERN ONLY	TOROIDAL ALMOST FULL EARTH	DESPUN EARTH COVER	DESPUN EARTH COV. AND 2 SPOTS STEERABLE	DESPUN EARTH COV. AND 2 HEMI	3–AXIS EARTH COV., 2 HEMI, 2 ZONE	3–AXIS DUAL POL. EARTH COV., 2 HEMI, 2 ZONE, 2 SPOTS	DESPUN DUAL POL. EARTH COV., 2 HEMI, 4 ZONE
Ku–BAND	N.A.	N.A.	N.A.	N.A.	N.A.	2 SPOTS STEERABLE	2 SPOTS STEERABLE	2 SPOTS STEERABLE
L–BAND	N.A.	N.A.	N.A.	N.A.	N.A.	EARTH COV.	N.A.	N.A.

Figure 1.8 Evolution of Intelsat satellites (Ref. 1).

with antennas of 14 to 19 m in diameter, are designed especially to operate within 14/11-GHz frequency bands.

Inmarsat system. The safety of life at sea remains a major problem for ship owners and seafarers. During 1968–1980, tankers alone accounted for 1004 serious casualties throughout the world. A major problem in preventing fatalities has been claimed to be the poor communications available for requesting medical assistance. A step toward improving maritime communications was taken in 1976 with the introduction of maritime satellite communications by the Marisat satellite systems under a partnership of Comsat General (holding 86%) with RCA, ITT, and Western Union. The satellites provided global coverage, instantaneous high-quality service, and interconnection with the worldwide public telecommunications networks. Besides emergency communications, offshore rigs may transmit oil well log and seismic data to land-based computers for rapid analysis, oil tankers may be diverted, crew morale can be improved by the ability to make private telephone calls to their homes, and port surveillance and navigation information can be provided.

The international Maritime Satellite Organization (Inmarsat) came into being on July 16, 1979, pursuant to the signing of its convention and operating agreement by 26 member states and signatories. The purpose of Inmarsat is to make provision for the space segment necessary for improving maritime communications for peaceful purposes. The Inmarsat system, which went into operation on February 1, 1982, has taken over the Marisat system, which provided similar services since 1976. Shore-to-ship communications are in the 6-GHz band from coastal earth stations to the satellite and in the 1.5-GHz band from the satellite to the ship. The ship-to-shore communications are in the 1.66-GHz band from ship to satellite and in the 4-GHz band from satellite to the coastal earth stations.

The initial Inmarsat space segment consisted of three Marisat satellites and a Marecs A satellite, Europe's first commercial satellite series dedicated to maritime communications. The three Marisat satellites provided up to 10 channels each for telephone, telex, facsimile, and data transmission. Marecs A has a 46-voice channel capacity. The space segment capacity was further increased by leasing maritime communications subsystems, each with a capacity of 30-voice channel capacity, on Intelsat V. In the future, multi-purpose satellites capable of providing aeronautical mobile communications in addition to the existing maritime service are planned.

The ground segment for maritime communications consists of two parts: a coast earth station and a ship earth station. A typical coastal earth station consists of an 11- to 14-m-diameter antenna and the associated radio-frequency (RF) equipment for interfacing with the international switched telecommunications networks. The ship earth station, approved by Inmarsat for use in its system, is called a Standard A terminal. It has a relatively high gain, narrow beamwidth, and a capacity for one voice/data channel and one telex channel. The ship earth station consists of two parts, the above-deck antenna shrouded in a glass-fiber randome and the below-deck hardware. The ship earth station, sized between 0.8 and 1.2 m in diameter, requires a complex stabilization system to keep the parabolic antenna pointed at the satellite at all times, in spite of all the ship's motion in the pitch, roll, and yaw directions. Inmarsat plans to introduce two new types of ship earth stations, referred to as Standard C and Standard D. The Standard D will have an antenna about 3 m in diameter and will enable several voice and telex channels to be used simultaneously. This terminal would be attractive to the offshore industry, which requires a larger voice and data communications capacity. The Standard C will be considerably smaller, for use on small ships such as those used in the fishing industry.

When Inmarsat began operations in February 1982, there were 1007 ships and oil drilling rigs equipped for satellite communications. The demand for satellite communications in the maritime community is growing more rapidly than expected. Since then, the number of ships and off-shore drilling rigs equipped with the ship earth stations has increased to more than 2,600 by July 1, 1984, more than doubling the figures of two years ago. Early forecasts of 4500 ship terminals by 1990 are now regarded as being too conservative.

Broadcast Satellites

A direct broadcast satellite receives TV and radio programs from the central earth stations and transmits them back, after very high signal power amplification, directly to homes for reception on parabolic antennas of less than 1 m diameter. The feasibility of direct broadcast services was demonstrated by ATS-6 in 1974 through experiments in India for educational purposes. In 1976, the Canadian experimental CTS satellite (Hermes) and in 1978 the Japanese Broadcast Satellite (BSE) were launched and then used for broadcasting. In the Soviet Union, broadcasting by means of satellites has been provided by the EKRAN system since 1976 and more recently by the Gorizont satellites. In India, DBS services were initiated by Insat in 1983. Initially, DBS service was considered to be suited only for developing countries and remote areas where terrestrial services were not adequately developed. However, this service has recently found wide acceptance in the United States and Europe, where terrestrial services are well developed.

For broadcasting satellites, the World Administrative Radio Conference of 1977 provided the basic regulations. These regulations include orbit locations, a channelization scheme, frequency and polarization assignments, antenna coverage areas, and maximum power level (EIRP). Based on these regulations, the governments of Germany and France are jointly developing TV broadcasting satellites. The satellites, TV-SAT for Germany and TDF-1 for France, use the same spacecraft bus and differ only as dictated by WARC requirements (antennas, frequencies, EIRP). Each spacecraft employs five channels, three of which are operated simultaneously. The spacecraft are three-axis stabilized with approximately 3 kW of power from the solar array. The transponder system consists of a high-powered TWTA, in the range of 230 to 260 W. The EIRP is 65.5 dBW for Germany and 62.9 dBW for France. The uplink and downlink frequencies are in the 17 to 18 GHz and 11 to 12 GHz bands, respectively. In the United States, several companies are developing broadcasting satellite systems.

The DBS satellites differ from conventional satellites, such as Intelsat satellites, mainly in terms of transmitted power. For conventional satellites, TWTAs typically operate at 10 W. However, on DBS satellites, TWTAs operate nominally at 200 W. This exceptional high power level influences spacecraft design, specifically temperature control. These satellites need large solar arrays to provide power to the TWTAs. During the solar eclipse periods, without batteries, the DBS satellite will become inoperational for periods of up to 72 minutes every day. The effect of eclipses, however, can be accommodated by proper orbital location of the satellites. If the satellite is positioned west of the service area such that the eclipse period would not begin until 2:00 A.M., the satellite can be considered fully operational, as few viewers watch TV at that time. Hence batteries are required only for housekeeping.

Military Satellites

Communications satellites have been found to be of great military use in the following applications: long-distance communications to isolated areas; high-data-rate transmission for intelligence applications; rapid extension of circuits into new areas; and mobile communications to moving platforms, such as aircraft, ships, and submarines. The design of a military satellite is driven by survivability against physical attack, electronic countermeasures, security for sending secret messages, and flexibility for mobile system and remote-area access. These design considerations result in an increase in complexity and cost.

Multi-Mission Satellites

In the previous sections, we have discussed dedicated satellites for each mission. However, some users have found it more cost-effective to combine several missions into one satellite, called a "multi-mission satellite." Insat 1, Indian National Satellite System, is one example in this class. It combines long distance telecommunications, meteorological, and direct TV broadcast services into one satellite. The configuration of INSAT 1 is shown in Fig. 1.9. The payload of INSAT 1 provides the following services:

(a) Twelve (12) transponders of 36 MHz bandwidth each, operating in 6/4 GHz frequency (C) band with total telecommunications capacity of approximately 4,300 two-way telephone circuits.

(b) Two (2) high-power transponders, operating in 6/2.5 GHz frequency (C/S) band, each capable of providing one direct TV broadcast channel

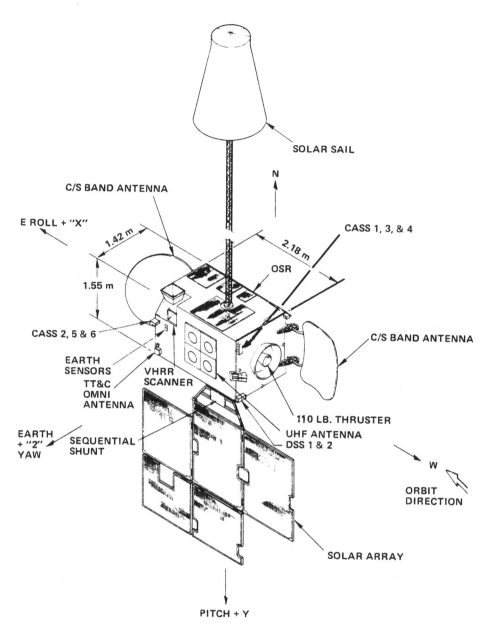

Figure 1.9 INSAT-1 spacecraft deployed. (Courtesy of Dept. of Space, Gov't. of India)

for direct reception/TV networking and five carriers for services, such as national and regional radio networks and disaster warning systems.

(c) A Very High Resolution Radiometer (VHRR) with visible (0.55–0.75 μm) and infra-red (10.5–12.5 μm) band channels with resolutions of 2.75 km and 11 km respectively and with half-hourly full coverage.

(d) A data channel with 400 MHz earth-to-satellite link for relay of meteorological, hydrological and oceanographic data from unattended land and ocean based, automatic data collection-cum-transmission platforms (DCPs).

The spacecraft is biased momentum three-axis-stabilized. INSAT 1A was launched on April 10, 1982 by a Delta 3910/PAM and INSAT 1B on August 30, 1983 by the STS/PAM-D. The combination of communications and meteorological missions has resulted in several unique design features for INSAT 1. In order to ensure a clear field of view (FOV) into cold space for the radiation cooler of the VHRR on the north side, the solar array is deployed only on the south side of the spacecraft. The large solar torque due to asymmetrical solar array configuration is offset by a solar sail on a long boom. The attitude control requirements for VHRR image stability are: short term stability of 0.0008° in a period of 2.2 seconds and long term stability of 0.0044° for a half-hour duration. The communication antenna pointing accuracy requirements are 0.2° for roll and pitch and 0.4° for yaw. In comparison to antenna pointing accuracy requirements, the image stability requirements are more demanding. Therefore, during imaging, all house-keeping operations, including stepping-up of the solar array and thruster firing, are inhibited. Housekeeping functions are performed during the 7-minute interval between successive full frame imaging operations. INSAT 1 is the first satellite to use a unified bi-propellant propulsion system, which combines the functions of apogee injection, orbit corrections, and attitude control.

1.3 LAUNCH VEHICLES

A typical flight profile used to obtain geosynchronous orbit is shown in Chapter 2 (Fig. 2.1). The satellite is first injected into a low-altitude circular parking orbit, the low altitude being about 200 km. The satellite, with its upper stages still attached, coasts in the parking orbit until it is above the equator, at which point the first of the upper stages, also called the perigee kick motor (PKM), injects the satellite into a geosynchronous transfer orbit. The transfer orbit is a highly elliptical orbit with maximum altitude equal to that of the geosynchronous orbit, 35,786 km. The inclination angle of the orbit depends on the latitude of the launch site and launch azimuth. A due-east launch, with azimuth equal to 90°, results in a transfer orbit inclination equal to the latitude of the launch site; any other azimuth angle of the launch will increase the inclination. Since a zero inclination angle is required for fully geostationary satellites, it is advantageous to have the launch site as close to the equator as possible. The satellite is injected into the geosynchronous orbit by the second upper stage, the apogee kick motor (AKM), which not only increases the transfer orbit apogee velocity to match the geosynchronous orbit velocity, but reduces the orbit inclination to zero.

For a geosynchronous orbit satellite, the launch vehicle selected will determine whether the PKM and AKM are part of the spacecraft or of the launch vehicle. Spacecraft launched by Titan IIIC will need neither a PKM nor an AKM; spacecraft launched by Delta, Atlas-Centaur, or Ariane will need an AKM; and spacecraft launched by the Space Shuttle will need both a PKM and an AKM. These upper stages can be an integral part of the spacecraft or developed separately and attached to the spacecraft for launch.

Launch vehicles impose several constraints on spacecraft designs. The most obvious constraints are the mass and volume available to the spacecraft. Other launch vehicle characteristics that influence spacecraft design include mechanical and electrical interfaces, radio link performance through the fairing (required for certain spacecraft ground tests while they are enclosed in the fairing), launch vibration environment, nutation angle at separation from spin-stabilized stages, and thermal environment. Therefore, spacecraft designers need to know much more about the launch vehicle than the available mass and volume. This additional information is usually available in the user's manual of the launch vehicle.

During the 1970s most commercial communications satellites were launched by either the Delta or Atlas-Centaur launch vehicles. Both of these U.S. launch vehicles are expendable (i.e., they are not recoverable). In 1982, NASA introduced a reusable launch vehicle, the Space Shuttle. The Space Shuttle has a significantly higher payload capability. However, since it only injects the spacecraft into a low-altitude parking orbit, upper stages are necessary as part of the Shuttle payload to inject the spacecraft into a geosynchronous transfer orbit. McDonnell Douglas Corporation (MDAC) has developed the PAM-D and PAM-A upper stages, which provide payload capabilities equivalent to the Delta and Atlas-Centaur, respectively. The Space Shuttle can launch several Delta- and Atlas-Centaur-sized space-craft on the same mission. The Space Shuttle normally provides a lower launch cost compared with that of expendable launch vehicles. Therefore, spacecraft in the late 1970s and early 1980s were designed to be compatible with both expendable launch vehicles (i.e., Delta or Atlas-Centaur) and the Space Shuttle with corresponding upper stages. In the future, NASA plans to discontinue support for expendable launch vehicles. The European Space Agency (ESA) introduced Ariane, an expendable launch vehicle, in 1979. The payload capability of Ariane I is similar to that of Atlas-Centaur. It can launch either one Atlas-Centaur-sized spacecraft or two Delta-sized spacecraft. With the withdrawal of NASA support for expendable launch vehicles, the spacecraft in the mid-1980s are designed to be compatible with the Space Shuttle and Ariane. Japan has also developed N-Vehicles, which are essentially a copy of the Delta launch vehicles.

Table 1.1 gives some key characteristics of the launch vehicles for geosynchronous orbit satellites. As shown in the table, Ariane and Delta have several versions. The payload capability of the Space Shuttle is given for several perigee stages. The inclination angle of the geosynchronous

TABLE 1.1 LAUNCH VEHICLE CHARACTERISTICS

Launch Vehicle	Date of First Launch	Transfer Orbit Inclination (deg)	Synchronous Transfer Orbit Payload (kg)	Synchronous Orbit Payload (kg)	Shroud Diameter (m)
Expendable					
N-Vehicle		30.0	250	132	
Delta 2914	1972	28.7	724	384	2.4
Delta 3914	1975	28.7	954	505	2.4
Delta 3910/PAM	1980	28.7	1154	612	2.4
Delta 3920/PAM	1982	28.7	1312	695	2.4
Atlas Centaur I	1966	27.1	1900	1007	3.0
Atlas Centaur II	1985	27.1	2000	1179	
Ariane I	1979	9.65	1700	1002	2.9
Ariane II		8.5	2000	1179	2.9
Ariane III	1984	8.5	2580	1521	2.9
Ariane IV		8.5	4200	2478	3.6
Reusable					
Shuttle	1981				4.57
STS/PAMD			1250	625	
STS/PAMD2			1840	867	
STS/PAMA			2000	960	4.57
STS/IUS				2120	4.57
STS/Centaur				4805	4.57

transfer orbit varies because of the use of different launch sites. As mentioned earlier, for nominal launch (launch azimuth equal to 90°), the inclination angle is equal to the latitude of the launch site. For U.S. launch vehicles, the launch site is the Eastern Test Range (ETR) in Florida at a latitude of 28.3°. Ariane is launched from French Guiana at a latitude of approximately 5°. The Japanese launch site is at about 30°N latitude. It should be noted that the information given in Table 1.1 is preliminary. A spacecraft designer should use the latest user's manuals of the launch vehicles and contact the launch vehicle contractors if further information is needed. The characteristics of the Space Shuttle and Ariane are discussed in more detail in the following sections.

Space Shuttle

The Space Shuttle, developed by NASA and shown in Fig. 1.10, consists of three main parts: the orbiter, which carries the payload, the crew, and the main engine; two solid rockets; and the external tank, which carries the fuel and the oxidizer.

An overview of a typical Shuttle mission is given in Fig. 1.11. During launch, the three main engines deliver 5,005 kN (1,125,000 lb) of thrust by burning liquid oxygen and liquid hydrogen from the external tank. The main

| | Standard Orbits | |
Altitude (nm)	Inc (deg)	Payload lb (Kg)
160	28.5	65,000 (29,484)
160	56	57,000 (25,855)
160	90	37,000 (16,783)
160	104	30,000 (13,608)

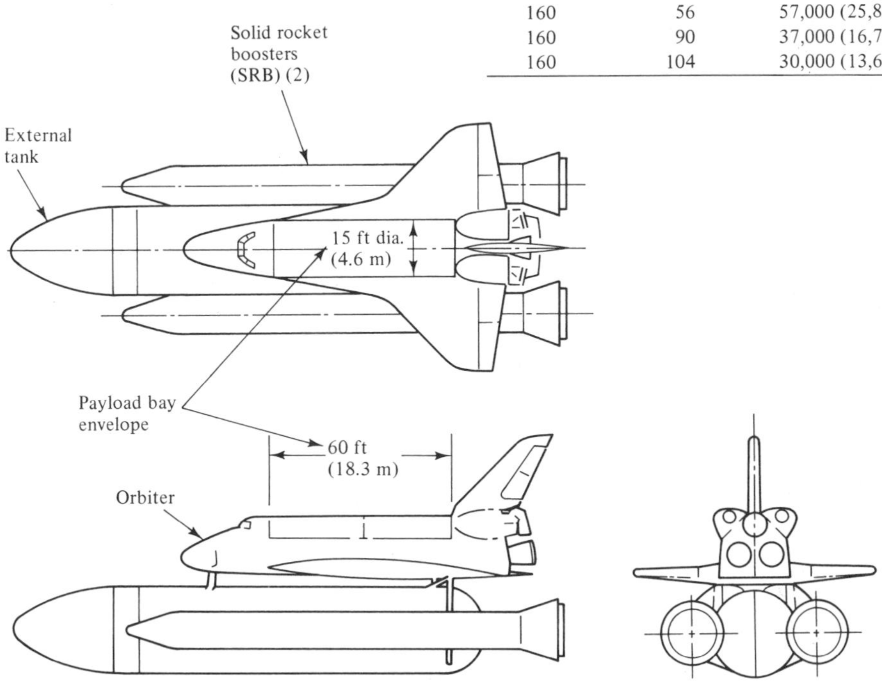

Figure 1.10 Space Shuttle flight system (Ref. 10).

engines are augmented by two solid rockets. The combined thrust is 28,584 kN (6,425,000 lb) at lift-off, almost the thrust developed by Saturn V. After solid rocket burnout, approximately 2 minutes after lift-off from the Kennedy Space Center (Eastern Test Range) and at an altitude of approximately 43 km, the spent rockets are jettisoned and recovered for reuse. The external tank is jettisoned when empty and 8 minutes after lift-off at an altitude of 62 nautical miles. The only part of the Shuttle system that is not reused is the external tank, which burns up as it reenters the atmosphere.

The orbiter is placed in its circular parking orbit by its orbital maneuvering system (OMS), a self-contained storable bipropellant liquid system. Two 26.7-kN (6,000-lb)-thrust liquid rocket engines burn for about 2 minutes for orbital insertion. The orbiter can remain in the parking orbit for a nominal mission of 7 days. The orbiter attitude is controlled to 0.5° and its attitude rate to 0.1° per second. The orbiter remains in the parking orbit to perform experiments and deploy satellites. After completion of the mission, the orbiter deorbits by firing two 26.7-kN (6,000-lb)-thrust liquid rocket motors of the OMS for about 95 seconds. This imparts a braking velocity of 121.9 m/s (400 ft/s), which starts the reentry into the atmosphere. Peak surface temperatures greater than 1,093.3°C (2,000°F) are experienced. The

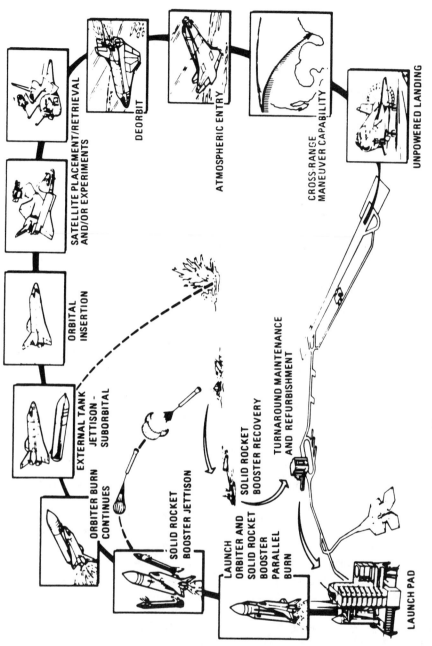

Figure 1.11 Typical Shuttle mission.

SATELLITE PLACEMENT/RETRIEVAL AND/OR EXPERIMENTS

DEORBIT

ATMOSPHERIC ENTRY

CROSS-RANGE MANEUVER CAPABILITY

UNPOWERED LANDING

ORBITAL INSERTION

EXTERNAL TANK JETTISON – SUBORBITAL

ORBITER BURN CONTINUES

SOLID ROCKET BOOSTER JETTISON

SOLID ROCKET BOOSTER RECOVERY

TURNAROUND MAINTENANCE AND REFURBISHMENT

LAUNCH ORBITER AND SOLID ROCKET BOOSTER PARALLEL BURN

LAUNCH PAD

heat is absorbed in special ceramic tiles that cover most of the surface of the orbiter. Of the 31,000 tiles used, 25,000 are of different sizes. From deorbit to landing, the orbiter coasts without power and then lands as a glider.

Payload capability. One of the main features of the Space Shuttle orbiter is its large payload volume. The allowable payload volume is 4.57 m (15 ft) in diameter by 18.29 m (60 ft) in length. Figure 1.12 shows a set of curves of shuttle payload performance to circular orbit for a KSC launch. The payload mass is plotted against the circular orbital altitude for various inclinations and additional OMS kits to extend the performance. As an example, consider curves for a 28.5° inclination, due east launch from KSC. Starting at an altitude of 100 nautical miles, the payload mass is constant at 29,484 kg (65,000 lb) until about 230 nautical miles. From the performance point of view, the payload mass would be much higher at 100 nautical miles and drop off monotonically to the value at 230 nautical miles, but the structural capabilities of the orbiter have been limited to a payload of 29,484 kg (65,000 lbs). After about 230 nautical miles, the performance drops off dramatically. The performance can be extended by adding one or more OMS kits. The mass for the OMS kits has been subtracted so that the payload weight shown is the payload available to the user. Since the Space

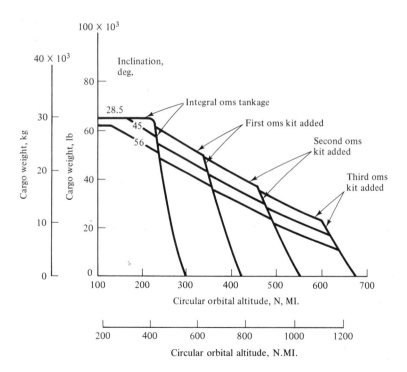

Figure 1.12 Space Shuttle payload performance to circular orbit (KSC launch, payload delivery only) (Ref. 10).

Shuttle returns to earth like an airplane, its center of gravity must be constrained for stability reasons.

In general, several payloads will share the Space Shuttle. To compute a shared user cost, one computes the ratio of the length used by the payload, which includes satellite, upper stage, cradle, and electronics, to the 18.29-m (60-ft) bay length. One also computes the ratio of the mass of the payload to the Shuttle capability, which is 29,484 kg (65,000 lb). The load factor is the greater of these two ratios. The charge factor is obtained by dividing the load factor by 0.75, which accounts for the assumption that on the average the shuttle will be only 75% full. For load factors equal to or greater than 0.75, the charge factor is unity. There is a minimum charge factor for small values of the load factor. The launch cost is the cost factor times the price for a dedicated launch.

Since the Space Shuttle puts the payloads in a low parking orbit, geosynchronous orbit payloads will need perigee motors to put them into the transfer orbit and apogee motors to put them into the geosynchronous orbit. Therefore, the total payload for the Space Shuttle will consist of perigee motor, apogee motor, and the spacecraft. Some spacecraft, such as Leasat, have been designed to be launched only by the Space Shuttle. These spacecraft have perigee and apogee motors as integral parts and they have a direct structural interface with the Shuttle. The Space Shuttle provides numerous attachment points along the side and bottom of the cargo bay for structural interfaces. The majority of communications satellites, however, are designed to be compatible for launch with the Shuttle and expendable launch vehicles due to launch cost and schedule considerations. For these spacecraft, apogee motors are developed as a part of the spacecraft and the perigee stages are developed separately. The perigee stage interfaces with the Shuttle and the spacecraft interfaces with the perigee stage. For these spacecraft, launch vehicle interface constraints are provided by the perigee stage and the selected expendable launch vehicle. Figure 1.13 shows an exploded view of a PAM-D. This perigee stage, developed by MDAC, provides an option for a satellite to be launched by either a Delta launch vehicle or by the Space Shuttle. The cradle assembly attaches the perigee motor/spacecraft to the orbital bay. The attachment is through four longeron fittings on the two sides of the cargo bay and the keel fitting along the bottom of the bay. This stage has been used by many satellites (e.g., SBS, ANIK-C). Similar perigee stages have been developed for Atlas-Centaur-and Ariane-sized spacecraft.

Ariane

Ariane, as shown in Fig. 1.14, is a three-stage expendable launch vehicle developed by the European Space Agency. The first stage consists of four Viking V engines, developing 2,485 kN thrust on the ground. The propellants (UDMH and N_2O_4) are contained in two identical steel tanks. The second stage consists of a single Viking IV engine developing a thrust

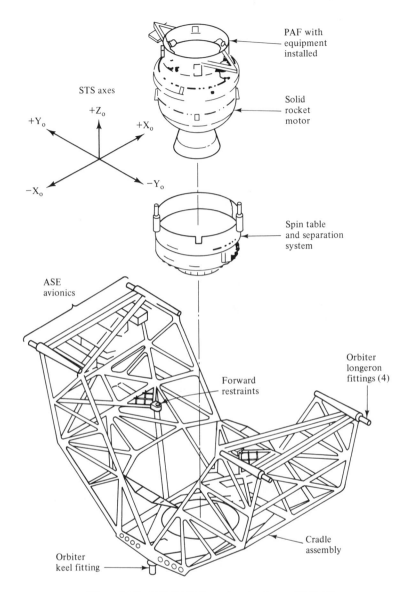

STS axes

$+Z_0$

$+Y_0$

$+X_0$

$-X_0$

$-Y_0$

PAF with equipment installed

Solid rocket motor

Spin table and separation system

ASE avionics

Orbiter longeron fittings (4)

Forward restraints

Cradle assembly

Orbiter keel fitting

Figure 1.13 STS PAM-D system hardware (Ref. 11).

of 797 kN in vacuum, the specific impulse being 295.2 s. The third stage consists of a single HM7-type engine developing a thrust of 61.8 kN in vacuum with a specific impulse of 440.6 s.

The Ariane launch vehicle can simultaneously place two independent satellites in orbit, as shown in Fig. 1.15. The structure used for this purpose is called SYLDA. SYLDA is basically a load-carrying structure comprising a conical adapter which carries the lower satellite enclosed inside a shell and the upper satellite on the top attachment. The shell consists of two

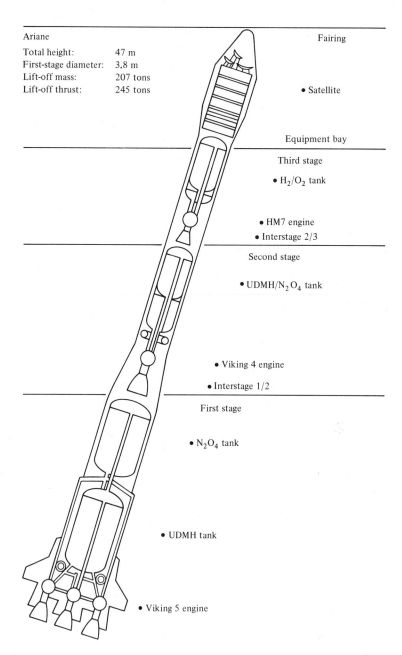

Ariane

Total height:	47 m
First-stage diameter:	3,8 m
Lift-off mass:	207 tons
Lift-off thrust:	245 tons

Fairing

• Satellite

Equipment bay

Third stage

• H_2/O_2 tank

• HM7 engine
• Interstage 2/3

Second stage

• $UDMH/N_2O_4$ tank

• Viking 4 engine

• Interstage 1/2

First stage

• N_2O_4 tank

• UDMH tank

• Viking 5 engine

Figure 1.14 Ariane (Ref. 12).

separate parts held together by a clamp band. Access to the inner (lower) satellite is via apertures in the shell. The two satellites are injected and separated in the following sequence. After engine cutoff, the third stage is oriented in the direction specified by the upper satellite, and spun up if required. The two connecting bolts are pyrotechnically cut, separating the upper satellite. Then the third stage orients the inner satellite to the required attitude and spin-up if necessary and separates the upper part of the support structure. The inner satellite is then separated a little later in the same way as the upper satellite.

Ariane II is an uprated version of the Ariane I. In this uprating, the increase has been in the thrust of the Viking engines, available mass of the third stage cryogenic propellants along with specific impulse of the engine using these propellants, and the size of the payload fairing. Ariane III is the same configuration as Ariane II except for the addition of two solid propellant boosters to the first stage for additional thrust augmentation. Ariane IV is the uprated version of Ariane II and III. The first stage is lengthened by seven meters to increase the propellant capacity. The first stage engines and the second and third stages are identical with Ariane II and III except for structural strengthening of the stages. There are several versions of Ariane IV achieved by the addition of various combinations of solids and liquid boosters to the basic vehicle. The launch capability of Ariane IV for synchronous transfer orbit varies between 1,900 and 4,200 kg (given in Table 1.1) depending on the selected version.

1.4 SPACECRAFT DESIGN CONSIDERATIONS

The spacecraft bus is designed to meet the following payload requirements:

- Provide support to equipment in a layout that minimizes signal losses and interconnections
- Provide the required electrical power within the specified voltage tolerances
- Provide temperature control within the limits imposed by the equipment
- Keep spacecraft attitude and orbit within allowable limits
- Provide telemetry and command services to permit ground monitoring and control of the transponder function
- Provide support to the total mass with adequate stiffness, alignment, and dimensional stability

For a communications payload, the requirements come from the following key parameters: operating frequency, coverage area, power of the signal, and number of channels. These parameters are finalized by a series of trade-offs between the spacecraft and earth station capabilities. The trend in the past decade has been to simplify the earth stations so that customers could have them located at their premises, eliminating the need for the terrestrial back-haul system. This trend has resulted in the spacecraft having higher transmitted power, higher operating frequencies, and higher antenna pointing accuracies.

For a communications satellite, approximately 87% of the total electrical power is used by the communications payload. Therefore, an electric power subsystem is highly dependent on the requirements of the communication

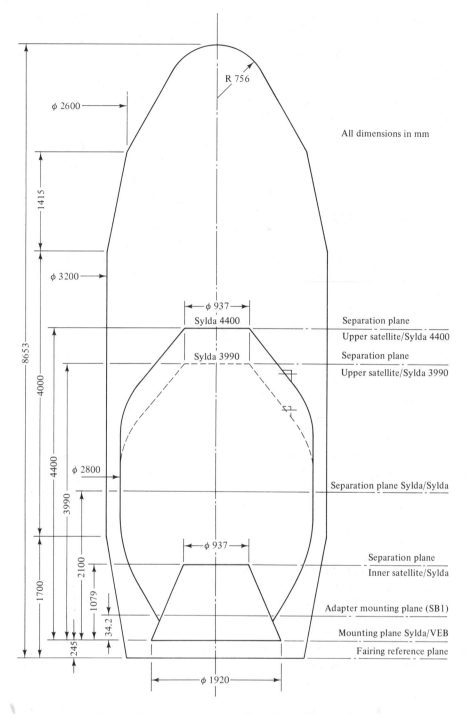

Figure 1.15 General configuration of SYLDA (Ref. 12).

payload. Direct-broadcast satellites provide very high signal power amplification in order that the signal can be received by antennas about 1 m in diameter, normally mounted on the roofs of houses. This results in the use of high-powered amplifiers, approximately 20 times higher than those used in conventional satellites, with the associated high electric power requirements. The requirements of communications payload operation during the eclipse period have a significant effect on the electric power subsystem. For the eclipse period, batteries must be carried to meet the communications equipment power demand when the solar array power is not available. A considerable mass saving can be achieved if the communication mission can accept a reduction in payload operation during eclipse, resulting in a much smaller battery size.

The efficiency of the power amplifiers in a communications satellite is approximately 30%. Approximately 70% of the electric power is therefore converted into heat, most of which has to be radiated into space by the thermal control subsystem. For direct-broadcast satellites, which use very high-powered amplifiers, thermal control is a difficult problem. For these satellites, active thermal control using heat pipes is commonly employed.

The communication mission determines the number and size of antennas. Antennas have a significant influence on spacecraft configuration. Launch vehicle shroud constraints require some of the antennas to be stowed during launch. Shadowing by antennas also influences the design of the thermal control subsystem and the solar array. The required coverage area and the communications link budget give the necessary antenna pointing accuracy. The antenna pointing error has contributions from attitude control errors, thermal distortion, spacecraft alignment errors, and orbit errors. The attitude control error limits are normally one-tenth of the antenna beamwidth, which is inversely proportional to the diameter of the reflector and the operating frequency. Therefore, the current trend for reflectors of larger diameters and the use of higher operating frequencies imposes requirements for tighter attitude control accuracies (low-attitude control errors). The requirement for north-south station keeping is also an important factor that influences spacecraft design. Marisat, the first maritime communications satellite, did not specify north-south station keeping for a lifetime of 7 years, resulting in a significant propellant mass saving.

Launch Vehicle

A launch vehicle imposes several constraints on the spacecraft design. The primary constraints are mass, size, electrical and mechanical interfaces, vibration, and thermal environment during launch. With the introduction of the Space Shuttle and liquid upper stages, the mass constraint is expected to be significantly relaxed.

Attitude Stabilization

The spacecraft configuration and subsystem designs are strongly influenced by the stabilization technique selected. There are two primary methods of spacecraft attitude stabilization: spin stabilization and three-axis stabilization.

In spin stabilization, gyroscopic stiffness is provided by all or part of the spacecraft body. There are two subsets of this class: single-spin stabilization and dual-spin stabilization. In single-spin stabilization, the entire spacecraft is spinning about an axis of the principal moment of inertia. Early communications satellites, such as Syncom and Intelsat I and II, were single-spin stabilized. Such stabilization is still generally used in the geosynchronous transfer orbit. The primary limitation of single-spin stabilization is that it cannot have earth-oriented antennas. This limitation is overcome in dual-spin stabilization, where the spacecraft bus rotates, providing gyroscopic stiffness, while the antennas, including the communications equipment in some cases, form a despun platform pointing toward the earth. The Intelsat III, IV, and VI satellite series uses this stabilization method.

Three-axis stabilization stabilizes the entire spacecraft, pointing toward the earth. There are several types of three-axis stabilization systems. The control torques along three axes can be provided by several combinations of momentum and reaction wheels and thrusters. Basically there are two main types of three-axis stabilization systems: a momentum wheel along the pitch axis and an all-reaction-wheel system. In the momentum wheel system, angular momentum along the pitch axis provides gyroscopic stiffness. The control torque along the pitch axis is provided by the change in the speed of the momentum wheel. The torque along the roll axis is provided by thrusters or a reaction wheel. Yaw error is controlled indirectly due to the interchange of yaw and roll error every quarter of the orbital period. In an all-reaction-wheel system, each axis has a reaction wheel and is controlled independently. Hence a yaw sensor is required for control of the yaw axis.

Figure 1.16 illustrates the main configuration differences between a dual-spin-stabilized spacecraft and a three-axis-stabilized spacecraft. A dual-spin-stabilized spacecraft consists of a spinning cylindrical drum covered with solar cells. The antennas are mounted on the despun platform with their boresights approximately normal to the spin axis. The despun platform is connected to the spinning section by a mechanical despin mechanism which provides the relative motion and transfers signal and power. In a three-axis-stabilized spacecraft, the entire spacecraft is fixed relative to the earth. It consists of two sun-tracking, deployed solar arrays which are stowed during launch. The antennas are mounted on the main spacecraft body with their boresights approximately parallel to the yaw axis.

Temperature control for a spin-stabilized spacecraft is much simpler than that for a three-axis-stabilized spacecraft. In a dual-spin spacecraft, the spinning solar array drum equalizes the solar radiation and provides a comfortable temperature, 10 to 25°C, for the internal equipment, except during eclipse, when the solar array drum temperature drops to as low as −85°C. The north face can radiate freely to space except for the obstruction

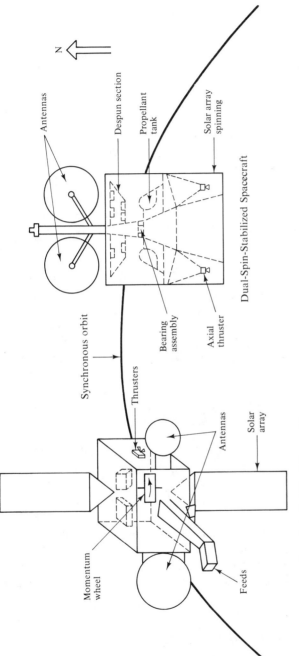

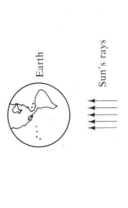

Figure 1.16 Spacecraft attitude stabilization.

caused by the antennas. The south face contains the apogee nozzle and is usually covered by a shield or multilayer insulating blanket. Thus, for Intelsat IV, almost all of the heat from the equipment is radiated by the north-face radiator. Recently, higher-powered spacecraft, such as SBS and Intelsat VI, use a radiator on the solar array drum.

For three-axis-stabilized spacecraft, the temperature control is independent of the solar array temperature, which varies between 60°C in sunlight and −160°C in eclipse. The spacecraft body rotates about the N-S axis at one revolution per sidereal day. Thus surfaces are normally covered with superinsulation blankets to avoid the extreme daily temperature variation. The north and south surfaces contain radiators because these surfaces are least affected by daily transient variations in solar incidence angle. The maximum seasonal solar incidence angle variation is ±23.5°.

A dual-spin-stabilized satellite has a high inherent gyroscopic stiffness, typically 2,000 N·m·s. The main disturbance torque is solar torque due to the offset between the solar pressure center and the spacecraft center of mass. Depending on the spacecraft configuration, the attitude is corrected every 3 to 60 days by ground command.

In a three-axis-stabilized spacecraft, attitude corrections are done more frequently. The angular momentum of the momentum wheel is significantly smaller than in dual-spin-stabilized spacecraft, 35 N·m·s for an Intelsat V momentum wheel versus 2,000 N·m·s for a typical dual-spin-stabilized spacecraft. The number of thrusters needed for a dual-spin-stabilized spacecraft are significantly lower than that for a three-axis-stabilized spacecraft. As an example, Intelsat IV, a dual-spin-stabilized spacecraft, has only six thrusters: two radial, two axial, and two spin-up thrusters. On the other hand, Intelsat V, a three-axis-stabilized satellite, has 20 thrusters. In addition, the duty cycle for thrusters on spin-stabilized spacecraft is well defined, normally by the spin rate of the spacecraft and the thruster firing angle. This is not the case for a three-axis-stabilized spacecraft.

One of the main limitations with spin-stabilized spacecraft is the available solar power due to both the geometric constraints of the launch vehicle on the diameter of the solar drum and to the requirement that the area should be approximately π times that of a sun-oriented solar array in a three-axis-stabilized spacecraft. A good figure of merit for a solar array is its power at the end of life divided by the total mass of the solar array, in W/kg. For a dual-spin-stabilized spacecraft, the typical figure of merit is in the range 9 to 11 W/kg, while an equivalent figure of merit for a three-axis-stabilized spacecraft is in the range 18 to 22 W/kg.

1.5 SPACECRAFT CONFIGURATIONS

This section provides a brief description of each subsystem of Intelsat V, a three-axis-stabilized spacecraft, and Intelsat VI, a dual-spin-stabilized spacecraft.

Intelsat V: Three-Axis Stabilization

Intelsat V, as shown in Fig. 1.17, is a three-axis-stabilized spacecraft. It weighs approximately 1,900 kg at launch and is designed to be compatible for launch by Atlas-Centaur, STS-PAM-A, or Ariane. Depending on the operational configuration, each satellite can carry up to 12,000 two-way telephone circuits and two color-TV transmissions. The propellant budget is summarized in Table 1.2 and the spacecraft mass breakdown is given in Table 1.3.

The communications subsystem provides an RF bandwidth capability of 2137 MHz. It is accomplished by extensive frequency reuse of 4 and 6 GHz by both spatial and polarization isolation and by introducing the 14/11-GHz bands to international traffic. Figure 7.14 shows the Intelsat V Atlantic Ocean antenna coverages. Spatial isolation is provided between the east and west 6/4- and 11/14-GHz beams. Polarization isolation is provided between the hemi and zone beams of the 6/4-GHz coverages by using right-circular polarization for the hemi beams and left circular polarization for the zone beams.

The spacecraft antennas consist of communications and telemetry/command antennas. The communications antennas consist of 4-GHz transmit and 6-GHz receive hemi/zone antennas, 14/11-GHz east and west spot antennas, the 4-GHz transmit, and the 6-GHz receive global antennas and 11-GHz beacon antennas. The hemi/zone transmit and receive antennas consist of single offset-fed parabolic reflectors of diameters 2.244 m and 1.54 m, respectively, illuminated by clusters of square feed horns. Each 14/11-GHz spot beam antenna consists of a nominal 1-m-diameter single offset-fed reflector illuminated by a conical corrugated feed horn. The two

TABLE 1.2 INTELSAT V PROPELLANT BUDGET

	Centaur Launch	
Maneuver	Magnitude	Fuel Weight (kg)
Transfer orbit		
Spin-up	45 r/min	1.5
Active nutation damping	10 min time constant	4.3
Reorientation	48°	1.2
Drift orbit		
AM dispersion correction, including coverage reorientation	65.8 m/s	28.5
Spin-down	45 r/min	1.4
Synchronous orbit		
N-S station keeping	347.5 m/s	106.0
E-W station keeping	29.0 m/s	11.7
Attitude maintenance	7 years	12.3
Residuals		2.0
Total fuel requirements without pressurant		168.9

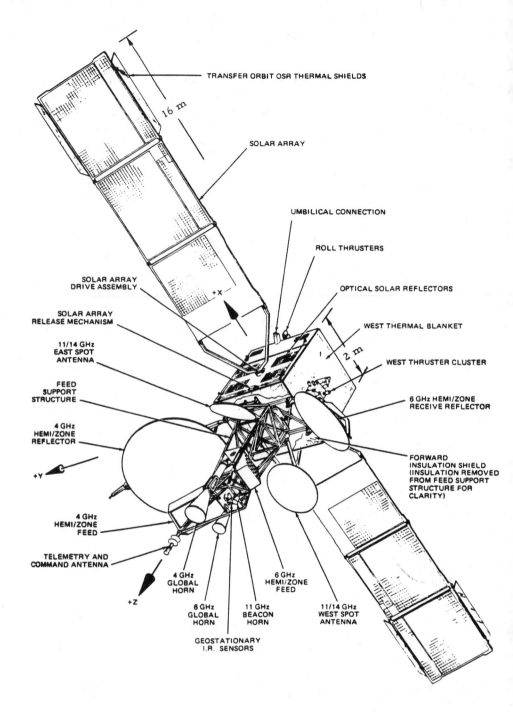

Figure 1.17 Intelsat V spacecraft configuration. (Courtesy of INTELSAT and FACC)

TABLE 1.3 SPACECRAFT MASS SUMMARY (KG)

Subsystem	Intelsat VI	Intelsat V	DOMSAT	TV-SAT
Antenna	309	59	32	57
Repeater	326	175	65	195
TT&C	80	28	27	29
Attitude control	70	73	23	52
Propulsion*	120	96	41	123
Electric power	330	142	124	249
Thermal control	52	26	18	85
Structures	280	157	90	178
Wire harness	99	40	19	55
Balance mass	23	15	3	5
Spacecraft dry	1689	~~750~~ 811	442	1028
Margin	90	24	9	131
Residual propellant	28		6	
End of life	1807	835	457	1159
Station keeping	420	173	90	7-year
BOL mass	2227	1008	547	propel-
Reorientation	10		23	lant
AMF propellant	1439	861	470	=1156
Separation mass	3676	1869	1040	2315

*Includes apogee motor inert mass.

(4 GHz transmit and 6 GHz receive) global antennas are circularly polarized conical horns. A switching network provides considerable interconnectivity between the various coverage areas as well as channel allocation between hemi, zone, and global coverages. The communications subsystem uses graphite/epoxy material extensively for antenna feeds, waveguides, contiguous output multiplexers, and input channel filters.

The communications subsystem consists of 27 independent transponder channels, of which 24 are at least 72 MHz wide. Three types of TWTAs are used. At 4 GHz, 4.5- and 8.5-W output TWTAs are used for zone and hemi/global use, respectively. At 11 GHz, 10-W-output-power TWTAs are used for the spot beam channels.

Intelsat V is spin stabilized during transfer orbit. The stabilization is provided by active nutation control using thrusters. In geosynchronous orbit, the spacecraft is three-axis stabilized. Pitch attitude is controlled by a fixed-momentum-wheel reaction torque. The roll attitude is controlled by thrusters. The yaw attitude is passively maintained because of roll–yaw coupling due to the angular momentum. The pitch and roll attitudes are determined by the infrared sensors. To reduce unbalanced solar pressure disturbance torque, a magnetic roll torque compensation scheme using a dipole aligned with the spacecraft yaw axis is included. This reduces the number of daily thruster firings around the roll axis from about 200 to about 17 during the solstices.

The station-keeping mode occurs during corrections for north-south or east-west station keeping. These corrections are implemented by firing thrusters in pairs. This also results in disturbance torques due to thrust misalignment and imbalance. In this mode, the momentum wheel is either operational or commanded to a preset speed. An active yaw control is also provided by using fine digital sun sensors and yaw thrusters. The expected pointing accuracy (3σ) is $\pm 0.12°$ in the roll and pitch axes and $\pm 0.33°$ in the yaw axis.

The Intelsat V propulsion system consists of a solid-propellant apogee motor and a combined catalytic/electrothermal monopropellant hydrazine subsystem. Two 22.2-N (5-lb$_f$) thrusters are used during geosynchronous transfer orbit for orientation correction and active nutation damping. Spacecraft spin/despin, east-west station keeping and pitch and yaw are performed by the 2.67N (0.6 lb$_f$) thrusters. These thrusters also serve as a backup to the 0.3-N (0.07 lb$_f$) electrothermal thrusters normally used for north-south station keeping, which, as shown in Table 1.2, require the largest portion of the propellant. Such thrusters can deliver an average specific impulse of about 304 s by heating hydrazine propellant to about 2,200°C prior to ejection. The mass saving of hydrazine over 7 years is on the order of 20 kg compared to the use of catalytic hydrazine thrusters, which deliver a specific impulse of about 230 s. The propellant is kept in two screen-type surface-tension propellant/pressurant tanks which are fed to two redundant sets of thrusters.

The electrical power system for Intelsat V is designed to accommodate a continuous spacecraft primary load of approximately 1 kW for a 7-year orbit lifetime. Primary power is provided by two sun-oriented solar array wings. During any period of insufficient solar array power, such as during eclipse, power is supplied by two nickel–cadmium batteries. Each solar array wing consists of a deployment mechanism, three rigid panels, and a solar array drive mechanism which provides a rotation of 1 revolution per day to keep the array pointed at the sun. The total panel area is 18.12 m^2 and is covered with 17,568 solar cells. During transfer orbit, the array is stowed in such a way that one outer panel per wing provides load support and battery charging.

Each battery consists of 28 series-connected 34-Ah nickel–cadmium cells. The allowable depth of discharge is less than 55%. The power control electronics consist of a power control unit (PCU) and shunt dissipators. The output of each solar array wing is independently regulated to 42 ± 0.5 V dc by use of a sequential linear partial shunt regulator.

The structural configuration, shown in Fig. 1.18, consists of three main elements: antenna module, communications module, and spacecraft bus. The spacecraft bus structure and communications module structure are U-shaped and when joined together form a rigid box of dimensions $1.65 \times 2.01 \times 1.77$ m. The box is additionally stiffened by a central tube which supports the apogee motor. Aluminum is the main material of these

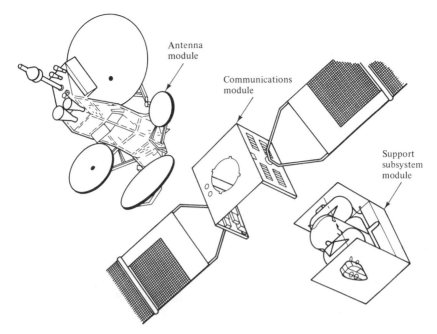

Figure 1.18 Intelsat V modular design. (Courtesy of INTELSAT and FACC)

structures, in contrast to the antenna module, which uses graphite/epoxy extensively in order to limit thermal distortion.

Thermal control of the Intelsat V spacecraft is accomplished by using conventional passive techniques. The passive thermal design is augmented with heater elements for components having relatively narrow allowable temperature ranges. High thermal dissipators, such as the TWTAs, are located on the north and south panels of the main body so that they may efficiently radiate their energy to space via heat sinks and optical solar reflector radiators. The east and west panels, antenna deck, and aft surfaces are covered with multilayer insulation to minimize the effect of solar heating on equipment temperature during a diurnal cycle. Thermal control for the antenna module tower is achieved by the use of a three-layer thermal shield. Thermal control of the antenna reflectors, positioners, feeds, and horns is obtained by using thermal coatings and insulation. Heater elements are used on the propellant tanks, lines, valves, apogee motors, and batteries to maintain temperature above the minimum allowable levels.

Intelsat VI: Dual-Spin Stabilization

Intelsat VI, as shown in Fig. 1.19, is a dual-spin-stabilized spacecraft. It is divided into spun and despun sections which are interconnected by a bearing and power transfer assembly. The spun section consists of a solar panel assembly, an aft thermal barrier, batteries, and a liquid bipropellant subsystem. The despun system contains mainly a communications subsystem.

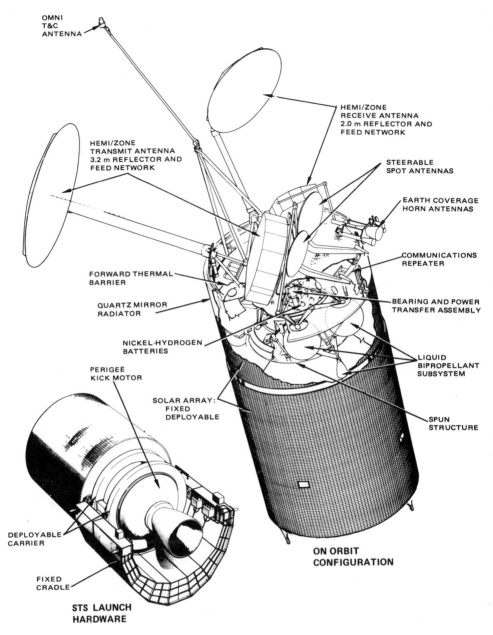

OMNI
T&C
ANTENNA

HEMI/ZONE
RECEIVE ANTENNA
2.0 m REFLECTOR AND
FEED NETWORK

HEMI/ZONE
TRANSMIT ANTENNA
3.2 m REFLECTOR AND
FEED NETWORK

STEERABLE
SPOT ANTENNAS

EARTH COVERAGE
HORN ANTENNAS

COMMUNICATIONS
REPEATER

FORWARD THERMAL
BARRIER

BEARING AND POWER
TRANSFER ASSEMBLY

QUARTZ MIRROR
RADIATOR

NICKEL-HYDROGEN
BATTERIES

LIQUID
BIPROPELLANT
SUBSYSTEM

PERIGEE
KICK MOTOR

SOLAR ARRAY:
FIXED
DEPLOYABLE

SPUN
STRUCTURE

DEPLOYABLE
CARRIER

ON ORBIT
CONFIGURATION

FIXED
CRADLE

STS LAUNCH
HARDWARE

Figure 1.19 Intelsat VI spacecraft configuration. (Courtesy of INTELSAT and HAC)

It weighs approximately 3,700 kg at launch and is designed to be compatible for launch by either the Space Shuttle or Ariane IV. Depending on the operational configuration, each satellite can carry up to 40,000 two-way telephone circuits and two color-TV transmissions. The total spacecraft mass breakdown is summarized in Table 1.3.

The communications subsystem is located in the despun section. The antenna coverage area for the Atlantic region is shown in Fig. 7.16. The antenna system consists of a 2-m-diameter receive (6 GHz) and a 3.2-m-diameter transmit (4 GHz) reflector; 1.12-m- and 1.0-m-diameter east and west spot beam steerable reflector antennas (14/11 GHz); and transmit and receive global horn antennas (6/4 GHz).

The repeater system consists of 50 distinct transponders operating over the 6/4-GHz and 14/11-GHz bands. The high-heat-dissipating units, traveling-wave-tube amplifiers, and solid-state power amplifiers are mounted on the outside surface of the rim shelf to provide close proximity to the mirrored radiator on the solar panel, for better heat transfer. The top of the circular shelf contains the output multiplexor and the beacon transmitters, and the bottom side contains the input multiplexers, switch matrices, receivers, and telemetry and command hardware.

Intelsat VI is spin stabilized in both the geosynchronous transfer orbit and geosynchronous orbit. The spin axis attitude is determined by rotor-mounted sun and earth sensors. Control of the spin axis attitude, the rotor spin rate, and the spacecraft orbit is accomplished by continuous or pulsed firing of selected spacecraft thrusters. Intelsat VI has an RF antenna tracker for initial alignment of the C-band hemi/zone antennas and precision closed-loop pointing of their beams. The expected nominal antenna pointing error, 3σ, is $\pm 0.12°$ in roll and pitch and $0.15°$ in yaw. By using an RF tracker, the roll and yaw errors of the C-band antennas (6/4 GHz) can be reduced to $\pm 0.05°$. The orbit tolerance is $\pm 0.10°$ in both longitude and latitude.

A liquid bipropellant system is used for both the apogee motor and on-orbit control. It uses nitrogen tetroxide (N_2O_4) as an oxidizer and monomethylhydrazine (MMH) as a fuel. The thruster system consists of four 22-N radial thrusters to provide east-west station keeping and spin-up/spin-down control. At the aft end of the spacecraft are two 22-N axial thrusters for north-south station keeping and attitude control, and two 490-N apogee thrusters for apogee boost and reorientation maneuvers. The eight propellant tanks and the thrusters are divided into two redundant groups which are arranged such that the closure of the latch valve in one group will allow completion of the mission by the other group.

The solar power is provided by a body-fixed solar panel and a cylindrical telescoping panel. The solar panel assembly is made of lightweight Kevlar honeycomb. On the fixed panel, K4-3/4 cells are used because of their lower solar absorptance, which results in lower temperatures for the bus subsystems. The telescopic panel uses K7 cells for higher power. The nickel–hydrogen batteries are mounted on a platform at the aft end of the spacecraft, permitting more efficient heat rejection. The power system is centrally preregulated by a dual-independent bus system. Under steady-state conditions the voltage is controlled between 28.0 and 30.0 V.

The Intelsat VI thermal control design is passive with radial heat rejection, augmented by heaters. The cylindrical radiator rejects approximately two-thirds of the total heat load. It consists of silvered-quartz mirrors bonded to a graphite substrate. Battery temperatures are maintained between -12 and $20°C$ during charge and nominally $20°C$ during overcharge. Heater elements are used on the batteries, liquid bipropellant subsystem, and antenna actuators to keep the temperatures above minimum allowable values.

The spun structure consists of a shear tube, an annular equipment shelf, and 12 struts to support the shelf. The propellant tanks are fastened to the spacecraft base ring for direct reaction of all axial load and 50% of lateral flight loads. Only 50% of the lateral loads affect the shear tube. The construction is all aluminum except for the beryllium struts. The despun structure is essentially a communications equipment platform which also supports the antenna system. The despun structure consists mainly of aluminum except for the graphite/epoxy antenna reflectors and their support structure, which are required for low thermal distortion.

1.6 MASS AND POWER ESTIMATION

The first step during a feasibility study for a mission is to identify the available launch vehicle for the mission. Next, the stabilization system of the spacecraft is selected and the mass and power requirements of the subsystem of the spacecraft bus, excluding the electric power subsystem, are estimated. The available mass for the payload and the electric power subsystem is then calculated by subtracting the estimated masses of subsystems from the total mass of the spacecraft, which is based on the payload capability of the launch vehicle. Trade-offs are then performed for the payload in terms of available mass versus available electric power to define the mission payload. This process is iterative. If the available mass/power ratio for the payload is inadequate, the launch vehicle, spacecraft configuration, and masses of the subsystems must be reevaluated. Therefore, an accurate prediction of masses and power requirements of the subsystems is very important during feasibility study of a spacecraft.

The mass of a spacecraft is significantly influenced by its functional, reliability, and cost requirements. The challenge to a spacecraft designer is to select the spacecraft configuration, technologies, and equipment to meet the spacecraft functional requirements with high reliability and low cost. The individual masses of the spacecraft subsystems will depend on the selected spacecraft technologies and redundancies in the equipment. This section provides a review of the available technologies for spacecraft subsystems and their influence on spacecraft mass.

Propulsion

A propulsion system normally consists of an apogee motor and a reaction control system that provides thrust for orbit corrections and attitude control. The propellant mass can be calculated by using the following equation:

$$M_p = M_i (1 - e^{-\Delta V / Ig}) \quad /e \, ft \tag{1.1}$$

where M_p = mass of the propellant, kg
      M_i = initial mass of the spacecraft, kg
      ΔV = required velocity change, m/s
      I = specific impulse of the propellant, s
      g = gravitational accelerations, m/s^2 ($9.81\ m/s^2$)

The performance of a thruster is rated by its specific impulse. The higher the specific impulse, the lower will be the propellant mass requirements for a given spacecraft mass and velocity change. The specific impulses for different propulsion systems are given in Table 3.2. A spacecraft designer normally has to make a choice between a solid motor and a bipropellant thruster for an apogee motor; between monopropellant, electrothermal, bipropellant, and ion thrusters for orbit corrections; and between monopropellant and bipropellant thrusters for attitude control. In the 1970s, the propellant systems generally consisted of solid motors for apogee injection and monopropellant hydrazine thrusters for orbit corrections/attitude control. In the 1980s, the trend is to use a unified bipropellant propulsion system for both apogee injection and orbit corrections/attitude control. The latter system not only provides a higher specific impulse, but also flexibility in the use of propellant between apogee injection and orbit corrections. Ion thrusters, although having a very high specific impulse, have not yet been used on commercial spacecraft because of reliability and high electric power requirement considerations.

The required changes in the velocity (ΔV) of a spacecraft for orbital maneuvers are determined after selecting the mission sequence. However, as a first approximation, apogee injection requirements for STS launched and Ariane launched spacecraft can be assumed to be 1.819 km/s and 1.514 km/s, respectively. The ΔV requirements for inclination and longitude station-keeping are given in Tables 2.4 and 2.5, respectively. The ΔV requirements for other orbit maneuvers and attitude control can be extrapolated from similar existing spacecraft designs. Knowing the spacecraft mass, ΔV requirements, and specific impulses of the thrusters, the total propellant mass can be calculated from Eq. (1.1).

The dry mass of a propulsion system is determined after the selection of tanks and thrusters. However, as a first approximation, the dry mass of the propulsion system can be determined by using the following empirical equation.

The dry mass of the apogee motor, M_{AI}, is given by:

RCS

$$M_{AI} = 0.07\ M_p \text{ (solid propellant)} \tag{1.2}$$

$$= 0.1\ M_p \text{ (bipropellant)}$$

where M_p is the mass of the propellant for the apogee motor. The dry mass for the hydrazine reaction control system, for orbit corrections and attitude control, can be estimated from the following empirical equations:

Three-axis-stabilization:

$$M_{RC} = (0.01 + 0.0115\ \sqrt{Y})\ M_{SC} \tag{1.3}$$

Dual-spin-stabilization:

$$M_{RC} = (0.006 + 0.007 \sqrt{Y})M_{SC} \qquad (1.4)$$

where Y = spacecraft life, years (between 5 and 10 years)
M_{RC} = dry mass of the reaction control system, kg
M_{SC} = beginning of life spacecraft mass, kg

The dry mass for a unified bipropellant propulsion system, which combines the functions of apogee injection, orbit corrections, and attitude control, can be estimated by the following equation:

$$M_{UB} = C_{UB}M_{PR} \qquad (1.5)$$

where M_{UB} = dry mass of the unified propulsion system, kg
C_{UB} = 0.084 for three-axis-stabilization
    = 0.054 for dual-spin stabilization
M_{PR} = total mass of the propellant, including apogee injection
    requirements, kg
    $= M_p + M_{RC}$
By combining the propellant mass and the dry mass, the total propulsion subsystem mass is calculated.

Electric Power

The mass of an electric power subsystem depends on several factors: electric power requirements of the spacecraft, selected solar cells, array structural design, batteries and their depth of discharge, power regulation system, and attitude stabilization.

A solar array consists of three parts: solar cells, array structure, and cover glasses. There are only two types of solar cells that can be considered for space use, namely silicon cells and gallium arsenide solar cells. Gallium arsenide solar cells have slightly higher efficiency than that of silicon cells. However, due to their higher cost and non-availability in production quantities, these cells have not yet been used on operational spacecraft. The silicon cells have been used exclusively in commercial spacecraft and are expected to continue to be used in the near future. The trend in the silicon solar cell technology is primarily directed toward decreasing the cell thickness from the current 200-300 μm to 50-100 μm, while improving efficiency and reducing radiation damage. In the 1970s, as shown in Fig. 6.36, the efficiency of silicon cells has improved by approximately 50 percent. The solar array mass structure constitutes a major part of solar array mass. A variety of design concepts, shown in Fig. 6.14, have been developed for solar array structures that exhibit lower mass by using light-weight advanced materials, high deployment reliability, and higher natural frequencies. The specific performance of a solar array is expressed in watts/kg, which is obtained by dividing the total power delivered by the solar array by its total mass. Since the solar array power output, as shown in Fig. 6.15, varies as a function of time in orbit, the specific performance is normally specified at the end of life (EOL) at summer solstice. The power output as a function of solar array mass is given in Fig. 6.37 for different solar array designs. The specific performances of sun tracking solar arrays used on three-axis-stabilized spacecraft are significantly higher than those solar arrays of spin-stabilized spacecraft.

Batteries, which provide power during eclipse, constitute approximately 40 percent of the electrical power subsystem mass for a spacecraft that requires the communication subsystem to be fully operational during eclipse. Since the eclipse period is a maximum of 1.2 hours' duration near local midnight, some spacecraft are designed to have only a fraction of communications subsystem on during the eclipse. The specific performance of a battery is expressed as watt-hr/kg, the total energy delivered by the battery divided by its mass. The specific performance for a battery will also depend on its allowable depth of discharge. Higher depth of discharge will increase specific performance but will reduce the battery life. There are only two types of batteries that can be considered at present for geo-synchronous spacecraft use, namely Ni-Cd and Ni-H$_2$ batteries. In the 1960s and 1970s, Ni-Cd batteries were exclusively used for communications satellites. The current trend is to use Ni-H$_2$ batteries because they allow a higher depth of discharge and a longer life, resulting in higher specific performance. The masses of Ni-Cd and Ni-H$_2$ batteries as a function of power load at the battery terminal during a 1.2-hr eclipse are given in Figs. 6.38 and 6.39, respectively. Knowing the power requirements during an eclipse, these figures can be used to estimate the mass of the battery.

The choice of the bus voltage level and range has a major effect on the design of a power regulation system and its mass. The regulated bus generally results in a higher power regulation system mass of the EPS subsystem and lower power regulation mass for the load equipment in comparison with an unregulated bus. Therefore, a trade-off between different power regulation methods should be based on the total power regulation mass.

The mass of an electric power subsystem (EPS) can be calculated only after performing a preliminary design, as discussed in Section 6.5. However, as a first approximation, EPS mass can be calculated by using specific masses of EPS subsystem equipment from Table 6.5.

Structure

The mass of a spacecraft structure is highly dependent on the spacecraft configuration. For a spacecraft with large deployable reflectors, the reflector support mass could be quite significant. Similarly, for a spacecraft with high power dissipation equipment, such as TWTAs, the thicknesses of the face skins of the honeycomb equipment panels may be increased and heat sinks/thermal doublers added to conduct the heat from the equipment to the radiator. The estimation of the spacecraft structure mass is complicated because there is no general agreement among spacecraft designers on whether the masses of reflector support structures, heat sinks, thermal doublers, and deployment mechanisms should be part of the mass of the spacecraft structure or the other subsystems. The conventional material for spacecraft structures is aluminum with the exception of antenna support structures that use graphite/epoxy because of low thermal distortion requirements. Studies have shown that a significant mass saving, approximately 30 percent, can be achieved by using advanced materials, such as beryllium, boron/ aluminum, and graphite/epoxy, instead of aluminum for primary spacecraft structures. The advanced materials, however, result in additional material and manufacturing cost. The specific performance of a spacecraft structure can be defined as the ratio between the spacecraft structure mass and the spacecraft mass at launch. The mass of a spacecraft structure can be accurately estimated only after finalization of the spacecraft configuration and then only by performing preliminary structural design. However, as a first approximation, the spacecraft structure mass can be estimated by the following empirical equation:

$$M_{ST} = C_{ST} \times M_{SP} \qquad (1.6)$$

where M_{ST} = structure mass, kg
$\quad\ C_{ST}$ = 0.087 for three-axis-stabilization
$\qquad\ \ $ = 0.097 for spin-stabilization
$\quad\ M_{SP}$ = spacecraft separation mass, i.e., spacecraft mass that the structure has to support during launch, kg

Thermal Control

The mass of a thermal control subsystem is highly dependent on the desired temperature limits of the spacecraft equipment. The thermal control design is passively achieved by the use of proper thermal coatings and insulation at the exterior surfaces of the spacecraft. Some equipment, such as Visible and Infrared Spin Scan Radiometer (VISSR) in a meteorological satellite, will require a special radiator cooler to keep the operating temperatures of infrared detectors at about 95 K ($-178°C$). Batteries, the temperature of which is kept within a very narrow range because of life considerations, are normally provided with heaters to control the lower limits of the temperature range. Propulsion systems are also provided with heaters to keep the temperature above the freezing temperature of the hydrazine. High thermal dissipation units, such as TWTAs, require heat sinks/doublers to spread the heat over the equipment panel and conduct it to the radiator. For very high thermal dissipation units, such as TWTAs for direct broadcast satellites, the mass of the required heat sinks and doublers, will be normally unacceptable. Therefore, for such applications, a combination of heat pipes, which provide high thermal conductivity in the panel, and louvers, which vary the effective radiator area of the spacecraft, will be generally more mass efficient. In general, the thermal control mass for a spacecraft with high thermal dissipation units and a smaller radiator area will be higher. As given in Table 1.3, the thermal control mass of TV-SAT that has high power TWTAs (230-260W) is approximately three times that of Intelsat V that has relatively low power TWTAs (10W). Thermal control mass also depends on the type of attitude stabilization. Thermal control design for spin-stabilized spacecraft is generally simpler than that for a three-axis-stabilized spacecraft. The thermal control mass of a spacecraft can be accurately estimated only after knowing the desired temperature limits of the spacecraft equipment and performing a trade-off study between different thermal control designs, such as heat sinks vs. heat pipes/louvers. However, as a first approximation, the thermal control mass can be estimated from the following empirical equation:

$$M_T = C_T \times W_D \tag{1.7}$$

where M_T = mass of thermal control, kg
C_T = 0.04 for three-axis stabilization
= 0.031 for dual-spin-stabilization
W_D = equipment power dissipation, W

If the equipment power dissipation value is not known, the thermal control mass can be estimated by the following empirical equation

$$M_T = C_{MT} \cdot M_{SC} \tag{1.8}$$

where C_{MT} = 0.032 for three-axis-stabilization
= 0.027 for dual-spin-stabilization

Attitude Control Subsystem

The mass of an attitude control subsystem is a function of the type of attitude stabilization, required attitude control accuracy, redundancies in actuators and sensors, and size and mass of the spacecraft. The complexity and mass of an attitude control subsystem increases with the increase in attitude control requirements and the size of the spacecraft. The mass of an attitude control subsystem can be divided into two categories. First, those that are almost independent of spacecraft mass, such as sensors and attitude control electronics and second, those that are highly dependent on spacecraft mass and size, such as momentum wheels and solar array drives. Some spacecraft designers consider solar array drives as part of attitude control subsystem while others consider it as part of electric power subsystems. The mass of an attitude control system can be calculated only after the selection of sensors and actuators. However, as a first approximation, the attitude control subsystem mass can be estimated by using the following empirical equations. For a three-axis-stabilized spacecraft, the attitude control system mass is given by:

$$M_{AC} = 65 + 0.022 \, (M_{SC} - 700) \tag{1.9}$$

In the above equation, the solar array drive mass is included in the attitude control system mass. For dual-spin stabilization, the attitude control system mass is given by:

$$M_{AC} = 31 + 0.027 \, (M_{SC} - 700) \tag{1.10}$$

where M_{AC} is the mass of the attitude control system.

The spacecraft mass summary for four geosynchronous spacecraft is given in Table 1.3. Intelsat V and TV-SAT are three-axis stabilized spacecraft, while Intelsat VI and DOMSAT are dual-spin stabilized spacecraft. It should be noted that the masses given in Table 1.3 are based on preliminary information. Actual spacecraft masses may be slightly different. The empirical formulas given in this section are based on 10 geosynchronous spacecraft designs. It is, however, more accurate to extrapolate subsystem masses from those of a spacecraft with similar functional and reliability requirements, if the mass data for such spacecraft are available. A procedure for preliminary estimation of the spacecraft masses of subsystems by extrapolation is demonstrated in the following example.

Example 1.1

Given

1. Launch vehicle: Ariane IV, mass at separation = 3440 kg
2. Three-axis stabilization
3. Solid apogee motor, $\Delta V = 1487$ m/s
4. Spacecraft design similar to that for Intelsat V
5. Life: 7 years

Determine the available mass and power for the communication payload.

Solution   The masses of the spacecraft subsystems are calculated by performing a preliminary analysis using empirical formulas and extrapolation from Intelsat V. For maneuvers where the required changes in the velocity and specific impulses of the propellant are known, the propellant mass is calculated by using Eq. (1.1). For apogee motor firing (AMF), $M_i = 3425$ kg (Table 1.4), $\Delta V = 1,487$ m/s, and $I = 285$ s. Substituting these parameters into Eq. (1.1), the propellant mass for AMF is 1,413 kg. Similarly, for station-keeping maneuvers, station repositioning, and deorbiting, the propellant masses are calculated and are given in Table 1.4. For maneuvers for which the required changes in the velocity are not accurately known, such as before AMF, post AMF, and attitude control, the propellant masses are extrapolated from Intelsat V. Table 1.4 gives a detailed breakdown of the propellant budget. From the table, the spacecraft mass at the beginning of life, M_{SC}, is 1,964 kg.

Propulsion subsystem. The dry mass of the propulsion subsystem is calculated from Eq. (1.3).

For the present example, with a catalytic hydrazine thruster, $Y = 7$, and $M_{SC} = 1,964$. Substituting these values in Eq. (1.3) gives us

$$M_{RC} = (0.01 + 0.0115 \sqrt{7}) \cdot 1,964$$

$$\approx 80 \text{ kg}$$

TABLE 1.4 PROPELLANT BUDGET

Maneuver	ΔV (m/s)	Specific Impulse (s)	Mass Change (kg)	Final Mass (kg)	Efficiency
Separation				3440	
Before AMF (spin-up, reorientation, and nutation damping)			15	3425	
AMF	1487	285	1413	2012	1.0
Post AMF (dispersion correction, spindown, station repositioning)			48	1964	
N-S station keeping	351	220	321.4	1642.6	0.91
E-W station keeping	13	220	10.0	1632.6	0.99
Station-repositioning	30	220	22.8	1609.8	0.99
Attitude control			19	1590.8	
Deorbit	7	220	5.2	1585.6	0.99
Pressurant			6.0	1579.6	
Margin propellant			8.6	1571	
			1869.0		

Apogee motor inert mass. The apogee motor inert mass is calculated by using Eq. (1.2).

For the present apogee motor,

$$M_p = 1413 \text{ kg}$$

and

$$M_{AI} = 0.07 \times 1413 \approx 99 \text{ kg}$$

Attitude control subsystem. It is assumed that the attitude control system is similar to that of Intelsat V, whose total attitude control subsystem mass is 73 kg. Momentum wheels are sized for steady-state yaw error. Assuming that the disturbance torque is twice that for Intelsat V and that the pointing accuracy is improved by a factor of 2, the required angular momentum will be four times that of Intelsat V. This will result in the momentum wheel mass being twice that for Intelsat V (i.e., 40 kg). It is also assumed that the solar array is approximately twice that for Intelsat V. The mass of the solar array drive is increased by a factor of 1.4 (i.e., 16.8 kg). The mass of the attitude control system, which is independent of spacecraft mass, is 41 kg. Hence the total attitude control subsystem mass is 98 kg.

Telemetry and command. The telemetry and command subsystem is assumed to be similar to that for Intelsat V. However, due to the increase in complexity of the communications payload, the Intelsat V mass is multiplied by a factor of 1.4. So the telemetry and command subsystem mass is $28 \times 1.4 = 39$ kg.

Structure. The mass of the structure subsystem will be significantly dependent on the spacecraft configuration, in particular the antenna configuration. However, as a first approximation, it can be extrapolated from Intelsat V. The coefficient C_{ST} in Eq. (1.6) for Intelsat V is 0.084. The spacecraft separation mass is 3,440 kg. Substituting these values in Eq. (1.6), the structural mass is 289 kg.

Thermal control. Thermal control mass is highly dependent on the communication payload heat dissipation. For the current spacecraft, thermal control mass is extrapolated from Intelsat V. For Intelsat V, the values of the coefficient C_{MT} in Eq. (1.8) is 0.0275. The beginning of life mass for the spacecraft is 1,964 kg. Substituting these values in Eq. (1.8), the thermal control mass of the spacecraft is 54 kg.

Electrical/mechanical integration. The electrical and mechanical integration masses are extrapolated from Intelsat V as follows by slightly reducing the coefficient values because of increased spacecraft mass:

$$M_E = 0.039 \times M_{SC}$$
$$= 0.039 \times 1,964 = 77 \text{ kg}$$
$$M_M = 0.014 \times M_{SC}$$
$$= 0.014 \times 1,964 \approx 28$$

Mass margin. The mass margin is assumed to be approximately 10% of the spacecraft dry mass. For this example, it is assumed to be 157 kg.

Electric power. The mass of the electrical subsystem will depend on the power requirements of the spacecraft. In the calculation, the following assumptions are made:

1. 10% margin for solar array
2. 5% margin for the equipment
3. Ni-H$_2$ batteries with 70% depth of discharge and spacecraft fully operational during eclipse periods.
4. Design similar to that for Intelsat V

The power for the spacecraft bus is extrapolated from Intelsat V by multiplying its housekeeping power, 211 W, by the factor 1.9 which is a ratio between the masses of the spacecraft and Intelsat V. Therefore,

$$\text{housekeeping power} = 1.9 \times 211 \cong 400 \text{ W}$$

Let P be the power required by the communications payload and the total power required be $P + 400$. Applying the assumed margins, the loads for the battery and the solar array are:

$$\text{battery load} = (P + 400)1.05$$
$$\text{solar array load} = (P + 400)1.05 \times 1.1$$

The mass of the electrical power subsystem can be estimated by using Table 6.5.

It is assumed that, similar to Intelsat V, the bus is partially regulated, i.e., the solar array voltage during sunlight is regulated by a shunt regulator and the battery voltage during eclipse period is unregulated. Assuming the power is in 2.5 kw range, the specific masses for the solar array, charge array, shunt, charge control, battery, and discharge regulator from Table 6.5 are 42, 6.6, 7.5, 1.5, 47.3, and 0.2 grams/watt, respectively. The total mass of the electric power subsystem, M_{EL}, is given by

$$M_{EL} = (P + 400) \times 1.05 \, (1.1 \times 42 + 1.1 \times 6.6 + 7.5 + 1.5$$
$$+ 47.3 + 0.2) \times 0.001$$
$$= (P + 400) \times 0.1155$$

The estimated masses of the spacecraft subsystems are given in Table 1.5. The allocated mass for the communications and electric power subsystems is 650 kg. Let the communication mass be M_c; then

$$M_c = 650 - (P + 400) \times 0.1155$$

The equation above provides a trade-off between the available mass versus power for the communication payload. As an example, for $P = 2,500$ W, the mass of the payload, M_c, is 315 kg and the mass of the electric power subsystem, M_{EL}, is 335 kg.

TABLE 1.5 SPACECRAFT MASS SUMMARY

Subsystem	Mass (kg)
Communications	
Antennas	650
Electric power	
Structure	289
Thermal	54
Propulsion	80
Telemetry and command	39
Attitude control	98
Electric integration	77
Mechanical integration	28
Apogee motor inert	99
Mass margin	157
Dry spacecraft mass	1571
Propellant/pressurant	456
Apogee motor expendable	1413
Spacecraft mass at separation	3440

1.7 COMMUNICATIONS SATELLITES TRENDS

The evolution of communications satellites over the past years in terms of mass and power is shown in Figs. 1.20 and 1.21, respectively. These are two categories of satellites, Intelsat for international and Domsat for domestic communications. The Intelsat requirements to handle high-capacity trunkline traffic from a limited number of earth stations over a widespread area has increased the spacecraft complexity in each successive generation. As the payload capability of the launch vehicle has increased, communications satellites have increased in mass to make full use of the existing launch

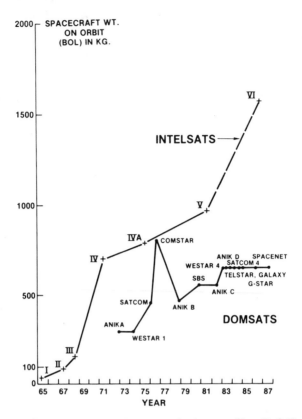

Figure 1.20 Mass trends in communications satellites (Ref. 7).

capability. In domestic satellites, the masses of the satellites have followed the payload capability of the Delta launch vehicle and have leveled off in the early 1980s. The power level, however, has kept on increasing, particularly for direct-broadcast satellites. There is also a trend toward longer lifetime; for example, from 1.5 years for Intelsat I to 10 years for Intelsat VI.

In the past, satellite masses have kept pace with launch vehicle capability. However, with the availability of the Space Shuttle, it is unlikely that the maximum capacity will be called for in the near term, except for special cases. Communications capacity demand will continue to increase. This, combined with the trend toward longer operational lifetimes and more and more satellites being launched, would lead to a serious physical crowding of the geosynchronous orbit, interference problems, and to saturating the capacity limit imposed by the available bandwidth. To avoid this problem, a number of space segment concepts are being developed, such as multi-missions, clusters, and platforms.

One approach to reducing crowding of the geosynchronous orbit by small dedicated satellites is to combine the payloads of a number of missions into a large multimission satellite. With the availability of the Space Shuttle and Ariane, launch mass is not the limitation. Several satellites have already

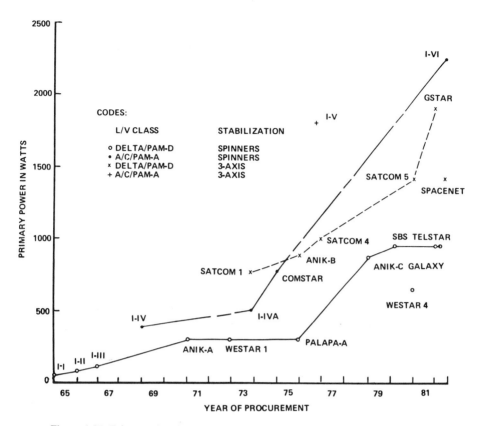

Figure 1.21 Primary electric power trends in communications satellites (Ref. 7).

taken this approach, including Insat-1, which has combined communications and meteorological payload capabilities. Another approach is to cluster the satellites at one longitude, provided that the frequencies of operation are compatible. The satellites can be controlled independently or their relative positions can be controlled. In the latter case, traffic may be exchanged between the satellites via intersatellite communications links. Another advantage of the cluster system is that the entire space segment does not have to be implemented at the same time. The system could be expanded as the traffic grows. A third approach is to use a geostationary platform, consisting of several elements which are assembled only after they have been separately put into space. It allows the same interconnectivity of traffic as for clusters but without requiring numerous intersatellite communication links, and it avoids the complexity of relative position control.

REFERENCES

1. J. N. Pelton, M. Perras, and A. Sinha, "Intelsat, The Global Telecommunications Network," Pacific Telecommunications Conference, Honolulu, Hawaii, Jan. 16–19, 1983.

2. R. J. Rush, J. T. Johnson, and W. Baer, "Intelsat V Spacecraft Design Summary," AIAA 7th Communications Satellite Conference, San Diego, Calif., Apr. 23, 1978.

3. S. B. Bennett and D. J. Braverman, "Intelsat VI Technology," International Astronautical Congress, 1982.

4. J. Schubert, D. E. Koelle, and H. Kellermeir, "The German TV-Broadcasting Satellite," XXX International Astronautical Congress, Munich, Germany, 1979.

5. R. L. Sackheim, D. E. Fritz, and H. Macklis, "The Next Generation of Spacecraft Propulsion," AIAA/SAE/ASME 15th Joint Propulsion Conference, Las Vegas, Nev., June 18–20, 1979.

6. D. H. Martin, "Communications Satellites 1958 to 1982," The Aerospace Corporation Report SAMSO-TR-79-078, September 1979.

7. R. Lovell and S. Fordyes, "A Figure of Merit for Competing Communications Satellite Designs," *Space Communications and Broadcasting,* Vol. 1, No. 1, Apr. 1983.

8. E. W. Ashford, "Future Configurations of Communications and Broadcast Satellites," *Space Communications and Broadcasting,* Vol. 1, No. 1, Apr. 1983.

9. Hughes Aircraft Company Geosynchronous Spacecraft Case Histories, January 1981.

10. Space Transportation System User Handbook, NASA, June 1977.

11. PAM-D User's Requirement Document, McDonnell Douglas Astronautics Company, MDC G6626F, July 1985.

12. Ariane User's Manual, *European Space Agency,* Issue 1980–Rev. 0.

2

ORBIT DYNAMICS

2.1 INTRODUCTION

A spacecraft will perform its mission only after it acquires the operational orbit and maintains it within allowable tolerances. A typical mission sequence for a geosynchronous satellite launched by the Space Shuttle is shown in Fig. 2.1. The sequence starts with the launch of the Shuttle from Cape Canaveral at a time compatible with the launch-window constraints. The Shuttle achieves a circular orbit with the altitude of about 296-km at an inclination angle of approximately 28.7°. After the spacecraft checkout, the orbiter is oriented to the proper attitude for the perigee motor firing. The spacecraft, together with the perigee motor, is separated from the orbiter approximately 45 minutes before the perigee motor firing. The perigee motor is fired at the equatorial crossing. This puts the satellite into a transfer orbit whose typical parameters are: perigee altitude of 296 km, apogee altitude of 35,786 km, and an inclination of 28.7°. Once the spacecraft is acquired at the ground station, the perigee motor case is ejected by a ground command.

To prepare for the proper attitude for apogee motor firing, the spacecraft attitude and orbit are determined by ground stations. Typically, four orbit periods in the transfer orbit provide ample time for several orbit and attitude determinations and to reorient the spacecraft from the injection attitude to the apogee motor firing (AMF) attitude. The apogee motor firing places the satellite into an almost geosynchronous orbit. For spin-stabilized spacecraft, the satellite spin axis is reoriented and antennas and solar arrays are deployed. The spacecraft drifts toward its final position in about 1 to 2 weeks, at which time station-acquisition maneuvers will be performed to circularize and synchronize the final orbit. Following station acquisition, the spacecraft is ready for operational use. The spacecraft orbit is perturbed by gravitational

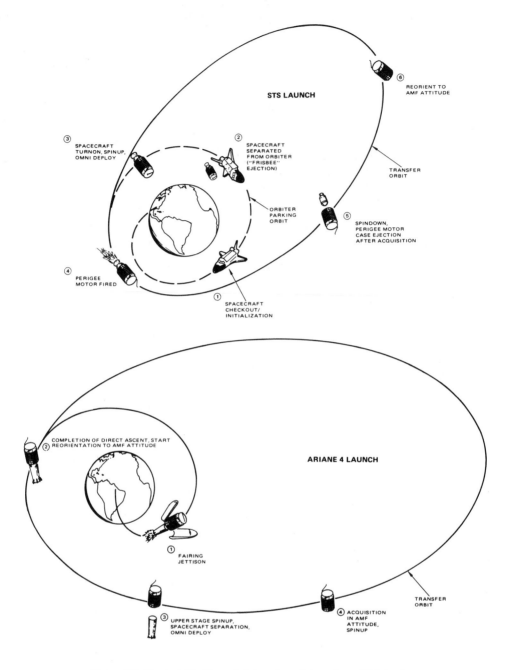

Figure 2.1 Typical Shuttle and Ariane launch transfer orbit injection.

effects of the sun and the moon, and the oblateness of the earth. To keep the orbital elements within allowable limits, orbital correction maneuvers are performed throughout the life of the spacecraft.

The orbital maneuvers influence the design of some of the spacecraft subsystems. In the propulsion subsystem, the propellant mass is determined on the basis of required changes in the velocity for the orbital maneuvers, as discussed earlier. To design these subsystems for meeting orbital requirements, a basic knowledge of orbital dynamics is necessary.

This chapter presents the basic principles of orbital dynamics. The section on the orbit equation discusses Kepler's laws, presents a derivation of the general orbit equation, and defines the orbital elements for an elliptic orbit. The ideal synchronous orbit is defined in the next section. The effects of inclination, incorrect radius, and eccentricity are also discussed. The section on orbit perturbation covers the perturbation of orbit parameters by the use of the variation-of-parameter method, and it discusses perturbation forces, orbital motion of the earth and moon around the sun, perturbation of orbital inclination due to the sun and moon and the oblateness of the earth, as well as longitudinal perturbations. The section on station keeping covers both inclination (north-south) and longitudinal (east-west) station keeping and station repositioning. In the final section, the velocity requirements for launching a satellite from Cape Canaveral are calculated and the launch windows for expendable-launch-vehicle and Shuttle-launched spacecraft are discussed. Solar eclipses, azimuth, and elevation of a satellite are also discussed.

2.2 ORBIT EQUATIONS

Newton's three laws of motion and law of gravitation form the foundation of orbital dynamics. The first law of motion states that in the absence of forces a body remains at rest or continues in uniform rectilinear motion. Newton's second law of motion states that the force acting on a mass is equal to the time rate change of its linear momentum, where the momentum is measured relative to an inertial frame. The third law states that the force exerted by a mass m_i on a mass m_j acts along the radius vector from m_i to m_j and is equal in magnitude but opposite in direction to the force exerted by m_j on m_i. The law of gravitation asserts that the force of attraction between two point masses is proportional to the product of the masses and inversely proportional to the square of the distance between them. For a satellite orbiting the earth, the gravitational force is

$$F = \frac{GMm}{r^2} = \frac{\mu_e m}{r^2} \tag{2.1}$$

where F = gravitational force, N
M = mass of the earth, kg
G = gravitational constant, $km^3\ kg^{-1}/s^2$
r = distance of the satellite from the center of mass of the earth, km

$$\mu_e = GM = 398{,}601.2 \text{ km}^3/\text{s}^2$$

m = mass of the satellite, kg

Motions of Planets and Satellites

The motions of planets and satellites are governed by Newton's second law in conjunction with Newton's gravitational law. Newton formulated his laws of motion on the basis of Kepler's three laws of planetary motion; Kepler, in turn, deduced his laws from observations of the motion of planets made by Tycho Brahe. Kepler's laws can be stated as follows:

First Law. The orbit of a planet around the sun is an ellipse, with the sun at a focus.

Second Law. The radius vector from the sun to a planet sweeps equal areas in equal time intervals.

Third Law. The squares of the orbit periods of the planets are proportional to the cubes of the semimajor axes of the ellipses.

Kepler's laws were deduced on purely geometric grounds and are only approximately correct. They describe the motion of a single particle about a fixed center. Using Newton's laws, one can derive similar laws for the motion of two particles about their barycenter. In the case of a system consisting of the sun and one of the planets, the mass of the sun is much larger than the mass of the planet, so that for all practical purposes the barycenter coincides with the center of the sun. Later in this section the nature of the approximation involved in Kepler's laws is discussed.

Although Kepler's laws were formulated for the motion of planets orbiting the sun, they can be used for the motion of any two-body system, such as the earth and the moon, the earth and an artificial satellite, the moon and an artificial satellite, and so on. Our interest is in the orbital motion of a satellite around the earth. Following is a derivation of Kepler's laws.

Newton's second law, using polar coordinates (Fig. 2.2) in conjunction with Newton's gravitational law, yields the following differential equations of motion (Ref. 1):

$$\ddot{r} - r\dot{\theta}^2 = -\frac{\mu}{r^2} \tag{2.2a}$$

$$r\ddot{\theta} + 2\dot{r}\dot{\theta} = \frac{1}{r}\frac{d}{dt}(r^2\dot{\theta}) = 0 \tag{2.2b}$$

Equation (2.2b) leads to the principle of conservation of angular momentum,

$$r^2\dot{\theta} = h = \text{const.} \tag{2.3}$$

where h can be identified as the angular momentum per unit mass, also known as the specific angular momentum.

The orbit equation represents a relation between r and θ, with the

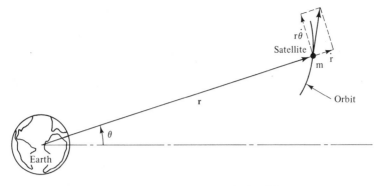

Figure 2.2 Satellite under gravitational force.

time t appearing only implicitly. To eliminate the time dependence from the equations of motion, we use Eq. (2.3) and consider

$$\dot{r} = \frac{dr}{dt} = \frac{dr}{d\theta}\frac{d\theta}{dt} = \frac{h}{r^2}\frac{dr}{d\theta} = -h\frac{d}{d\theta}\left(\frac{1}{r}\right) = -h\frac{du}{d\theta} \tag{2.4}$$

where $u = 1/r$. Similarly,

$$\ddot{r} = \frac{d\dot{r}}{dt} = \frac{d\dot{r}}{d\theta}\frac{d\theta}{dt} = \frac{h}{r^2}\frac{d}{d\theta}\left(-h\frac{du}{d\theta}\right) = -h^2 u^2 \frac{d^2 u}{d\theta^2} \tag{2.5}$$

Substituting Eqs. (2.3) and (2.5) into Eq. (2.2a), we obtain the orbit differential equation

$$\frac{d^2 u}{d\theta^2} + u = \frac{\mu}{h^2} \tag{2.6}$$

The general solution of Eq. (2.6) is the orbit equation

$$u = \frac{\mu}{h^2} + c \cos (\theta - \theta_0) \tag{2.7}$$

where c and θ_0 are constants of integration. The constant θ_0 can be made equal to zero by measuring θ from a given direction. On the other hand, the constant c can be expressed in terms of the system total energy E, which is constant by virtue of the fact that this is a conservative system. The total energy per unit mass can be shown to be

$$E = T + U = \frac{1}{2}V^2 - \frac{\mu}{r} = \frac{1}{2}(\dot{r}^2 + r^2 \dot{\theta}^2) - \frac{\mu}{r}$$
$$= \frac{h^2}{2}\left[\left(\frac{du}{d\theta}\right)^2 + u^2\right] - \mu u \tag{2.8}$$

where T and U are the kinetic and potential energy per unit mass and V is the magnitude of the velocity vector $\mathbf{V}$. The fact that the potential energy is negative, $U = -\mu/r$, is due to using the reference for measuring the potential energy at infinity (Ref. 1). Introducing Eq. (2.7) into Eq. (2.8), we obtain

$$c = \frac{\mu}{h^2} \sqrt{1 + \frac{2Eh^2}{\mu^2}} \tag{2.9}$$

so that the orbit equation can be rewritten in the form

$$r = \frac{h^2}{\mu} \cdot \frac{1}{1 + e \cos \theta} \tag{2.10}$$

where

$$e = \sqrt{1 + 2\frac{Eh^2}{\mu^2}} \tag{2.11}$$

is the orbit eccentricity (Ref. 1). Equation (2.10) represents a general conic section. The particular type of conic section depends explicitly on the eccentricity e and implicitly on the total energy E. The various possibilities are:

$$\text{Circle: } e = 0 \quad \left(E = -\frac{\mu^2}{2h^2} \right)$$

$$\text{Ellipse: } 0 < e < 1 \quad \left(-\frac{\mu^2}{2h^2} < E < 0 \right)$$

$$\text{Parabola: } e = 1 \quad (E = 0)$$

$$\text{Hyperbola: } e > 1 \quad (E > 0)$$

Note that the total energy E can be negative. This is due to the fact that the reference for the potential energy was taken as infinity, as indicated earlier. Because the values of E given here are for comparison purposes only, the reference is immaterial. The circle and the ellipse are closed orbits and the parabola and hyperbola are open orbits. For planets orbiting the sun, the eccentricity is close to zero, so that the orbits of the planets around the sun are ellipses of very low eccentricity, which is the essence of Kepler's first law. A typical ellipse is shown in Fig. 2.3, where a and b denote the

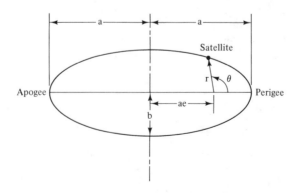

Figure 2.3 Parameters of an ellipse.

semimajor and semiminor axes, respectively. The points closest to and farthest from the focus are known as perihelion (perigee) and aphelion (apogee), for orbits around the sun (earth).

To prove Kepler's second law, consider the element of area Δa enclosed by the radii r and $r + \Delta r$ separated by the angle $\Delta \theta$. For small angles $\Delta \theta$, we can write

$$\Delta A \cong \frac{1}{2} r \cdot r \, \Delta \theta = \frac{1}{2} r^2 \, \Delta \theta \qquad (2.12)$$

Dividing Eq. (2.12) through by Δt, letting $\Delta t \to 0$, and recalling Eq. (2.3), we obtain

$$\lim_{\Delta t \to 0} \frac{\Delta A}{\Delta t} = \frac{dA}{dt} = \lim_{\Delta t \to 0} \frac{1}{2} r^2 \frac{\Delta \theta}{\Delta t} = \frac{1}{2} r^2 \dot{\theta} = \frac{h}{2} = \text{const.} \qquad (2.13)$$

which proves Kepler's second law.

Next, let us consider the orbital period P. From analytical geometry, the area of the ellipse A equals $\pi a \times b$, so that recalling that the area rate is constant, we obtain

$$P = \frac{A}{dA/dt} = \frac{2\pi ab}{h} \qquad (2.14)$$

Letting $r = r_p$ at $\theta = 0$ and $r = r_a$ at $\theta = \pi$ in Eq. (2.10) in sequence, we can write

$$a = \frac{1}{2}(r_p + r_a) = \frac{1}{2} \frac{h^2}{\mu} \left(\frac{1}{1 + e} + \frac{1}{1 - e} \right) = \frac{h^2}{\mu} \frac{1}{1 - e^2} \qquad (2.15)$$

Moreover,

$$b = a\sqrt{1 - e^2} \qquad (2.16)$$

Substituting h from Eq. (2.15) into Eq. (2.14) and using Eq. (2.16), we obtain

$$P = 2\pi \sqrt{\frac{a^3}{\mu}} \qquad (2.17)$$

from which we conclude that

$$\frac{a^3}{P^2} = \frac{\mu}{4\pi^2} = \text{const.} \qquad (2.18)$$

which proves Kepler's third law.

As mentioned earlier, Kepler's laws are only approximately correct. Indeed, whereas Eq. (2.18) indicates that the ratio a^3/P^2 is constant, and the same for all planets, from the two-body problem one concludes   1t this is only approximately true, as $\mu = G(M_s + M_p)$, where M_s is the mass of the sun and M_p is the mass of the planet (Ref. 3). Hence the ratio differs slightly from planet to planet. The developments noted above, of course, are equally valid for the motion of satellites around the earth.

In placing a satellite in orbit or in changing orbits, a certain velocity must be achieved in a certain position. For this reason it is useful to derive an expression relating the velocity to the radial distance. To this end, we turn our attention to the total energy [Eq. (2.8)]. At apogee the only component of velocity is in the transverse direction, or

$$V_a = r_a\dot{\theta} = \frac{h}{r_a} = \frac{\mu}{h}(1 - e) \tag{2.19}$$

so that, letting $v = v_a$ and $r = r_a$ in Eq. (2.8), we obtain

$$E = \frac{1}{2}\left(\frac{\mu}{h}\right)^2 (1 - e)^2 - \left(\frac{\mu}{h}\right)^2 (1 - e) = -\frac{1}{2}\left(\frac{\mu}{h}\right)^2 (1 - e^2) = -\frac{\mu}{2a} \tag{2.20}$$

Comparing Eqs. (2.8) and (2.20), we conclude that the orbital velocity is

$$V = \sqrt{\mu\left(\frac{2}{r} - \frac{1}{a}\right)} \tag{2.21}$$

The foregoing equations describe the orbit on the assumption that the plane of the orbit is given. In satellite dynamics, there is often the added task of determining the position of the orbital plane relative to an inertial space. This task is described in the following section.

Elliptical Orbit

To define the position of a satellite in space, it is necessary first to define the orbit relative to an inertial space and then the position of the satellite in the orbit. Figure 2.4 shows an inertial coordinate system commonly used to describe the motion of earth satellites. The origin of the coordinate system is at the mass center of the earth. The direction of the first axis is along the line of the vernal equinox, that is, the intersection of the earth equatorial plane and the earth orbital plane about the sun, where the latter is known as the ecliptic plane. The second axis is in the equatorial plane and perpendicular to the first axis. The third axis is normal to both axes and along the north pole. Spherical coordinates are ordinarily used to define the position of a satellite as follows: r the radial distance of the satellite from the origin O, the right ascension α being the angle in the equatorial plane from the vernal equinox eastward to the projection of the radius vector $\mathbf{r}$ on the equatorial plane, and the declination δ the angle from the equatorial plane of the radius vector $\mathbf{r}$.

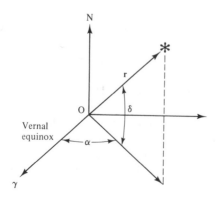

Figure 2.4 Inertial coordinate system for satellites.

Five orbital elements are necessary to define an elliptic orbit, and one orbital element to define the position of a satellite in that orbit. The classical orbital elements, shown in Fig. 2.5, are as follows:

a = semimajor axis

e = eccentricity

i = angle between the orbit plane and the equatorial plane, inclination angle (2.22)

Ω = right ascension of the ascending node

ω = angle between the ascending node and the perigee

T = time when the satellite passes the perigee, which determines the position of the satellite in the orbit

The ascending node is a point in the orbit where the satellite crosses the equatorial plane and moves from south to north.

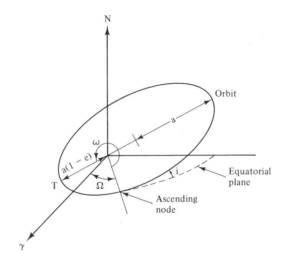

Figure 2.5 Orbit elements of an elliptic orbit.

Knowing the six orbital elements, described in the preceding section, the position and the velocity of the satellite at some specified time can be determined. The method is, however, complicated by the fact that in an elliptic orbit the angular velocity is not constant. As can be concluded from the conservation of angular momentum, the angular velocity is smallest at apogee and largest at perigee. It is frequently convenient to express the position and velocity of a satellite in an elliptic orbit in terms of an angle E subtending at the geometric center of the orbit and called eccentric anomaly, instead of the angle θ, which is known as the true anomaly (Fig. 2.6). To define the eccentric anomaly, an auxiliary circle is drawn by using the major axis of the ellipse as the diameter. The position of the satellite on the ellipse is then projected onto the circle. The eccentric anomaly E

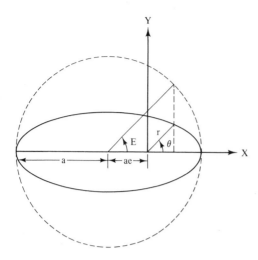

Figure 2.6 Eccentric anomaly of an elliptic orbit.

is the angle between the projected point and the perigee subtending at the geometric center of the ellipse.

In terms of the eccentric anomaly, the position and velocity of the satellite are (Ref. 2)

$$X = a (\cos E - e)$$
$$Y = a \sqrt{1 - e^2} \sin E$$
$$r = a (1 - e \cos E) \quad\quad (2.23)$$
$$\dot{X} = -a \sin E \, \dot{E}$$
$$\dot{Y} = a \sqrt{1 - e^2} \cos E \, \dot{E}$$

The angular momentum h can be expressed in terms of E as follows:

$$h = |\mathbf{r} \times \mathbf{v}| = X\dot{Y} - Y\dot{X}$$
$$= \dot{E} \, a^2 \sqrt{1 - e^2} \, (1 - e \cos E) \quad\quad (2.24)$$

Combining Eq. (2.24) and Eq. (2.15), we obtain

$$\dot{E} (1 - e \cos E) = \sqrt{\frac{\mu}{a^3}} = n \qu\quad (2.25)$$

or

$$\dot{E} = \frac{1}{r} \sqrt{\frac{\mu}{a}} \qu\quad (2.26)$$

where n is known as the mean angular velocity. Integrating Eq. (2.25), we can write

$$E - e \sin E = n(t - t_o) = M \qu\quad (2.27)$$

where t_o is a constant of integration and represents the time of passage of the perigee. The quantity $M = n(t - t_o)$ is called the mean anomaly. Substituting Eq. (2.26) into Eqs. (2.23), we have

$$\dot{X} = -\frac{\sqrt{\mu a}}{r} \sin E$$

$$\dot{Y} = \frac{\sqrt{\mu a (1 - e^2)}}{r} \cos E$$

(2.28)

Hence, to determine the position and velocity of a satellite at a given time, the mean anomaly must first be calculated. Next, the eccentric anomaly is calculated by using Eq. (2.27). This involves the solution of a transcendental equation. Having the eccentric anomaly, the position and velocity of a satellite are obtained by using Eqs. (2.23) and (2.28), respectively. It should be noted that the position and velocity are calculated in the orbit coordinate system. The properties of elliptic orbits are summarized in Table 2.1.

TABLE 2.1 PROPERTIES OF ELLIPTIC ORBITS

Perigee distance:	$r_p = a(1 - e)$	Orbital period:	$P = 2\pi \dfrac{a^{3/2}}{\mu^{1/2}}$
Apogee distance:	$r_a = a(1 + e)$	Velocity:	$V^2 = \mu\left(\dfrac{2}{r} - \dfrac{1}{a}\right)$
Semiminor axis:	$b = a\sqrt{1 - e^2}$ $= \sqrt{r_a r_p}$	Mean velocity:	$V_m = \sqrt{\dfrac{\mu}{a}}$
		Velocity at perigee:	$V_p = \sqrt{\dfrac{2\mu r_a}{(r_a + r_p)r_p}}$
Eccentricity:	$e = \dfrac{r_a - r_p}{r_a + r_p}$	Velocity at apogee:	$V_a = V_p \dfrac{r_p}{r_a}$
Radial distance:	$r = a(1 - e \cos E)$		
Mean anomaly:	$M = E - e \sin E$ $= M_0 + n(t - t_0)$	Mean angular velocity:	$n = \dfrac{\sqrt{\mu}}{a^{3/2}}$

2.3 GEOSYNCHRONOUS ORBIT

In an ideal geosynchronous orbit, a satellite moves around the earth in the equatorial plane so that the period of revolution of the satellite is exactly equal to the period of the rotation of the earth about its own axis. Hence, to an observer on the earth, the satellite appears to be stationary. Because the observer moves in a circle around the earth axis with sideral period of 23 h 56 min 4.09 s, the satellite orbit must also be circular and of the same period. The 24-hr day is from sunrise to sunrise, while the sideral day is from the rising of a star to the next rising of the same star. The radius of the orbit is calculated by using Eq. (2.17), as follows:

$$a = \left(\frac{\mu_e P^2}{4\pi^2}\right)^{1/3} = \left(\frac{398{,}601.2 \times 86{,}164.09^2}{4\pi^2}\right)^{1/3}$$

$$= 42{,}164.16 \text{ km}$$

(2.29)

where a is the orbit radius measured in kilometers and P is the orbit period measured in seconds. The location of the satellite is usually specified by the longitude λ of the subsatellite point, which is a point where the line joining the satellite and mass center of the earth intersects the surface of the earth.

The orbit elements of the ideal geosynchronous orbit are as follows:

$$a = 42{,}164 \text{ km} \qquad e = 0 \qquad i = 0 \tag{2.30}$$

For a circular equatorial orbit, the angles Ω and ω are immaterial. The position of a satellite is specified by the longitude of the subsatellite point instead of the time T.

Effect of Imperfections on a Geosynchronous Orbit

Even if a satellite is injected into the ideal geosynchronous orbit with the orbital elements given by Eq. (2.30), the gravitational forces due to the sun and the moon, as well as the differential gravity due to the oblateness of the earth, tend to change the orbital elements. In practice, the geosynchronous orbit is allowed to deviate slightly in one or more orbital parameters from the ideal geosynchronous orbit values depending on the functional requirements of the satellite.

Circular orbit in equatorial plane with incorrect radius ($e = 0$, $i = 0$, $a \neq 42{,}164$ km). An inaccuracy in the radius of a geosynchronous orbit results in an incorrect orbit period, as shown by Eq. (2.17). For a larger radius, $a > 42{,}164$ km, the orbit period is greater than the sideral period, so the satellite will drift westward. Conversely, for a smaller radius, $a < 42{,}164$ km, the satellite will drift eastward. The drift rate can be calculated by using the following mean angular velocity equation:

$$n = \sqrt{\frac{\mu}{a^3}} \tag{2.31}$$

Incrementing Eq. (2.31), the longitude drift rate with respect to the initial ideal geosynchronous orbit is

$$\Delta n = -\frac{3}{2}\frac{n_s}{a_s}\Delta a = -\frac{3}{2}\frac{360°/\text{day}}{42{,}164 \text{ km}}\Delta a \tag{2.32}$$

$$= -\frac{0.013°/\text{day}}{\text{km}}\Delta a$$

where the subscript s denotes the ideal geosynchronous orbit parameters.

Inclined circular orbit with correct radius ($i \neq 0$, $e = 0$, $a = 42{,}164$ km). In an inclined circular orbit with the correct orbital period, the subsatellite point oscillates about a position on the earth. The oscillation is mainly in the latitude. Figure 2.7 shows the inclined orbit. At a time t

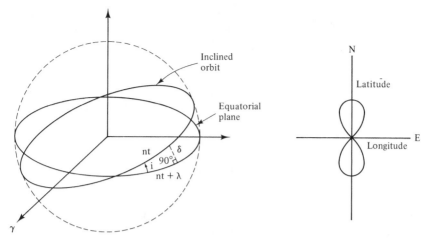

Figure 2.7 Inclined circular orbit with correct radius.

from the ascending node, the latitude of the subsatellite point can be determined by using spherical trignometry from the formula

$$\sin \delta = \sin i \sin nt \tag{2.33}$$

For small i and δ,

$$\delta \cong i \sin nt \tag{2.34}$$

The change in the longitude of the subsatellite point due to the inclination can also be calculated by using spherical trigonometry as follows:

$$\cos i = \tan (nt + \lambda) \cot nt \tag{2.35}$$

or

$$1 - \cos i = - \frac{\sin \lambda}{\sin nt \cos (nt + \lambda)} \tag{2.36}$$

For small i and λ,

$$\lambda = - \frac{i^2}{4} \sin 2nt \tag{2.37}$$

Hence, due to the inclination of the orbit, the subsatellite point performs a figure-eight motion, as shown in Fig. 2.7. For small inclinations, the north-south latitude motion is predominant.

Elliptic orbit in equatorial plane with correct radius ($e \neq 0$, $i = 0$, $a = 42,164$ km). Figure 2.8 shows an elliptic orbit of a satellite. Using Kepler's second law, it can be shown that the angular velocity of the satellite is largest at perigee, smallest at apogee, and equal to earth's angular velocity (i.e., no drift) at points A and B ($E = \pi/2$ and $3\pi/2$). The subsatellite point is at the center position at apogee and perigee and maximum eastward and westward at about points A and B, respectively. The amplitude $\Delta\lambda$ of the

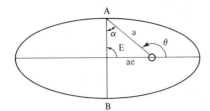

Figure 2.8 Elliptical orbit in equatorial plane with correct radius.

drift can be determined by calculating it at point A $(E = \pi/2)$,

$$\Delta\lambda = \theta - M \Big|_{E = \pi/2} \tag{2.38}$$

Using Eq. (2.27), we obtain

$$M = E - e \sin E \Big|_{E = \pi/2} = \frac{\pi}{2} - e \tag{2.39}$$

From Figure 2.8,

$$\sin \alpha = \frac{ae}{a} = e$$

For small e and α,

$$\alpha = e$$

and

$$\theta = \frac{\pi}{2} + \alpha = \frac{\pi}{2} + e \tag{2.40}$$

Substituting Eqs. (2.39) and (2.40) into Eq. (2.38), we obtain

$$\Delta\lambda = 2e$$

Similarly, it can be shown that the drift at point B $(E = 270°)$ is $-2e$ ($2e$ westward). Hence the presence of eccentricity in the intended circular orbit results in east-west oscillation of the subsatellite point with an amplitude equal to $2e$.

2.4 ORBIT PERTURBATIONS

In Section 2.2 the equations of motion of a satellite around the earth were derived by assuming that the earth was spherical and by neglecting the disturbance forces due to the gravitational attraction of the sun and the moon, solar radiation pressure, and so on. In reality, these effects exist and they tend to perturb the orbital elements of the satellite. For a communications satellite with small beam antennas, it is necessary to keep the orbital elements within certain allowable limits. Hence it becomes necessary to correct these orbital elements periodically during the lifetime of the satellite. To calculate the fuel required for the correction, it is necessary to predict the rate of change of the orbital elements due to the above-mentioned perturbing forces.

It is convenient to use the variation-of-parameters method to determine the influence of perturbing forces on the rate of change of the orbital elements. The analysis is simplified by using the orbital coordinate system $i_r i_\theta i_z$, as shown in Fig. 2.9. The transformation from the inertial coordinate system IJK to the orbital coordinate system can be carried out in three stages: a rotation Ω about the K axis, transfoming the IJK system to $I_1 J_1 K_1$; a rotation i about I_1, transforming $I_1 J_1 K_1$ to $I_2 J_2 K_2$; and a rotation θ about K_2, transforming $I_2 J_2 K_2$ to the orbital coordinate system $i_r i_\theta i_z$. Combining the transformation matrices associated with the three rotations leads to

$$\begin{Bmatrix} i_r \\ i_\theta \\ i_z \end{Bmatrix} = \begin{bmatrix} T \end{bmatrix} \begin{Bmatrix} I \\ J \\ K \end{Bmatrix} \tag{2.41}$$

where the transformation matrix is

$$\begin{bmatrix} T \end{bmatrix} = \begin{bmatrix} \cos \Omega \cos \theta & \sin \Omega \cos \theta & \sin i \sin \theta \\ -\sin \Omega \cos i \sin \theta & +\sin \theta \cos \Omega \cos i & \\ -\cos \Omega \sin \theta & -\sin \Omega \sin \theta & \sin i \cos \theta \\ -\sin \Omega \cos i \cos \theta & +\cos \Omega \cos i \cos \theta & \\ \sin \Omega \sin i & -\cos \Omega \sin i & \cos i \end{bmatrix} \tag{2.42}$$

The angular rate ω of the orbital coordinate system with respect to the inertial coordinate system is given by

$$\omega = \begin{bmatrix} \cos \theta & \sin \theta & 0 \\ -\sin \theta & \cos \theta & 0 \\ 0 & 0 & 1 \end{bmatrix} \begin{bmatrix} 1 & 0 & 0 \\ 0 & \cos i & \sin i \\ 0 & -\sin i & \cos i \end{bmatrix} \begin{Bmatrix} 0 \\ 0 \\ \dot\Omega \end{Bmatrix}$$

$$+ \begin{bmatrix} \cos \theta & \sin \theta & 0 \\ -\sin \theta & \cos \theta & 0 \\ 0 & 0 & 1 \end{bmatrix} \begin{Bmatrix} \dfrac{di}{dt} \\ 0 \\ 0 \end{Bmatrix} + \begin{Bmatrix} 0 \\ 0 \\ \dot\theta \end{Bmatrix} \tag{2.43}$$

$$= \begin{Bmatrix} \dot\Omega \sin i \sin \theta + \dfrac{di}{dt} \cos \theta \\ \dot\Omega \sin i \cos \theta - \dfrac{di}{dt} \sin \theta \\ \dot\Omega \cos i + \dot\theta \end{Bmatrix}$$

The perturbing force per unit mass of the satellite is resolved into three components along the orbital coordinate axes—f_r, f_θ, and f_z—as shown in Fig. 2.9. The force f_z tends to rotate the orbital plane, resulting in change in i and Ω. The forces f_r and f_θ tend to change the orbit element a, e and ω.

This section covers (1) the effects of perturbing forces on orbit parameters a, i, and Ω; (2) perturbation forces; (3) orbital motion of the earth and the moon around the sun; (4) perturbation of the orbital inclination due to the sun and moon and the oblateness of the earth; and (5) longitudinal perturbation due to the triaxiality of the earth.

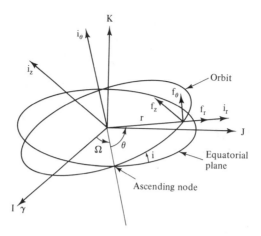

Figure 2.9 Orbit coordinate system.

Perturbation of Orbit Parameters

Variation of the semimajor axis. From Eq. (2.20), the semimajor axis a is related to the orbital energy E per unit mass by

$$E = -\frac{\mu}{2a} \tag{2.44}$$

The perturbation forces will cause a change in the orbital energy, which in turn will result in change of the semimajor axis. Taking the differential of Eq. (2.44) yields

$$dE = \frac{\mu}{2a^2}\, da = (\mathbf{f} \cdot \mathbf{v})dt \tag{2.45}$$

$$\frac{da}{dt} = \frac{2a^2}{\mu}\,(\mathbf{f} \cdot \mathbf{v}) \tag{2.46}$$

Variation in the inclination and right ascension of the node. The out-of-plane force f_z rotates the orbital plane, as shown in Fig. 2.10. Impulse

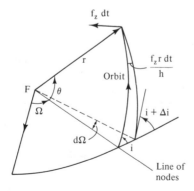

Figure 2.10 Orbit perturbation due to out-of-plane force.

Orbit Dynamics   Chap. 2

$f_z \, dt$ rotates the orbital plane through an angle of magnitude $f_z \, r \, dt/h$ about the radius vector. Here r is the radius and h the angular momentum of the orbit, $v_\theta \, r$. Hence

$$di = \frac{f_z \, r \, dt}{h} \cos \theta \qquad (2.47)$$

or

$$\frac{di}{dt} = \frac{r \cos \theta}{h} f_z \qquad (2.48)$$

and

$$\sin i \, d\Omega = \frac{f_z \, r \, dt}{h} \sin \theta \qquad (2.49)$$

or

$$\frac{d\Omega}{dt} = \frac{r \sin \theta}{h \sin i} f_z \qquad (2.50)$$

The expressions for the rate of change of e and ω are

$$\dot{e} = \frac{r}{h} [\sin \theta^* \, (1 + e \cos \theta^*) f_r + (e + 2 \cos \theta^* + e \cos^2 \theta^*) f_\theta] \quad (2.51)$$

$$\dot{\omega} = \frac{r}{he} [- \cos \theta \, (1 + e \cos \theta^*) f_r + \sin \theta^* \, (2 + e \cos \theta^*) f_\theta] \quad (2.52)$$

The angle θ^* is measured from the perigee of the orbit.

Perturbation forces. The perturbation forces are defined as those forces which change the orbital parameters of the satellite with respect to the earth. The satellite will experience perturbation forces due to the gravitational effects of the sun and the moon. The perturbation force from a perturbing body is the difference between the gravitational force due to the perturbing body at the satellite and the gravitational force the satellite would experience if it were at the center of the earth. From Fig. 2.11, the perturbing force per unit mass of a satellite is

$$f_p = \mu_p \frac{r_p \mathbf{i}_p - r \mathbf{i}_r}{|r_p \mathbf{i}_p - r \mathbf{i}_r|^3} - \frac{\mu_p \mathbf{i}_p}{r_p^2} \qquad (2.53)$$

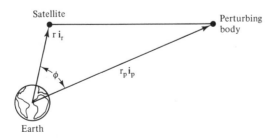

Figure 2.11 Perturbation forces.

where μ_p = gravitational constant of the perturbing body

r_p = distance from the earth mass center to the perturbing body

i_p = unit vector from the earth to the perturbing body

r = distance from the earth mass center to the satellite

i_r = unit vector from the earth to the satellite

The subscript p is to be replaced by s if the perturbing body is the sun, and by l if the perturbing body is the moon. If $r \ll r_p$, then

$$\mathbf{f}_p = \frac{\mu_p}{r_p^2} \left(\frac{r}{r_p} \right) [3(\mathbf{i}_r \cdot \mathbf{i}_p)\mathbf{i}_p - \mathbf{i}_r] \tag{2.54}$$

The unit vectors $\mathbf{i}_r$ and $\mathbf{i}_p$ can be written in terms of the inertial coordinates system IJK as

$$\mathbf{i}_r = (\cos \Omega \cos \theta - \sin \Omega \cos i \sin \theta)\mathbf{I} + (\cos \theta \sin \Omega \tag{2.55}$$
$$+ \cos \Omega \cos i \sin \theta)\mathbf{J} + (\sin i \sin \theta)\mathbf{K}$$

$$\mathbf{i}_p = (\cos \Omega_p \cos \theta_p - \sin \Omega_p \cos i_p \sin \theta_p)\mathbf{I} \tag{2.56}$$
$$+ (\cos \theta_p \sin \Omega_p + \cos \Omega_p \cos i_p \sin \theta_p)\mathbf{J} + (\sin i_p \sin \theta_p)\mathbf{K}$$

where Ω, i, and θ are the orbital elements of the satellite and Ω_p, i_p, and θ_p are the orbital elements of the perturbing body, as defined in Fig. 2.9.

The component f_z of the perturbing force in the direction normal to the satellite orbit plane is

$$f_z = \mathbf{f}_p \cdot \mathbf{i}_z$$
$$= \frac{3\mu_p}{r_p^3} r (\mathbf{i}_r \cdot \mathbf{i}_p)(\mathbf{i}_p \cdot \mathbf{i}_z) \tag{2.57}$$

where

$$\mathbf{i}_z = (\sin \Omega \sin i)\mathbf{I} + (- \cos \Omega \sin i)\mathbf{J} + \cos i \, \mathbf{K} \tag{2.58}$$

Substituting Eq. (2.57) into Eqs. (2.48) and (2.50), we obtain

$$\frac{di}{dt} = \frac{3\mu_p r^2}{h \, r_p^3} \cos \theta \, (\mathbf{i}_r \cdot \mathbf{i}_p)(\mathbf{i}_p \cdot \mathbf{i}_z) \tag{2.59}$$

$$\frac{d\Omega}{dt} = \frac{3\mu_p r^2}{h \, r_p^3} \frac{\sin \theta}{\sin i} (\mathbf{i}_r \cdot \mathbf{i}_p)(\mathbf{i}_p \cdot \mathbf{i}_z) \tag{2.60}$$

The perturbation of the satellite orbit can be visualized better by studying the motion of the satellite orbit pole with respect to the perturbing orbit pole. The orbit pole is defined as a line normal to the orbit plane. The right ascension α and declination δ of the orbit pole are expressed in terms of the right ascension of the ascending node Ω and the orbit inclination i of the orbit as follows:

$$\alpha = \Omega - 90° \tag{2.61}$$
$$\delta = 90° - i \tag{2.62}$$

The poles of the satellite orbit (W) and the perturbing body orbit (W_p) are shown in Fig. 2.12 on the celestial sphere. The angle between the poles,

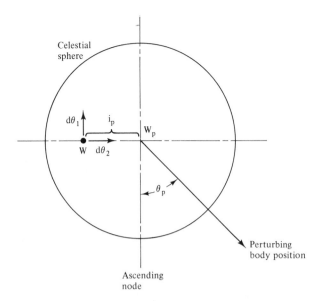

Figure 2.12 Motion of the satellite orbit pole.

i_p, is the difference of the inclination angles of the satellite and the perturbing body orbits; θ_p is the relative position of the perturbing body from the ascending node of its orbit.

The general motion of the satellite orbit pole W is composed of two rotations: one rotation, $d\theta_1$, always perpendicular to the plane containing W and W_p and the other, $d\theta_2$, along the plane containing W and W_p. Neglecting the periodic terms, where the period is equal to the satellite orbit period, the general motion is given by

$$\frac{d\theta_1}{dt} = \frac{3}{4} \frac{\mu_p \, r^2}{h \, r_p^3} \cos i_p \, (1 - \cos 2 \, \theta_p) \tag{2.63}$$

$$\frac{d\theta_2}{dt} = \frac{3}{4} \frac{\mu_p \, r^2}{h \, r_p^3} \sin i_p \sin 2 \, \theta_p \tag{2.64}$$

In this motion, $\dot{\theta}_2$ is periodic with perturbing orbit period and $\dot{\theta}_1$ has both secular and periodic terms. Over a long period, the net motion is rotation of the satellite orbit pole around the perturbing body orbit pole.

Orbital Motions of the Earth and the Moon Around the Sun

The earth revolves around the sun in the counterclockwise direction in a nearly circular orbit, $e = 0.016726$, with a period of approximately 365 days and a semimajor axis of 149.6×10^6 km. The earth spins around its own axis (an imaginary line passing through the two poles) from west to east (counterclockwise) with a period of 23 h 56 min 4.09 s. The earth axis is inclined from the normal of the orbital plane (ecliptic plane) by an angle 23.45° and is always directed toward a fixed point in the sky near the pole star.

The moon revolves around the earth in an almost circular orbit, $e = 0.0549$, with a semimajor axis of 384,400 km and a period of 27.3 days. The orbital plane is inclined to the ecliptic plane at an average angle of 5°8′, varying from 4°59′ to 5°18′. Due to the fact that the moon spins around its axis with the same period and direction as those of its orbit around the earth, it keeps the same face pointed toward the earth. The combined earth–moon system revolves around the sun in the same direction as that in which the moon is revolving around the earth. Thus it takes 2 days and 5 hours longer, or 29.5 days, for the moon to return to the same position relative to the sun (i.e., from full moon to full moon). The apparent motions of the sun and the moon with respect to the earth are given in the celestial sphere of Fig. 2.13.

The angle between the ecliptic plane (plane of the apparent motion of the sun) and the equatorial plane is 23.45°. The angle between the ecliptic plane normal and the lunar plane normal is 5.15° and the lunar plane normal precesses around the ecliptic plane normal with a period of 18.6 years. The earth's axis precesses clockwise around the ecliptic pole with a half-angle of 23.45° and a period of 25,000 years. Hence the vernal and autumnal equinoxes revolve clockwise with the same period.

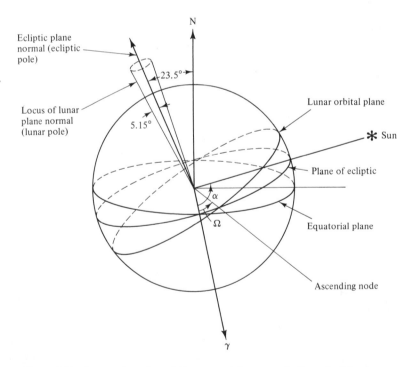

Figure 2.13 Apparent motion of the sun and the moon in the celestial sphere.

The angular position of the sun in the celestial sphere for a particular month and day varies little from year to year. Hence the values of the right ascension and declination given in Table 2.2 for 1969 can be used.

TABLE 2.2 SUN'S POSITION

Date		Right Ascension	Declination
Jan.	1	280°15′	−23°07′
	15	296°30′	−21°12′
Feb.	1	314°30′	−17°13′
	15	328°30′	−12°49′
Mar.	1	341°45′	−7°44′
	15	354°45′	−2°18′
Apr.	1	10°15′	4°23′
	15	23°0′	9°37′
May	1	38°0′	14°57′
	15	51°30′	18°46′
June	1	68°45′	22°0′
	15	83°15′	23°18′
July	1	99°45′	23°08′
	15	114°00′	21°36′
Aug.	1	131°00′	18°08′
	15	144°15′	14°11′
Sept.	1	160°00′	8°26′
	15	172°45′	3°11′
Oct.	1	187°00′	−3°02′
	15	199°45′	−8°22′
Nov.	1	216°00′	−14°18′
	15	230°00′	−18°23′
Dec.	1	247°00′	−21°45′
	15	262°15′	−23°15′

The right ascension of the ascending node of the lunar orbit measured in the ecliptic plane from the vernal equinox is given by

$$\Omega = 178.78 - 0.05295\, t \quad \text{(degrees)} \tag{2.65}$$

where t is the number of days from January 1, 1960.

Using spherical trignometry in conjunction with Fig. 2.14, the lunar orbit plane inclination i_l and the right ascension of the ascending node Ω_l are given by the equations

$$\cos i_l = \cos i_s \cos I_l - \sin i_s \sin I_l \cos \Omega \tag{2.66}$$

$$\sin \Omega_l = \frac{\sin I_l \sin \Omega}{\sin i_l} \tag{2.67}$$

where i_s = inclination angle of the plane of ecliptic, 23.45°

I_l = angle between the lunar orbit plane and ecliptic plane, 5.15°.

The inclination i_l of the lunar orbital plane varies between 18.3 and 28.6°. The right ascension of the ascending node oscillates between 13 and −13°. The values of i_l and Ω_l are given in Table 2.3 from 1979 to 2003.

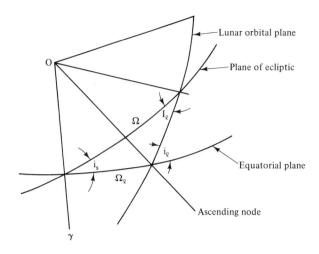

Figure 2.14 Orbital elements of moon orbit.

TABLE 2.3 INCLINATION DRIFT RATES

Date January 1	Ω moon (Deg.)	i_l (Deg.)	Ω_l (Deg.)	$\dfrac{di}{dt}\bigg\| \text{moon}$ (Deg./Yr.)	$\dfrac{di}{dt}\bigg\| \text{Total}$ (Deg./Yr.)
1978	190.660	18.412	−3.013	.480	.749
1979	171.332	18.375	2.459	0.480	0.749
1980	152.004	19.049	7.417	0.491	0.760
1981	132.623	20.300	10.975	0.513	0.782
1982	113.295	21.902	12.769	0.542	0.811
1983	93.966	23.627	12.910	0.574	0.843
1984	74.638	25.279	11.694	0.607	0.876
1985	55.257	26.703	9.447	0.636	0.905
1986	35.929	27.776	6.489	0.657	0.926
1987	16.601	28.422	3.089	0.671	0.940
1988	357.273	28.595	−0.511	−0.674	0.943
1989	337.892	28.284	−4.088	0.668	0.937
1990	318.563	27.511	−7.388	0.652	0.921
1991	299.235	26.330	−10.171	0.628	0.897
1992	279.907	24.830	−12.155	0.598	0.867
1993	260.526	23.135	−13.023	0.565	0.834
1994	241.198	21.423	−12.436	0.533	0.802
1995	221.870	19.898	−10.138	0.506	0.775
1996	202.542	18.793	−6.132	0.487	0.756
1997	183.161	18.311	−0.903	0.479	0.748
1998	163.833	18.557	4.504	0.483	0.752
1999	144.505	19.476	8.993	0.498	0.767
2000	125.177	20.888	11.875	0.523	0.792
2001	105.796	22.568	13.006	0.554	0.823
2002	86.468	24.287	12.581	0.587	0.856
2003	67.139	25.865	10.929	0.619	0.888

Perturbation of the Orbital Inclination

Perturbation due to the sun. The equations for the perturbing body are changed to the equations for the sun by changing the subscript p to s in Eqs. (2.56), (2.58), and (2.59) and noting that Ω_s is zero for the sun. Substituting Eqs. (2.55), (2.56), and (2.58) into Eq. (2.59) and averaging the right side of the equation with respect to θ (i.e., neglecting the periodic terms with the satellite orbital period), we obtain

$$\frac{di}{dt} = \frac{3}{4}\frac{\mu_s}{h}\frac{r^2}{r_s^3}\{\sin \Omega \cos \Omega \sin i \sin^2 i_s + \sin \Omega \cos i \cos i_s \sin i_s$$

$$+ \sin 2\,\theta_s\,[-\cos 2\Omega \sin i \cos i_s + \cos \Omega \cos i \sin i_s] \qquad (2.68)$$

$$+ \cos 2\theta_s\,[\sin \Omega \cos \Omega \sin i + \sin \Omega \cos \Omega \sin i \cos^2 i_s$$

$$- \sin \Omega \cos i \sin i_s \cos i_s]\}$$

Equation (2.68) contains periodic terms $\cos 2\theta_s$ and $\sin 2\theta_s$ with the period of 182.5 days (half of the sun orbit period). Considering only the secular terms, Eq. (2.68) reduces to

$$\frac{di}{dt} = \frac{3}{4}\frac{\mu_s}{h}\frac{r^2}{r_s^3}(\sin \Omega \cos \Omega \sin i \sin^2 i_s + \sin \Omega \cos i \sin i_s \cos i_s) \qquad (2.69)$$

For geosynchronous orbit $i \approx 0$, so that letting $\sin i = 0$ and $\cos i = 1$, Eq. (2.69) simplifies to

$$\frac{di}{dt} \approx \frac{3}{4}\frac{\mu_s}{h}\frac{r^2}{r_s^3}\sin \Omega \sin i_s \cos i_s \qquad (2.70)$$

It should be noted that the rate of change of the inclination depends on the right ascension of the satellite ascending node. The common strategy for the inclination stationkeeping, as discussed later, is to let the satellite orbit drift from the orbit inclination at the allowable limit and with Ω in the neighborhood of 270°. At $\Omega = 270°$, the rate of change of the inclination becomes

$$\frac{di}{dt} = -\frac{3}{4}\frac{\mu_s}{h}\frac{r^2}{r_s^3}\sin i_s \cos i_s \qquad (2.71)$$

Substituting in Eq. (2.71), the values of the parameters

$$\mu_s = 1.32686 \times 10^{11} \text{ km}^3/\text{s}^2$$

$$r = 42{,}164 \text{ km}$$

$$r_s = 1.49592 \times 10^8 \text{ km}$$

$$h = 129{,}640 \text{ km}^2/\text{s}$$

$$i_s = 23.45°$$

we obtain

$$\left.\frac{di}{dt}\right|_{\Omega=270°} = -0.269°/\text{yr} \qquad (2.72)$$

Hence the inclination decreases initially due to the effect of the sun at the rate of 0.269°/yr until it reaches zero. After it reaches zero, Ω changes to the neighborhood of 90°. At $\Omega = 90°$

$$\left.\frac{di}{dt}\right|_{\Omega=90°} = 0.269°/\text{yr} \tag{2.73}$$

Hence, after the inclination reaches zero, it increases again at the rate of 0.269°/yr. After it again reaches the allowable limit, the right ascension is changed from 90° to 270° by a station-keeping maneuver. It should be noted that in the discussion above, only the sun perturbation is considered. Similarly, substituting Eqs. (2.55), (2.56), and (2.58) into Eq. (2.60) and retaining only secular terms, we obtain

$$\frac{d\Omega}{dt} = \frac{3}{4}\frac{\mu_s}{h\sin i}\frac{r^2}{r_s^3}(-\sin^2\Omega \sin i \cos i - \cos^2 i_s \cos^2\Omega \sin i \cos i$$

$$+ \sin i_s \cos i_s \cos \Omega \cos 2i + \sin^2 i_s \sin i \cos i) \tag{2.74}$$

From Eq. (2.74), $\dot\Omega$ goes to infinity at $i = 0$. For a small inclination, this problem can be eliminated by selecting the reference plane to be the ecliptic plane, where i is never zero. However, this may not be desirable, as the position of the sun and the moon are referred to the equatorial plane and the gravitational perturbations arising from the nonspherical shape of the earth are most conveniently computed in the equatorial system. By using the transformation

$$I_x = \sin i \cos \Omega, \quad I_y = \sin i \sin \Omega \tag{2.75}$$

the singularity in the low-inclination case can be eliminated.

Perturbation due to the moon. For the moon, we proceed in the same way as for the sun. However, we note that for the moon Ω_l is not identically zero. Changing the subscript p for the perturbating force to l in Eqs. (2.56), (2.57), (2.59), and (2.60), substituting Eqs. (2.55), (2.56), and (2.58) into Eq. (2.59), and averaging the right side of the equation with respect to θ, we obtain

$$\frac{di}{dt} = \frac{3}{4}\frac{\mu_l}{h\,r_l^3}r^2\{\sin(\Omega-\Omega_l)\cos(\Omega-\Omega_l)\sin i \sin^2 i_l$$

$$+ \sin(\Omega-\Omega_l)\cos i \sin i_l \cos i_l + \sin 2\theta_l[-\cos 2(\Omega-\Omega_l)$$

$$\times \sin i \cos i_l + \cos(\Omega-\Omega_l)\cos i \sin i_l] + \cos 2\theta_l \tag{2.76}$$

$$\times [\sin(\Omega-\Omega_l)\cos(\Omega-\Omega_l)\sin i$$

$$+ \sin(\Omega-\Omega_l)\cos(\Omega-\Omega_l)$$

$$\times \sin i \cos^2 i_l - \sin(\Omega-\Omega_l)\cos i \sin i_l \cos i_l]\}$$

Equation (2.76) contains secular terms and periodic terms with a period of 14 days (half the moon orbital period). As mentioned earlier, the angle between the ecliptic plane normal and the lunar plane normal is 5.15° and the lunar plane normal precesses around the ecliptic plane normal with a period of approximately 18 years. Hence the lunar orbital inclination decreases from a maximum of 28.60° to a minimum of 18.30° in approximately 9 years and increases back to the maximum. The angle Ω_l varies from approximately $-13°$ to $+13°$. However, at the maximum and minimum inclination angles, Ω_l is zero. Considering only the secular terms and assuming almost geosynchronous orbit, $i = 0$, $\sin i = 0$, and $\cos i = 1$, the inclination drift rate can be approximated by

$$\frac{di}{dt} \approx \frac{3}{4} \frac{\mu_l}{h \, r_l^3} r^2 \sin (\Omega - \Omega_l) \sin i_l \cos i_l \tag{2.77}$$

Substituting for parameters in Eq. (2.77) the values

$$\mu_l = 4.9028 \times 10^3 \text{ km}^3/\text{s}^2$$

$$r = 42{,}164 \text{ km}$$

$$r_l = 3.844 \times 10^5 \text{ km}$$

$$h = 129{,}640 \text{ km}^2/\text{s}$$

$$i_l = 18.30 \text{ to } 28.60°$$

we have

$$\left. \frac{di}{dt} \right|_{i_l = 18.3°, \, \Omega = 90°, \, \Omega_l = 0} = 0.4780°/\text{yr} \tag{2.78}$$

$$\left. \frac{di}{dt} \right|_{i_l = 28.60°, \, \Omega = 90°, \, \Omega_l = 0} = 0.674°/\text{yr} \tag{2.79}$$

Similarly, using Eqs. (2.55), (2.56), (2.58), and (2.60), the expression for $d\Omega/dt$ resulting from perturbations due to the moon can be obtained.

Perturbation due to the oblateness of the earth. The earth is not perfectly spherical, but bulges around the equator. The polar and equatorial diameters are 12,713 km and 12,756.3 km, respectively. This oblateness results in a perturbing force with the components (Ref. 2)

$$f_r = -\frac{3\mu_e J_2 R_e^2}{2r^4} (1 - 3 \sin^2 i \sin^2 \theta) \tag{2.80}$$

$$f_\theta = -\frac{3\mu_e J_2 R_e^2}{r^4} \sin^2 i \sin \theta \cos \theta \tag{2.81}$$

$$f_z = -\frac{3\mu_e J_2 R_e^2}{r^4} \sin i \cos i \sin \theta \tag{2.82}$$

where R_e is the radius of the earth. Substituting Eq. (2.82) into Eqs. (2.48) and (2.50), we obtain

$$\frac{di}{dt} = -\frac{3\mu_e J_2 R_e^2}{h \, r^3} \sin i \cos i \sin \theta \cos \theta \tag{2.83}$$

$$\frac{d\Omega}{dt} = -\frac{3\mu_e J_2 R_e^2}{h \, r^3} \cos i \sin^2 \theta \tag{2.84}$$

Averaging Eqs. (2.83) and (2.84) with respect to θ over the satellite orbit period, we can write

$$\frac{di}{dt} = 0 \qquad (2.85)$$

$$\frac{d\Omega}{dt} = -\frac{3}{2}\frac{\mu_e}{h}\frac{J_2}{r^3} R_e^2 \cos i \qquad (2.86)$$

Hence the oblateness of the earth does not change the inclination angle i but does change Ω. Substituting in Eq. (2.86) the values

$$\mu_e = 398{,}601.2 \text{ km}^3/\text{s}^2$$
$$J_2 = 1.0823 \times 10^{-3}$$
$$R_e = 6.3782 \times 10^3 \text{ km}$$
$$r = 42{,}164 \text{ km}$$
$$h = 129{,}640 \text{ km}^2/\text{s}$$

we obtain

$$\frac{d\Omega}{dt} = 4.9 \cos i \qquad \text{deg/year} \qquad (2.87)$$

For near synchronous orbit, $i \approx 0$, so that

$$\frac{d\Omega}{dt} = 4.9°/\text{yr} \qquad (2.88)$$

Using Eqs. (2.51) and (2.52), the expressions for $\dot{e}$ and $\dot{\omega}$ due to interactions from the sun, the moon, and the oblateness of the earth can be determined.

Combined effect. The effect of the solar–lunar gravitational forces and the earth's oblateness on the satellite orbit normal is calculated by combining linearly the equations of the rates of change of the inclination and the right ascension, derived in the preceding section. For an initial synchronous orbit, the rate of change of the inclination is 0.269°/yr due to the sun; between 0.674° and 0.478°/yr due to the moon, depending on lunar orbit inclination; and zero due to earth's oblateness. The combined rate of change of the inclination is between 0.943° and 0.747°/yr. The orbit correction, called north-south station keeping, is generally required to maintain the orbit inclination within the allowable limits. In case an initial geosynchronous orbit is allowed to drift without any orbit correction, the inclination will gradually increase to the maximum value of approximately 15° in about 27.5 years, as shown in Fig. 2.15.

The perturbation effects over a long period can be understood better by considering the motion of the satellite orbit normal. The gravitational forces due to the sun and the moon cause the satellite orbit normal to rotate about the ecliptic pole and the lunar pole, respectively. The earth's oblateness causes the satellite orbit normal to rotate around the north pole. The combined effect is that the orbit normal of an initial geosynchronous

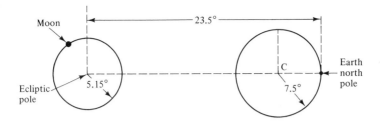

Figure 2.15 Combined effect on a satellite orbital inclination.

orbit rotates around the center point at a distance of 7.5° from the pole axis, and the full rotation around this point takes approximately 55 years.

Longitudinal Perturbations

The longitudinal drift acceleration $\ddot{\lambda}$ for a circular orbit is obtained by using Eqs. (2.31) and (2.46) as follows:

$$\ddot{\lambda} = \frac{360}{2\pi} \frac{dn}{dt} = -\frac{360}{2\pi} \frac{3}{a^2 n} (\mathbf{V} \cdot \mathbf{f})° / s^2 \tag{2.89}$$

Averaging the drift acceleration over the orbital period, we obtain

$$\ddot{\lambda}_{av} = -\frac{360}{2\pi} \frac{3}{a^2 n} \frac{1}{2\pi} \int_0^{2\pi} (\mathbf{V} \cdot \mathbf{f}) \, d\theta \tag{2.90}$$

From Eq. (2.90) it is clear that if the perturbing force is inertially fixed, the net effect over the orbital period is zero, because for a circular orbit the velocity magnitude remains constant but the direction changes by 360° over the orbital period. The sun, which moves only 0.986°/day, contributes very little to the drift acceleration. The moon, which moves approximately 13.2°/day, contributes to the longitude drift acceleration term with a period equal to one-half of the rotational period (27.3 days). The major contributor to the longitudinal drift acceleration is the ellipticity of the earth at the equator. The equatorial section of the earth is shown in Fig. 2.16. Due to the bulge, the gravitational force is not purely radial, but is has a component along or opposite to the velocity direction. For a geosynchronous satellite, the relative motion between the earth and the satellite is zero. Hence the lateral component of the gravitational force, which is always toward the nearest bulge, increases or decreases the energy of the orbit continuously. There are two stable longitudes S_1 and S_2 and two unstable longitudes US_1 and US_2, where the gravitational forces are radial and the longitude drift accelerations are zero. However, if the satellite longitude is near US_1 or US_2, the satellite will drift to the nearest stable point (S_1 or S_2). For example, if a satellite longitude is between S_1 and US_1, the lateral force component is along the velocity and this results in a negative longitude acceleration, as can be seen from Eq. (2.89). The satellite moves westward, toward the stable point S_1.

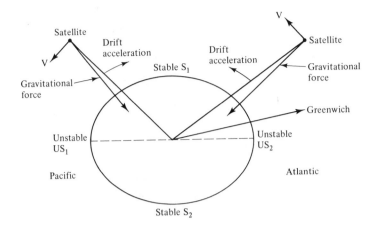

Figure 2.16 Longitudinal perturbation due to earth's ellipticity.

Figure 2.17 shows the drift acceleration for a geosynchronous satellite. The dashed line shows the drift acceleration due to second-order gravity effects only. Equilibrium points are 90° apart. The solid line shows the drift acceleration by including fourth-order gravity effects. The more complete gravity field results in equilibrium points that are not 90° apart. Considering only the second-order gravity effects, the longitudinal drift acceleration is given by the equation

$$\ddot{\lambda} = -0.00168 \sin 2(\lambda - \lambda_s) \quad \text{deg/day}^2 \quad (2.91)$$

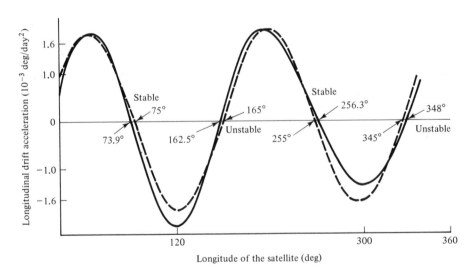

Figure 2.17 Longitudinal drift acceleration due to earth's ellipticity.

where

$$\ddot{\lambda} = \text{longitudinal drift acceleration, deg/day}^2$$
$$\lambda = \text{longitude of the satellite, deg}$$
$$\lambda_s = \text{stable longitude, 75°E and 255°E}$$

2.5 STATION KEEPING

The geosynchronous orbit of a satellite is defined by three elements: $a = 42,164$ km, $e = 0$, and $i = 0$. These orbital elements, however, tend to deviate from the ideal values because of perturbing forces, such as the gravitational forces of the sun, and the moon, and the nonspherical shape of the earth, as discussed in the preceding section. The orbital elements are kept within the allowable limits by performing orbital corrections periodically. This process is called station keeping. An error in the inclination mainly causes latitude (north-south) motion of the subsatellite point. The maneuvers to correct the inclination are called inclination (or north-south) station keeping. An error in the semimajor axis (a) results in longitude drift of the subsatellite point. The maneuvers to correct semimajor axis a are called longitude station-keeping (or east-west) maneuvers. The eccentricity is generally corrected indirectly, by performing longitude correction at apogee to increase the semimajor axis and at perigee to decrease the semimajor axis, as shown in Fig. 2.18.

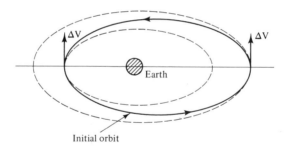

Figure 2.18 Indirect eccentricity correction.

The objective of station keeping is to maintain the orbital elements within the allowable limits for a given period and with minimum fuel and minimum number of maneuvers.

Inclination Station Keeping (North-South)

Inclination station keeping is better understood in terms of the motion of the pole of a satellite orbital plane. The right ascension α and declination δ of the orbital pole are expressed in terms of the right ascension of the ascending node and the orbit inclination as follows:

$$\alpha = \Omega - 90 \tag{2.92}$$

$$\delta = 90 - i \tag{2.93}$$

Figure 2.19 shows a typical correction cycle. The orbital pole is initially at point A, where the inclination is at the allowable inclination limit and Ω is in the neighborhood of 270°. Due to the perturbing forces of the sun, the moon, and the oblateness of the earth, the pole traces the path shown in Fig. 2.19. Initially, the inclination decreases until it reaches approximately zero and then it increases back to the allowable inclination limit to point B, where the right ascension is approximately 90°. At point B a maneuver is performed to move the pole to point C. It should be noted that the inclination remains the same before and after the maneuver. However, the right ascension of the ascending node of the orbit is changed by approximately 180°. The magnitude of ΔV required for the maneuver can be determined by considering the angular change in the angular momentum direction. If θ is the angular change in the angular momentum, which is equal to the great circle arc connection points B and C on the celestial sphere, the magnitude ΔV of the velocity change $\Delta \mathbf{V}$ is, as shown in Fig. 2.20,

$$\Delta V = 2V \sin \frac{\theta}{2} \tag{2.94}$$

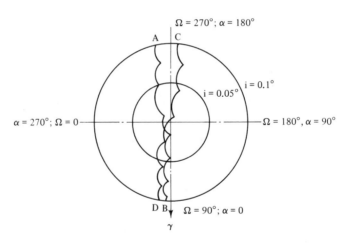

Figure 2.19 Inclination station keeping.

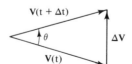

Figure 2.20 Velocity vector diagram for inclination correction.

where V = magnitude of the velocity of the synchronous satellite, 3.074 km/s

θ = angular change of the angular momentum

Using spherical trigonometry, the angle θ can be expressed in terms of the right ascension and declination of point B (the position before the maneuver) and point C (the pole position after the maneuver) as follows:

$$\theta = \cos^{-1} [\sin \delta_B \sin \delta_C + \cos \delta_B \cos \delta_C \cos (\alpha_B - \alpha_C)] \quad (2.95)$$
$$= \cos^{-1} [\cos i_B \cos i_C + \sin i_B \sin i_C \cos (\Omega_B - \Omega_C)]$$

To maximize the time interval between the maneuvers, the maneuver should bring the orbital node to such a position that the future trace of the orbital pole passes through the north pole ($i = 0$). Such a node is called an optimum node and is dependent on the location of the moon and the sun. The optimum node and the interval between the maneuvers are computed by means of the equations for di/dt and $d\Omega/dt$ derived earlier.

For the purpose of spacecraft design, we are interested in the total ΔV required during the design lifetime, the average ΔV required per maneuver, and the average time interval between the maneuvers. These parameters can be calculated using certain simplifying assumptions.

Assuming that the change in the right ascension of the orbital node caused by the maneuver is 180°, the velocity increment can be obtained by the equation

$$\Delta V = 6.148 \sin i \quad \text{km/s} \quad (2.96)$$
$$\cong 107.3 \, i \quad \text{m/s}$$

where i is the allowable inclination limit measured in degrees.

As calculated earlier, for a nearly synchronous orbit with Ω in the neighborhood of 270° or 90°, the inclination drift rate is 0.269°/yr due to the gravitational force of the sun; between 0.674° and 0.478°/yr due to the moon, depending upon its orbit inclination; and zero due to the earth's oblateness. The total inclination drift rate varies from maximum of 0.943°/yr to a minimum of 0.747°/yr, with a period of 9.3 years. Table 2.3 gives the inclination drift rate from 1971 to 2003. The design life of a communications satellite is normally 7 years. Hence the average inclination drift rate for the design life of a satellite can be assumed to be 0.8475°/yr. Assuming an average inclination drift rate of 0.8475°/yr, the average time interval T between the maneuvers is given by

$$T = \frac{2i}{0.8475} \times 365 = 861.4 \, i \quad \text{days} \quad (2.97)$$

Table 2.4 gives the average ΔV required for maneuver and the time interval between maneuvers for several inclination limits.

TABLE 2.4 INCLINATION STATION KEEPING

Inclination Limit (deg)	ΔV per Maneuver (m/s)	Average Time between Maneuvers (days)
0.1	10.7	86.14
0.5	53.65	430.7
1.0	107.30	861.4
2.0	214.56	1,722.8
3.0	321.76	2,584.2

Longitudinal Station Keeping (East-West)

Considering second-order effects only, the longitudinal drift acceleration $\ddot{\lambda}$ due to the ellipticity of the earth at the equator is

$$\ddot{\lambda} = -0.00168 \sin 2(\lambda - \lambda_s) \quad \text{deg/day}^2$$

where λ = satellite longitude (2.98)

λ_s = stable longitude, 75°E and 255°E

A typical east-west correction cycle is shown in Fig. 2.21. Let us assume that λ_0 is a desired longitude of the satellite, $\Delta\lambda$ is the allowable longitude deviation, and that the longitude drift acceleration $\ddot{\lambda}$ is negative. The maneuver is performed at the longitude west limit, $\lambda_0 - \Delta\lambda$, to impart to the satellite the drift rate $\dot{\lambda}_0$. The magnitude of $\dot{\lambda}_0$ is such that the satellite longitude drift rate is zero when it reaches the east limit. At this point the drift is reversed and the drift rate becomes $-\dot{\lambda}_0$ when the satellite reaches the longitude west limit $\lambda_0 - \Delta\lambda$. At the west limit, the maneuver is performed to impart the satellite the drift rate $\dot{\lambda}_0$ and then the cycle is repeated. For positive drift acceleration, the maneuver is performed at the longitude east limit, $\lambda_0 + \Delta\lambda$, and the cycle is reversed.

The drift rate $\dot{\lambda}_0$ can be calculated from the correction cycle characteristic

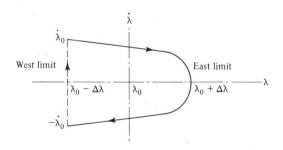

Figure 2.21 Longitudinal station keeping.

that the drift rate changes from $\dot{\lambda}_0$ to zero due to the drift acceleration $\ddot{\lambda}$ in the 2 $\Delta\lambda$ drift. Hence

$$0 = \dot{\lambda}_0^2 - 2\,|\,\ddot{\lambda}\,|\,2\,\Delta\lambda \tag{2.99}$$

or

$$\dot{\lambda}_0 = 2(|\,\ddot{\lambda}\,|\,\Delta\lambda)^{1/2} \tag{2.100}$$

The time interval between successive maneuvers can be calculated from the correction cycle characteristic that the drift rate changes from $\dot{\lambda}_0$ to $-\dot{\lambda}_0$ during the time interval T, or

$$-\dot{\lambda}_0 = \dot{\lambda}_0 - |\,\ddot{\lambda}\,|\,T \tag{2.101}$$

from which we obtain

$$T = \frac{2\dot{\lambda}_0}{|\ddot{\lambda}|} \tag{2.102}$$

Substituting $\dot{\lambda}_0$ from Eq. (2.100) into Eq. (2.101), we can write

$$T = 4\left(\frac{\Delta\lambda}{|\ddot{\lambda}|}\right)^{1/2} \tag{2.103}$$

The maneuver changes the drift rate from $-\dot{\lambda}_0$ to $\dot{\lambda}_0$.

The velocity change ΔV required per maneuver is calculated by using Eqs. (2.31) and (2.21) as follows:

$$n = \frac{\sqrt{\mu_e}}{a^{3/2}} \tag{2.104}$$

$$V^2 = \mu_e\left(\frac{2}{r} - \frac{1}{a}\right) \tag{2.105}$$

Incrementing Eq. (2.104), the longitude drift rate change Δn in rad/s is given by

$$\Delta n = -\frac{3}{2}\frac{\sqrt{\mu_e}}{a^{5/2}}\Delta a \tag{2.106}$$

Incrementing Eq. (2.105) and keeping r constant, because the radius of the satellite can be assumed to change very little during the maneuver, we obtain

$$2V\,\Delta V = \frac{\mu_e}{a^2}\Delta a \tag{2.107}$$

Substituting Δa from Eq. (2.107) into Eq. (2.106), we can write

$$\Delta n = -\frac{3}{a^{1/2}}\frac{V}{\sqrt{\mu_e}}\Delta V \tag{2.108}$$

or

$$\Delta V = -\frac{a^{1/2}\sqrt{\mu_e}}{3V}\Delta n = -\frac{a^2 n}{3V}\Delta n \tag{2.109}$$

For a synchronous satellite,

$$a = 42,164 \text{ km}$$

$$n = \frac{2\pi}{23.9 \times 3600} \text{ rad/s}$$

$$V = 3.074 \text{ km/s}$$

Inserting these parameters into Eq. (2.109), we obtain

$$\Delta V = 14,019.24 \, \Delta n \quad \text{km/s} \tag{2.110}$$

$$= 2.83 \, \Delta\dot\lambda \quad \text{m/s} \tag{2.111}$$

where $\Delta\dot\lambda$ is the longitude drift rate change in deg/day. The velocity increment per maneuver required to change the drift rate by $2\dot\lambda_0$ is given by

$$\Delta V|_{\text{maneuver}} = 5.66 \, \dot\lambda_0 \quad \text{ms}^{-1}/\text{maneuver} \tag{2.112}$$

$$= 11.32 \, (|\ddot\lambda| \, \Delta\lambda)^{1/2} \, \text{ms}^{-1}/\text{maneuver}$$

The velocity change required per year is

$$\Delta V|_{\text{year}} = 5.66 \, \dot\lambda_0 \frac{365}{T} \quad \text{ms}^{-1}/\text{yr} \tag{2.113}$$

Substituting $\dot\lambda_0$ from Eq. (2.100) and T from Eq. (2.103) into Eq. (2.113) and using Eq. (2.98), we obtain

$$\Delta V|_{\text{year}} = 1,032.95 \, |\ddot\lambda| \quad \text{ms}^{-1}/\text{yr} \tag{2.114}$$

$$= 1.74 \sin 2(\lambda - \lambda_s) \quad \text{ms}^{-1}/\text{yr}$$

As discussed in Section 2.5, the drift acceleration due to the moon is periodic with a period of 13.6 days. The drift acceleration consists of the secular acceleration due to the earth ellipticity at the equator and periodic acceleration due to the moon. For a larger longitude tolerance, the time between maneuvers spans several lunar cycles, so that the lunar effects can be neglected. However, for small longitude tolerance the lunar effects should not be neglected and numerical techniques should be used to maximize the time interval between successive maneuvers.

As discussed earlier, the velocity required for longitudinal station keeping depends on the longitude of the satellite. Considering the longitudinal drift acceleration due to the ellipticity of the earth only, the drift acceleration and velocity requirements are zero for the satellite at equilibrium longitudes. Due to the small difference between the semimajor and semiminor axes of the earth's equator, there is some uncertainty in the exact longitude of the equilibrium points. In Fig. 2.17, the dashed line shows the drift acceleration and equilibrium points due to second-order effects only and the solid line includes fourth-order effects. The maximum velocity change ΔV required for station keeping is 1.74 ms^{-1}/yr for $\lambda - \lambda_s = 45°$. The maximum velocity increment required per maneuver and the minimum time interval between successive maneuvers for different longitude tolerances are given in Table 2.5. The values were obtained by means of Eqs. (2.112) and (2.103).

TABLE 2.5 LONGITUDE STATION KEEPING

Longitude Tolerance	$\Delta V_{\max}/$ Maneuver (m/sec)	Minimum Time Interval Between Maneuvers (days)
±0.1	0.15	31
±0.2	0.21	43
±0.5	0.33	69
±1.0	0.46	97
±2.0	0.66	138
±3.0	0.80	169

Station Repositioning

A synchronous satellite is sometimes required to move from one station (longitude) to a different station (longitude). This can be accomplished in two maneuvers. The first maneuver imparts the satellite longitudinal drift rate by changing its orbit velocity. This results in an elliptic orbit. When the satellite reaches the neighborhood of the desired longitude and at a synchronous radius, the second orbit velocity correction is made to bring it back to the synchronous orbit. The velocity required depends on the number of days allowed for the station repositioning. Let us assume that the longitude of a satellite is to be changed in n days. The necessary drift rate $\Delta\dot\lambda$ is given by

$$\Delta\dot\lambda = \frac{\Delta\lambda}{n} \quad \text{deg/day} \tag{2.115}$$

Using Eq. (2.111), the velocity change ΔV_1 required for the first maneuver is

$$\Delta V_1 = 2.83\,\Delta\dot\lambda \quad \text{(m/s)} = 2.83\,\frac{\Delta\lambda}{n} \quad \text{(m/s)} \tag{2.116}$$

The velocity change ΔV_2 required by the second maneuver to correct the orbital velocity to the synchronous orbital velocity is equal and opposite to the velocity change ΔV_1 given by Eq. (2.116). The total velocity change ΔV required for station repositioning is

$$\Delta V = \Delta V_1 + \Delta V_2 = 5.66\,\frac{\Delta\lambda}{n} \quad \text{m/s} \tag{2.117}$$

The reason for adding the two velocity changes, although they are in opposite directions, is that the expenditure of fuel is proportional to velocity changes and the expenditure is the same regardless of the direction of the velocity changes.

Example 2.1

A spacecraft was injected into a geosynchronous orbit on July 1, 1979. The design life of the spacecraft is 10 years. The operational longitude is 325°E. The allowable inclination and longitude tolerances are 0.1°. Determine the average time interval between orbital correction maneuvers for north-south and east-west station keeping and the total ΔV required over 10 years.

Solution *North-south station keeping.* The average inclination drift rate per year from July 1, 1979 to July 1, 1989 from Table 2.3 is

$$= (0.760 + 0.782 + 0.811 + 0.843 + 0.876 + 0.905$$
$$+ 0.926 + 0.940 + 0.943 + 0.937)/10$$
$$= 0.8723°/\text{yr}$$

The average time interval between N/S station keeping maneuvers will be, from Eq. (2.97)

$$T = \frac{2 \times 0.1}{0.8723} \times 365.25 = 83.74 \text{ days}$$

The total number of maneuvers will be

$$\frac{\text{total inclination drift}}{2 \times \text{inclination tolerance}} = \frac{0.8723 \times 10}{2 \times 0.1} \approx 44$$

From Eq. (2.96), the total ΔV required is

$$\text{number of maneuvers} \times \Delta V \text{ per maneuver} = 44 \times 6.148 \sin 0.1°$$
$$= 0.472 \text{ km/s}$$

East-west station keeping. For the operational longitude 325°E, from Eq. (2.98), the longitudinal drift acceleration $\ddot{\lambda}$ is

$$\ddot{\lambda} = -0.00168 \sin 2 (325 - 255)$$
$$= 1.08 \times 10^{-3} \text{ deg/day}^2$$

The time interval between east-west maneuvers is given from Eq. (2.103) as

$$T = 4 \left(\frac{\Delta\lambda}{|\ddot{\lambda}|} \right)^{1/2}$$
$$= 4 \left(\frac{0.1}{1.08 \times 10^{-3}} \right)^{1/2} \text{ days}$$
$$= 38.5 \text{ days}$$

ΔV required per year from Eq. (2.114) is

$$\Delta V = 1.74 \sin 2(\lambda - \lambda_s) \quad \text{ms}^{-1}/\text{yr}$$
$$= 1.74 \sin 2 (325 - 255) = 1.118 \text{ ms}^{-1}/\text{yr}$$

Therefore ΔV required during the design life of 10 years of the satellite is 11.18 ms^{-1}/yr.

2.6 LAUNCH

A typical sequence of the major events involved in injecting a satellite into the synchronous orbit by an expendable launch vehicle is shown in Fig. 2.22. The satellite is launched eastward to be in the same direction as the earth's rotation, to minimize the parking orbit inclination to the latitude of the launch site and to maximize the assistance from the earth's rotational velocity. The parking orbit is a circular orbit with the altitude of about 185.2 km (100 nautical miles), and the inclination is equal to the latitude

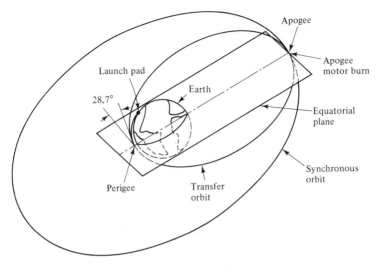

Figure 2.22 Sequence of major events.

of the launch site. The launch vehicle with the satellite coasts in the parking orbit for about 15 minutes to reach the equatorial plane. At the equatorial crossing, the final stage of the launch vehicle is fired to place the satellite into the elliptical transfer orbit with the apogee at synchronous orbit altitude, inclination equal to the launch site latitude, and both apogee and perigee in the equatorial plane. Next, the satellite is separated from the launch vehicle.

The satellite is left in the transfer orbit for several orbital periods so as to obtain accurate orbital data, reorient the spacecraft for correct apogee motor attitude, and wait for the apogee to be close to the operational longitude. For a satellite launched from Cape Canaveral, the first apogee of the transfer orbit is located at 85°E, the second at 285°E, the third at 124°E, and so on. At the selected apogee, the apogee motor is fired to make the orbit circular and simultaneously reduce its inclination to zero. To achieve this in one maneuver, the apogee of the transfer orbit must be in the equatorial plane. This explains why the launch vehicle and the spacecraft coast in the parking orbit until they reach the equatorial plane.

Velocity Requirements for a Satellite Launched from Cape Canaveral

The latitude of Cape Canaveral is 28.7°. Hence the parking orbit and the transfer orbit will have normally a 28.7° inclination angle. The orbital parameters are shown in Fig. 2.23. The velocity requirements for these orbits are calculated by using the orbit equations from Table 2.1 as follows:

The parking orbit velocity V_p is

$$V_p = \sqrt{\frac{\mu_e}{a}} = \sqrt{\frac{398,601.2}{6563.4}} = 7.79 \text{ km/s} \qquad (2.118)$$

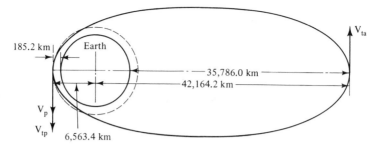

Figure 2.23 Orbital parameters for parking orbit and transfer orbit.

The orbit period τ_p for the parking orbit is

$$\tau_p = 2\pi \frac{a^{3/2}}{\mu^{1/2}} = 2\pi \sqrt{\frac{(6,563.4)^3}{398,601.2}} = 1.47 \text{ hr} \tag{2.119}$$

The transfer orbit velocity V_{tp} at perigee is

$$V_{tp} = \sqrt{\frac{2\mu_e r_a}{(r_a + r_p)r_p}} = \sqrt{\frac{2 \times 398,601.2 \times 42,164.2}{(42,164.2 + 6,563.4) \times 6,563.4}} \tag{2.120}$$
$$= 10.25 \text{ km/s}$$

Hence the velocity change required to transfer the satellite from the parking orbit to the transfer orbit without a plane change is

$$\Delta V_{tp} = V_{tp} - V_p = 2.46 \text{ km/s} \tag{2.121}$$

The transfer orbit period τ_t is given by

$$\tau_t = 2\pi \frac{a^{3/2}}{\mu^{1/2}} = 2\pi \sqrt{\frac{\left(\dfrac{r_a + r_p}{2}\right)^3}{\mu_e}}$$
$$= 2\pi \sqrt{\frac{\left(\dfrac{42,164.2 + 6,563.4}{2}\right)^3}{398,601.2}} = 10.51 \text{ hr} \tag{2.122}$$

The transfer orbit velocity V_{ta} at the apogee is

$$V_{ta} = V_{tp}\frac{r_p}{r_a} = 10.25 \times \frac{6,563.4}{42,164.2} = 1.596 \text{ km/s} \tag{2.123}$$

The synchronous orbit velocity V_s is

$$V_s = \sqrt{\frac{\mu_e}{a}} = \sqrt{\frac{398,601.2}{42,164.2}} = 3.075 \text{ km/s} \tag{2.124}$$

The velocity change ΔV_s provided by the apogee motor to transfer the satellite from the transfer orbit to synchronous orbit can be determined from the velocity vector diagram shown in Fig. 2.24. At apogee, the transfer orbit velocity is 1.596 km/s and the orbit plane is inclined at 28.7° relative

Orbit Dynamics   Chap. 2

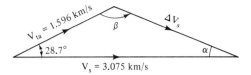

Figure 2.24 Velocity vector diagram at apogee burn.

to the equatorial plane. The apogee motor is required to provide a velocity change ΔV_s so that the satellite attains synchronous orbit velocity. From Fig. 2.24,

$$\Delta V_s = \sqrt{(1.596 \sin 28.7)^2 + (3.075 - 1.596 \cos 28.7)^2} \qquad (2.125)$$

$$= 1.842 \text{ km/s}$$

$$\alpha = \tan^{-1}\left(\frac{1.596 \sin 28.7}{3.075 - 1.596 \cos 28.7}\right) = 24.59° \qquad (2.126)$$

$$\beta = 180° - (28.7° + 24.59°) = 126.71° \qquad (2.127)$$

Hence the apogee motor is required to provide a velocity change of 1.842 km/s at an angle 24.59° with respect to the equatorial plane.

Launch Windows

Launch windows are those periods of time during which the spacecraft and mission constraints are satisfied. The spacecraft constraints are basically solar power, thermal control, and attitude control requirements during the transfer orbit. Electrical power is required for apogee motor ignition, heaters, telemetry/command, and so on. Solar panels normally supply this power except during the eclipse periods, when the power is supplied by batteries. The temperature of the spacecraft, such as that of the apogee motor and other propulsion elements, cannot be allowed to fall below a certain temperature. To satisfy the solar power and thermal control requirements, the sun angle, as defined in Fig. 2.25, is maintained near 90° and the eclipse duration is not allowed to exceed a certain time limit. The solar angle acceptable limits and eclipse duration limits are determined from the solar power and thermal control requirements of the spacecraft. For Intelsat IV, the sun angle acceptable limits are $90° \begin{Bmatrix} +10° \\ -25° \end{Bmatrix}$. Since the sun line is fixed in inertial space (except for approximately 1°/day relative motion) and the attitude at injection into the transfer orbit is fixed in earth coordinates, the sun angle varies with lift-off time. Attitude determination is required in the transfer orbit to plan and execute attitude maneuvers to reorient the spacecraft spin axis for apogee boost. These constraints depend on the sensors used for the transfer orbit. The sensor requirements for the transfer orbit are discussed in Chapter 3.

For a nearly geosynchronous orbit in which some orbit inclination is acceptable, north-south station keeping may not be required if the right ascension of the ascending node at the start of life is optimal (near 270°).

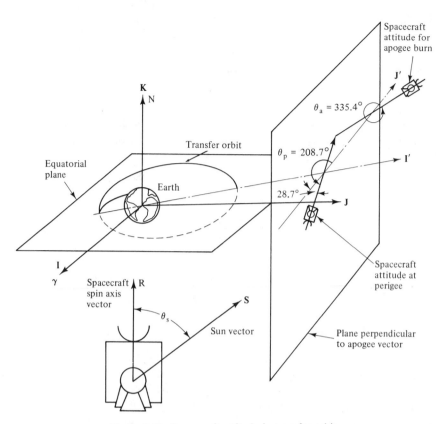

Figure 2.25 Spacecraft attitude in transfer orbit.

For example, if a 3° orbit inclination is acceptable, then for a satellite with a 7-year life, north-south station keeping is not required if the right ascension of the ascending node at the beginning of the life is near 270°. For such satellites, the launch date and the time may be further constrained by the allowable limits on the right ascension of the transfer orbit ascending node.

Sun Angle

The sun angle θ_s, as shown in Fig. 2.25, is defined as the angle between the spacecraft spin axis vector (positive on the antenna side) and the sun vector. The sun vector $\mathbf{S}$ is given in inertial coordinates by

$$\mathbf{S} = \cos \alpha_s \cos \delta_s \mathbf{I} + \sin \alpha_s \cos \delta_s \mathbf{J} + \sin \delta_s \mathbf{K} \qquad (2.128)$$

where

α_s = right ascension of the sun in the celestial sphere

δ_s = declination of the sun

The spacecraft spin axis vector $\mathbf{R}$ is defined in terms of the coordinate system I' J' K' by

$$\mathbf{R} = \cos \theta \, \mathbf{J}' + \sin \theta \, \mathbf{K}' \qquad (2.129)$$

where

$\mathbf{I'}$ = unit radial vector toward the apogee of the transfer orbit

$\mathbf{K'} = \mathbf{K}$

$\mathbf{J'} = \mathbf{K'} \times \mathbf{I'}$

θ = angle between $\mathbf{J'}$ axis and spacecraft spin axis $\mathbf{R}$
where $\mathbf{R}$ is in a plane perpendicular to $\mathbf{I'}$

It should be noted that the apogee of the transfer orbit is assumed to be in the equatorial plane. The coordinated system $\mathbf{I'\,J'\,K'}$ is obtained from the inertial coordinate system $\mathbf{I\,J\,K}$ by rotating it about the $\mathbf{K}$-axis by an angle equal to the right ascension of the apogee of the transfer orbit. The unit vectors $\mathbf{I'}$, $\mathbf{J'}$, and $\mathbf{K'}$ are related to $\mathbf{I}$, $\mathbf{J}$, and $\mathbf{K}$ as follows:

$$\begin{Bmatrix} \mathbf{I'} \\ \mathbf{J'} \\ \mathbf{K'} \end{Bmatrix} = \begin{bmatrix} \cos \alpha_a & \sin \alpha_a & 0 \\ -\sin \alpha_a & \cos \alpha_a & 0 \\ 0 & 0 & 1 \end{bmatrix} \begin{Bmatrix} \mathbf{I'} \\ \mathbf{J'} \\ \mathbf{K'} \end{Bmatrix} \tag{2.130}$$

where α_a is the right ascension of the apogee. Substituting Eq. (2.130) into Eq. (2.129), we obtain

$$\mathbf{R} = -\cos \theta \sin \alpha_a \, \mathbf{I} + \cos \theta \cos \alpha_a \, \mathbf{J} + \sin \theta \, \mathbf{K} \tag{2.131}$$

Using Eqs. (2.128) and (2.131), we can write

$$\cos \theta_s = \mathbf{S} \cdot \mathbf{R} = -\cos \alpha_s \cos \delta_s \cos \theta \sin \alpha_a \tag{2.132}$$
$$+ \sin \alpha_s \cos \delta_s \cos \theta \cos \alpha_a + \sin \theta \sin \delta_s$$

The right ascensions of the apogee and the perigee are calculated as follows. The right ascension of the launch site at the time of lift-off is

$$\alpha_l = \alpha_s + 180 + \lambda_l + \frac{360}{24} T_l \tag{2.133}$$

where

α_s = right ascension of the sun

T_l = Greenwich time of the launch

λ_l = longitude of the launch, east positive, west negative

The right ascension α_p of the perigee is

$$\alpha_p = \alpha_s + \lambda_p + \frac{360}{24} (T_l + t) + 180 \tag{2.134}$$

where λ_p = longitude of subsatellite point of the perigee

t = time from the lift-off to the perigee injection

The right ascension α_a of the apogee is

$$\alpha_a = \alpha_p + 180 = \alpha_s + \lambda_p + \frac{360}{24} (T_l + t) \tag{2.135}$$

For a satellite launched from Cape Canaveral by a Delta launch vehicle

$$\lambda_l = -80.56° \qquad t = 0.39 \text{ hr} \qquad \lambda_p = -6.4° \tag{2.136}$$

Substituting the parameters from Eq. (2.136) into Eq. (2.135), we obtain

$$\alpha_a = \alpha_s + \frac{360}{24} T_l - 0.55$$

$$\cong \alpha_s + \frac{360}{24} T_l$$

(2.137)

For a given lift-off time, the sun angle is calculated by substituting the parameters corresponding to that time in Eq. (2.132). Figure 2.26 shows the launch windows satisfying the sun angle constraints $90 \pm 35°$ for the spacecraft attitude at perigee after the separation, $\theta_p = 208.7°$, and at apogee motor burn, $\theta_a = 335.4°$, as a function of the launch date. The launch windows exist around Greenwich mean time (GMT) noon and GMT midnight. The noon windows exhibit long eclipse periods around transfer orbit apogee (more than 2 hours around equinoxes), which are undesirable. Midnight launches have short eclipses (less than 1/2 hour), if any, due to the fact that they occur at perigee. Hence GMT noon launches are considered only in rare instances when insufficient midnight launch windows occur. It should be noted that in the case of Ariane and Atlas/Centaur launches, the injection attitude can be selected by the user, so the relationship between the sun angle and lift-off time becomes more complex.

Launch windows for shuttle. For the shuttle, perigee motor firing can occur at equatorial crossings referred to as injection opportunities (IO)

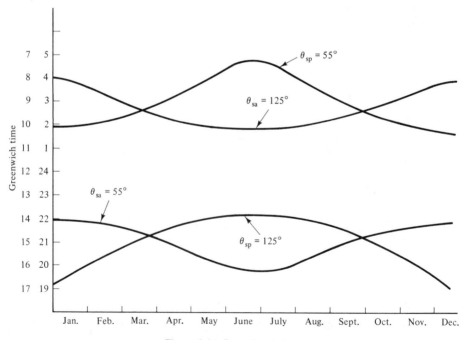

Figure 2.26 Launch windows.

at either ascending or descending nodes of the parking orbit. The IOs must occur near local noon or midnight at the injection point in order to satisfy sun angle constraints, i.e., to have sun angle near 90°. Figure 2.27 shows the case where all descending node IOs occur near local noon and all ascending node IOs near local midnight. In order to launch into the parking orbit shown in Fig. 2.27 from Kennedy Space Center, the launch must be around noon GMT or at dawn at the launch site. If the inertial orientation of the parking orbit shown in Fig. 2.27 were advanced by 180°, then all ascending node IOs would occur near local midnight, and lift-off from KSC must occur around midnight GMT or at evening at the launch site. Therefore, there are two opportunities every day, one near dawn at launch site and one at evening.

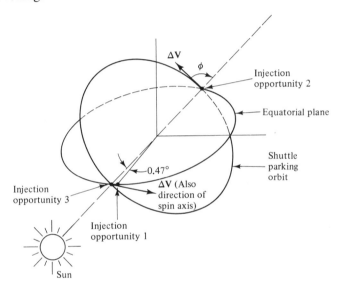

Figure 2.27 Injection opportunities for noon launches.

The parking orbit nodes regress at a rate of 7.5°/day due to the oblateness of the earth's mass distribution. This results in successive local-noon IOs or local-midnight IOs which are separated by 0.47°. Consequently, there is a reduction in allowable launch window intervals as dwell time in the parking orbit increases.

Local-noon IOs are limited by large eclipse periods during the transfer orbit. As both noon and midnight GMT launches provide local-noon and local-midnight IOs, the noon GMT launch window is preferred because of the launch of the Shuttle during daylight. For perigee injection, local-midnight IOs are preferred since local-noon IOs are limited by large eclipse periods.

Solar Eclipses

A geosynchronous satellite experiences two periods of solar eclipses centered around vernal and autumnal equinoxes. The longest eclipses, of about 70 minutes, occur at equinoxes, around March 21 and September 23. In this section the equations for eclipse periods are derived for circular orbits.

In a general eclipse case, the projection of the earth's shadow on the satellite orbit plane is an ellipse whose semiminor axis is the radius of the earth, r_e, and whose semimajor axis is $r_e/\sin \delta$, where δ is the angle between the sun's rays and the orbit plane, as shown in Fig. 2.28. An eclipse occurs only if the semimajor is greater than the satellite orbit radius, R (i.e., $\sin \delta < r_e/R$). For a geostationary spacecraft, this means that sun eclipses will occur only when the sun's declination, δ, is less than 8.7°. The equation of the ellipse is

$$\left(\frac{x}{r_e/\sin \delta}\right)^2 + \left(\frac{y}{r_e}\right)^2 = 1 \tag{2.138}$$

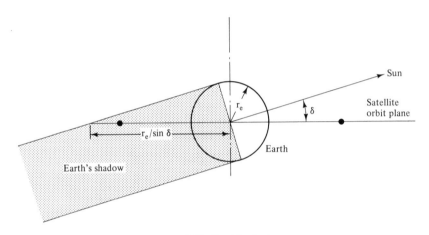

Figure 2.28 Earth's shadow.

The satellite enters the eclipse, as shown in Fig. 2.29, at $x = R \cos \theta$ and $y = R \sin \theta$. Substituting these values in the equations above, the angle θ is given by

$$\theta = \cos^{-1} \frac{(1 - r_e^2/R^2)^{1/2}}{\cos \delta} \tag{2.139}$$

the total eclipse period T_e is

$$T_e = \frac{P}{\pi} \cos^{-1} \frac{(1 - r_e^2/R^2)^{1/2}}{\cos \delta} \tag{2.140}$$

where P is the period of the satellite orbit. For the case of $\delta = 0$ (i.e., at equinox for geosynchronous orbit), the angle θ is given directly from Fig. 2.29 as

$$\theta_e = \sin^{-1} r_e/R \tag{2.141}$$

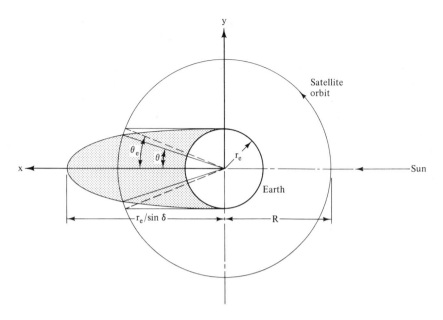

Figure 2.29 Earth's shadow in orbit plane.

and

$$T_e = \frac{P}{\pi} \sin^{-1} \frac{r_e}{R} \qquad (2.142)$$

For synchronous orbit, T_e is 1.16 hours or approximately 70 minutes.

In geosynchronous orbit, the eclipses occur near local midnight for a total of about 90 days per year in the spring and fall, centered around March 21 and September 23, respectively. Eclipses lasting about 70 minutes occur around these dates and those lasting longer than 1 hour occur about 50 days per year. Spring eclipses begin in late February or early March and end in about mid-April. Fall eclipses begin about September 1 and end about mid-October.

Azimuth and Elevation of a Satellite

In the previous sections, the satellite location was determined in the coordinate systems whose origins were at the center of the earth. For communications satellites, however, it is necessary to determine the satellite location with respect to the earth stations on the earth's surface. The location of a satellite with respect to an earth station is expressed in terms of azimuth, elevation, and slant range.

In the new coordinate system, the origin is at the observer's (earth station) position and the fundamental plane is the local horizon, which is a tangent plane to the observer's meridian at the observer's position. The location of an object (satellite) is defined by two angles: azimuth and elevation.

Azimuth angle A of the object is the angle from the north to the object's meridian in the horizontal planes. Elevation angle, h, is the angular elevation of the object above the horizon plane.

By plotting the north pole, the object's position and observer's zenith point directly overhead on the celestial sphere, as shown in Fig. 2.30, and using spherical trignometry, the expressions for azimuth and elevation of the object are determined. In Fig. 2.30, $\Delta\lambda$ is the difference between the longitudes of the object and the observer, ϕ is the observer's latitude, δ is the object's latitude, and A is the azimuth of the object. The angle between the normal to the horizon plane, the radial vector to the observer, and the radial vector to the object (satellite) is $90 - h_\infty$, where h_∞ is the elevation of the object. From the law of cosines applied to the spherical triangle in Fig. 2.30, the elevation angle h_∞ is given by

$$\sin h_\infty = \sin \delta \sin \phi + \cos \delta \cos \phi \cos \Delta\lambda \qquad (2.143)$$

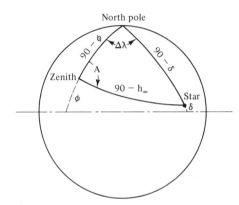

Figure 2.30 Azimuth of a satellite from an earth station.

The azimuth A can be calculated from the equation

$$\tan A = \frac{\sin \Delta\lambda}{\cos \phi \tan \delta - \sin \phi \cos \Delta\lambda} \qquad (2.144)$$

For communications satellites that are not infinitely far away, the observed elevation from an earth station, as shown in Fig. 2.31 is h and is given by

$$\tan h = \frac{\sin h_\infty - R/r}{\cos h_\infty} \qquad (2.145)$$

The range is the distance from the satellite to an earth station, a point on the earth. From Fig. 2.31, the range, ρ, is given by

$$\rho = \frac{r - R \sin h_\infty}{\cos (h_\infty - h)} \qquad (2.146)$$

A graph of the azimuth and elevation for geostationary satellites as a function of earth station latitude and the difference in the longitude between the earth station and the satellite is given in Fig. 2.32. To use the graph, find the intersection of the two appropriate contour lines for earth station latitude, and the longitude difference between the earth station and the satellite, and read off the elevation and azimuth from the proper scales.

Orbit Dynamics   Chap. 2

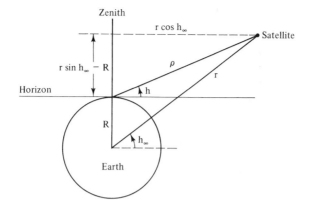

Figure 2.31 Elevation of a satellite, h, from an earth station.

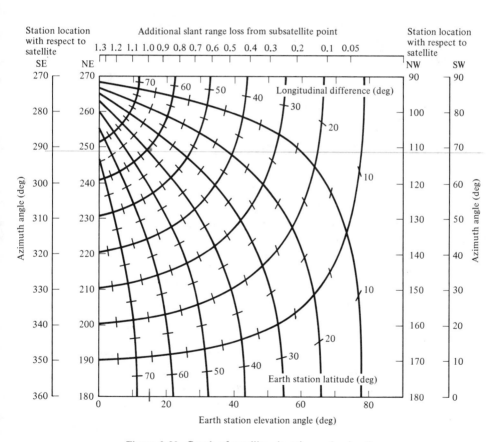

Figure 2.32 Graph of satellite elevation and azimuth.

PROBLEMS

2.1. A spacecraft was injected into a synchronous orbit on January 1, 1978. The design life of the spacecraft is 7 years. The operational longitude is 60°E. The allowable inclination and longitude tolerances are ±0.1°. Determine the average time interval between orbital correction maneuvers for north-south and east-west station keeping and the total ΔV required over 7 years.

2.2. A satellite is injected by the Shuttle in a 296-km circular parking orbit with an inclination of 28.5°. Determine (a) the parking orbit velocity, (b) the parking orbit period, (c) the ΔV provided by the perigee motor to transfer the satellite from the parking orbit to the transfer orbit, (d) the transfer orbit period, and (e) the ΔV provided by the apogee motor to transfer the satellite from the transfer orbit to the synchronous orbit.

2.3. A satellite is launched from Cape Canaveral by a Delta launch vehicle at 7 P.M. on January 1, 1980. The time from launch to the transfer orbit injection (perigee motor fire) is 0.39 hours. Determine (a) the right ascension of the perigee, (b) the right ascension of the apogee, (c) the sun angle at perigee, and (d) and the sun angle at apogee.

2.4. A spare satellite has to be moved from longitude 63°E to 335°E to replace an operational satellite. The maximum allowable time for repositioning is 30 days. Determine the ΔV required for repositioning the satellite.

2.5. The longitude and latitude for the Etam earth station are 280.3°E and 39.28°N, respectively. By using Fig. 2.32, determine the azimuth and elevation from the Etam earth station of a satellite whose longitude is 340°E.

REFERENCES

1. W. T. Thomson, *Introduction to Space Dynamics*, Wiley, New York, 1961.

2. M. H. Kaplan, *Modern Spacecraft Dynamics and Control*, Wiley, New York, 1976.

3. R. H. Battin, *Astronautical Guidance*, McGraw-Hill, New York, 1964.

4. P. R. Escobal, *Methods of Orbit Determination*, Wiley, New York, 1965.

5. R. M. Baker, Jr., *Astrodynamics: Applications and Advanced Topics*, Academic Press, New York, 1967.

6. L. Meirovitch, *Methods of Analytical Methods*, McGraw-Hill, New York, 1970.

7. G. E. Cook, *Luni-Solar Perturbations of the Orbit of an Earth Satellite*, Royal Aircraft Establishment, Tech Note G.W. 582, July 1961.

8. R. R. Allan, *Perturbation of a Geostationary Satellite, The Ellipticity of the Equator*, Royal Aircraft Establishment, Technical Note Space 43.

9. J. S. Tyler, J., and E. W. Onstead, "Launch Constraint for a Synchronous Communications Satellite Mission," *Journal of Spacecraft and Rockets*, Vol. 6, No. 8, Aug. 1980.

3

ATTITUDE DYNAMICS AND CONTROL

3.1 INTRODUCTION

Communications satellites require high pointing accuracies for the antennas so that they may provide the desired coverage on the earth's surface. This requirement is achieved through an attitude control system which maintains the spacecraft attitude, its orientation in space, within the allowable limits. It consists of sensors for attitude determination and actuators to provide corrective torques. The disturbance torques are from many sources, such as the solar pressure, the gravity gradient, and misalignments of the spacecraft thrusters. The attitude control system also provides the desired spacecraft attitude for orbital maneuvers.

A block diagram of an attitude control system is shown in Fig. 3.1. The basic elements of an attitude control system are spacecraft attitude dynamics, attitude sensors, control laws, actuators, and disturbance torques. Spacecraft attitude dynamics predicts the motion of the spacecraft body as a result of disturbance and control torques. Orbit dynamics, discussed in Chapter 2, deals with the motion of the center of mass of a spacecraft. Attitude dynamics, on the other hand, is concerned with the rotational motion of the spacecraft body about its center of mass. The sensors provide information on the attitude of the spacecraft. The sensor outputs are used by control laws to determine the required control torques. The control torques are provided by actuators, which could be a combination of momentum wheels, reaction wheels, and thrusters. To design a control system, it is also necessary to be able to estimate disturbance torques. This chapter provides a basic understanding of these elements of the attitude control systems for both spin-stabilized and three-axis-stabilized spacecraft.

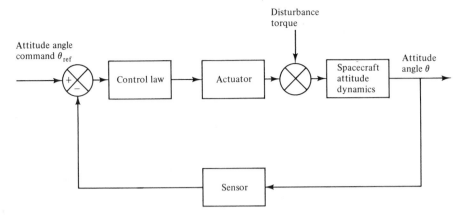

Figure 3.1 Block diagram of an attitude control system.

3.2 RIGID-BODY DYNAMICS

The attitude dynamics of a spacecraft is based primarily on rigid-body dynamical equations. These equations, however, are modified for flexibility in the cases of flexible spacecraft requiring high-attitude accuracies. The attitude motions are normally represented in the spacecraft body fixed coordinate systems, which are rotating and accelerating. Hence the subject of relative motion and transformation of coordinates plays an important role in attitude dynamics. This section deals with the general motion of a particle in a rotating coordinate system and derives the equations of motion of a rigid body.

Derivative of a Vector in a Rotating Coordinate System

In many situations it is convenient to refer a vector to a rotating coordinate system. In such cases, the time derivative of the vector consists of two parts: one due to the rate of change of the vector relative to the moving axes, and the other due to the rotation of the axes. Let XYZ be an inertial system and xyz a set of axes rotating with the angular velocity $\boldsymbol{\omega}$ relative to XYZ as shown in Fig. 3.2. Then, if $\mathbf{i}$, $\mathbf{j}$, and $\mathbf{k}$ are unit vectors along the x, y, and z axes, respectively, the vector $\mathbf{r}$ can be written in the form

$$\mathbf{r} = x\mathbf{i} + y\mathbf{j} + z\mathbf{k} \tag{3.1}$$

The time derivative of $\mathbf{r}$ is simply

$$\dot{\mathbf{r}} = \dot{x}\mathbf{i} + x\dot{\mathbf{i}} + \dot{y}\mathbf{j} + y\dot{\mathbf{j}} + \dot{z}k + z\dot{\mathbf{k}} \tag{3.2}$$

It can be shown (Ref. 1) that the time derivatives of the unit vectors are

$$\dot{\mathbf{i}} = \boldsymbol{\omega} \times \mathbf{i} \qquad \dot{\mathbf{j}} = \boldsymbol{\omega} \times \mathbf{j} \qquad \dot{\mathbf{k}} = \boldsymbol{\omega} \times \mathbf{k} \tag{3.3}$$

Moreover, introducing the notation

$$\dot{\mathbf{r}}_{\text{rel}} = \dot{x}\mathbf{i} + \dot{y}\mathbf{j} + \dot{z}\mathbf{k} \tag{3.4}$$

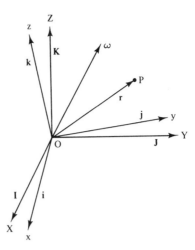

Figure 3.2 Derivative of vector in a rotating coordinate system.

where $\dot{\mathbf{r}}_{rel}$ is recognized as the time rate of change of $\mathbf{r}$ relative to xyz, Eq. (3.2) is reduced to

$$\dot{\mathbf{r}} = \dot{\mathbf{r}}_{rel} + \boldsymbol{\omega} \times \mathbf{r} \tag{3.5}$$

General Motion of a Particle

This approach can be used to develop expressions for the general motion of a particle. In the more general case, the frame xyz is capable not only of rotation relative to XYZ, but also of translation, as shown in Fig. 3.3. Denoting the origin of xyz by C, the position vector of a particle P relative to XYZ can be written in the form

$$\mathbf{R} = \mathbf{R}_c + \mathbf{r} \tag{3.6}$$

where $\mathbf{R}_c$ is the radius vector from O to C and $\mathbf{r}$ is the position vector of P relative to xyz. Differentiating Eq. (3.6) with respect to time, we obtain

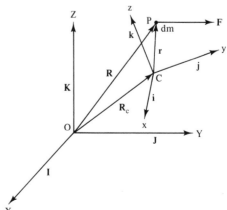

Figure 3.3 General motion of a particle.

the absolute velocity of P in the form

$$V = \dot{R} = \dot{R}_c + \dot{r} \tag{3.7}$$

Noting that $\dot{R}_c = V_c$ is the velocity of the origin C and recalling Eq. (3.5), we obtain the absolute velocity of P as follows:

$$V = V_c + V_{rel} + \omega \times r \tag{3.8}$$

where $V_{rel} = \dot{r}_{rel}$ is the velocity of P relative to xyz.

The absolute acceleration of P can be obtained by differentiating Eq. (3.8), so that

$$a = \dot{V} = \dot{V}_c + \frac{d}{dt}(V_{rel}) + \dot{\omega} \times r + \omega \times \dot{r} \tag{3.9}$$

But $\dot{V}_c = a_c$ is the acceleration of C. Moreover, recalling Eq. (3.4), we can write

$$\frac{d}{dt}(V_{rel}) = \ddot{x}i + \dot{x}\dot{i} + \ddot{y}j + \dot{y}\dot{j} + \ddot{z}k + \dot{z}\dot{k}$$
$$= a_{rel} + \omega \times V_{rel} \tag{3.10}$$

where $a_{rel} = \ddot{r}_{rel}$ is the acceleration of P relative to xyz. Inserting Eq. (3.10) into Eq. (3.9) and recalling Eq. (3.5), we obtain the absolute acceleration of P in the form

$$a = a_c + a_{rel} + 2\omega \times V_{rel} + \dot{\omega} \times r + \omega \times (\omega \times r) \tag{3.11}$$

where $2\omega \times V_{rel}$ is known as the Coriolis acceleration.

Momentum of a Rigid Body

The linear momentum of a particle is defined as the product of its mass and its velocity. Let dm be a differential element of mass in a rigid body and V the velocity of the element (Fig. 3.3); the linear momentum of a rigid body can be written in the form

$$P = \int_m V \, dm \tag{3.12}$$

where m is the mass. Substituting Eq. (3.8) into Eq. (3.12) and assuming that axes xyz are embedded in the rigid body, so that $V_{rel} = 0$, we obtain

$$P = \int_m (V_c + \omega \times r) \, dm = mV_c + \omega \times \int_m r \, dm \tag{3.13}$$

Equation (3.13) can be simplified significantly by choosing the origin C of the rotating coordinate system to coincide with the center of the mass (cm) of the rigid body, as in this case $\int_m r \, dm = 0$ by definition. It follows that

$$P = mV_c \tag{3.14}$$

or, the linear momentum of a rigid body is equal to the product of its mass and the velocity of its mass center.

The angular momentum of a particle dm about point C is defined as the moment about C of the linear momentum, or

$$\mathbf{h}_c = \mathbf{r} \times \mathbf{V} \, dm \tag{3.15}$$

Using the same approach as for the linear momentum, the angular momentum of the rigid body can be written in the form

$$\mathbf{H}_c = \int_m \mathbf{r} \times \mathbf{V} \, dm = \int_m \mathbf{r} \times (\mathbf{V}_c + \boldsymbol{\omega} \times \mathbf{r}) \, dm$$

$$= \left(\int_m \mathbf{r} \, dm \right) \times \mathbf{V}_c + \int_m \mathbf{r} \times (\boldsymbol{\omega} \times \mathbf{r}) \, dm \tag{3.16}$$

Once again letting the origin C of the frame xyz coincide with the mass center of the rigid body, Eq. (3.16) reduces to

$$\mathbf{H}_c = \int_m \mathbf{r} \times (\boldsymbol{\omega} \times \mathbf{r}) \, dm \tag{3.17}$$

Writing the angular velocity vector $\boldsymbol{\omega}$ by components

$$\boldsymbol{\omega} = \omega_x \mathbf{i} + \omega_y \mathbf{j} + \omega_z \mathbf{k} \tag{3.18}$$

and recalling Eq. (3.1), Eq. (3.17) yields

$$\begin{aligned} \mathbf{H}_c = &\; [I_{xx}\omega_x - I_{xy}\omega_y - I_{xz}\omega_z]\mathbf{i} \\ &+ [-I_{xy}\omega_x + I_{yy}\omega_y - I_{yz}\omega_z]\mathbf{j} \\ &+ [-I_{xz}\omega_x - I_{yz}\omega_y + I_{zz}\omega_z]\mathbf{k} \end{aligned} \tag{3.19}$$

where

$$I_{xx} = \int_m (y^2 + z^2) \, dm \qquad I_{yy} = \int_m (x^2 + z^2) \, dm$$

$$I_{zz} = \int_m (x^2 + y^2) \, dm \tag{3.20}$$

are the moments of inertia of the body about axes x, y, z, respectively, and

$$I_{xy} = \int_m xy \, dm, \qquad I_{xz} = \int_m xz \, dm, \qquad I_{yz} = \int_m yz \, dm \tag{3.21}$$

are the products of inertia. The axes about which products of inertia are zero are called the principal axes of moment of inertia.

It should be noted that for a coordinate system fixed in the body, the moments and products of inertia are constant quantities. Equation (3.19) can be written in the matrix form

$$\mathbf{H}_c = \begin{Bmatrix} H_x \\ H_y \\ H_z \end{Bmatrix} = \begin{bmatrix} I_{xx} & -I_{xy} & -I_{xz} \\ -I_{xy} & I_{yy} & -I_{yz} \\ -I_{xz} & -I_{yz} & I_{zz} \end{bmatrix} \begin{Bmatrix} \omega_x \\ \omega_y \\ \omega_z \end{Bmatrix} \tag{3.22}$$

Kinetic Energy of a Rigid Body

The kinetic energy of a rigid body is defined as

$$T = \frac{1}{2} \int_m \mathbf{V} \cdot \mathbf{V} \, dm \tag{3.23}$$

Using Eq. (3.8) with $V_{rel} = 0$, Eq. (3.23) reduces to

$$T = \frac{1}{2} m \mathbf{V}_c \cdot \mathbf{V}_c + \mathbf{V}_c \cdot (\boldsymbol{\omega} \times \int_m \mathbf{r}\, dm) + \frac{1}{2} \int_m (\boldsymbol{\omega} \times \mathbf{r}) \cdot (\boldsymbol{\omega} \times \mathbf{r})\, dm \quad (3.24)$$

and if C is the center of mass of the rigid body, the first integral will be zero and

$$T = \frac{1}{2} m V_c^2 + \frac{1}{2} [I_{xx}\omega_x^2 + I_{yy}\omega_y^2 + I_{zz}\omega_z^2$$
$$- 2\,\omega_x\omega_z I_{xz} - 2\omega_y\omega_z I_{yz} - 2\omega_x\omega_y I_{xy}] \quad (3.25)$$

where V_c is the magnitude of $\mathbf{V}_c$. The first term in Eq. (3.25) represents the kinetic energy of the body as if it were in pure translation and the remaining terms constitute the kinetic energy of the body as if it were in pure rotation about C. The kinetic energy of rotation, T_{rot}, can be expressed in the matrix form

$$T_{rot} = \frac{1}{2} [\omega_x \quad \omega_y \quad \omega_z] \begin{bmatrix} I_{xx} & -I_{xy} & -I_{xz} \\ -I_{xy} & I_{yy} & -I_{yz} \\ -I_{xz} & -I_{yz} & I_{zz} \end{bmatrix} \begin{Bmatrix} \omega_x \\ \omega_y \\ \omega_z \end{Bmatrix} \quad (3.26)$$

Equations of Motion

The equations of motion are based on Newton's second law, which states that the rate of change of the linear momentum of a particle is equal to the force acting on the particle. It should be noted that in applying Newton's second law one must measure the motion relative to an inertial coordinate system.

Let us consider a particle of mass, m, moving under the action of a force, $\mathbf{F}$, as shown in Fig. 3.3. According to Newton's second law,

$$\mathbf{F} = \frac{d}{dt}(m\mathbf{V}) = m\dot{\mathbf{V}} = m\mathbf{a} \quad (3.27)$$

The angular momentum of the particle relative to an arbitrary point C is

$$\mathbf{h}_c = \mathbf{r} \times m\mathbf{V} \quad (3.28)$$

The time derivative of the angular momentum is

$$\dot{\mathbf{h}}_c = \dot{\mathbf{r}} \times m\mathbf{V} + \mathbf{r} \times m\dot{\mathbf{V}} \quad (3.29)$$

so that, using Eqs. (3.7) and (3.27), we obtain

$$\dot{\mathbf{h}}_c = \dot{\mathbf{r}} \times m(\dot{\mathbf{R}}_c + \dot{\mathbf{r}}) + \mathbf{r} \times \mathbf{F} \quad (3.30)$$
$$= -\dot{\mathbf{R}}_c \times m\dot{\mathbf{r}} + \mathbf{r} \times \mathbf{F}$$

By definition, however, the moment of the force $\mathbf{F}$ about point C is

$$\mathbf{M}_c = \mathbf{r} \times \mathbf{F} \quad (3.31)$$

Substituting Eq. (3.30) into Eq. (3.31), we obtain

$$\mathbf{M}_c = \dot{\mathbf{h}}_c + \dot{\mathbf{R}}_c \times m\dot{\mathbf{r}} \quad (3.32)$$

The force and moment equations of the motion, Eqs. (3.27) and (3.32), were derived for a single particle, but they can be extended to the motion of rigid bodies. Recognizing that for rigid bodies there is no motion relative to the frame xyz and using Eq. (3.11), we can write the force equation of the motion

$$\mathbf{F} = \int_m \mathbf{a} \, dm = \int_m [\mathbf{a}_c + \boldsymbol{\omega} \times \mathbf{r} + \boldsymbol{\omega} \times (\boldsymbol{\omega} \times \mathbf{r})] \, dm \qquad (3.33)$$

where $\mathbf{F}$ is the resultant of the external forces acting on the rigid body and m is the total mass of the body. Note that mutual attraction forces between any two particles in the rigid body are internal to the body and cancel out in pairs. If C coincides with the center of the mass of the rigid body, so that $\int_m \mathbf{r} \, dm = 0$, Eq. (3.33) reduces to

$$\mathbf{F} = m\mathbf{a}_c \qquad (3.34)$$

or, a rigid body moves as if it were a single particle of mass equal to the total mass, concentrated at the center of mass C and acted on by the resultant $\mathbf{F}$ of the external forces.

The moment equation of motion about C can be obtained in an analogous manner. Using Eq. (3.32), we can write the moment equation about C in the form

$$\mathbf{M}_c = \dot{\mathbf{H}}_c + \int_m \dot{\mathbf{R}}_c \times \dot{\mathbf{r}} \, dm \qquad (3.35)$$

where $\mathbf{M}_c$ is the resultant of the external moments about C and $\mathbf{H}_c$ is the angular momentum of the rigid body about C. Using Eq. (3.5) and recalling that for a rigid body $\dot{\mathbf{r}}_{\text{rel}} = 0$, we can write

$$\int_m \dot{\mathbf{R}}_c \times \dot{\mathbf{r}} \, dm = \dot{\mathbf{R}}_c \times \left(\boldsymbol{\omega} \times \int_m \mathbf{r} \, dm \right) \qquad (3.36)$$

But if C is the mass center of the body $\int_m \mathbf{r} \, dm = 0$, Eq. (3.35) reduces to

$$\mathbf{M}_c = \dot{\mathbf{H}}_c \qquad (3.37)$$

or, the moment of the external forces about the center of the mass of a rigid body is equal to the time rate of change with time of the angular momentum of the body about the center of the mass.

It will be more convenient to express the angular momentum in terms of components along axes x, y, z in the form

$$\mathbf{H}_c = H_x \mathbf{i} + H_y \mathbf{j} + H_z \mathbf{k} \qquad (3.38)$$

Because x, y, z are rotating axes, we can use Eq. (3.5) for the time derivative of a vector expressed in terms of rotating axes and rewrite Eq. (3.37) in the form

$$\mathbf{M}_c = \dot{\mathbf{H}}_{\text{crel}} + \boldsymbol{\omega} \times \mathbf{H}_c \qquad (3.39)$$

Equations (3.34) and (3.39) can be written by components as follows:

$$F_x = m\ddot{x}_c \qquad (3.40a)$$

$$F_y = m\ddot{y}_c \qquad (3.40b)$$

$$F_z = m\ddot{z}_c \tag{3.40c}$$

$$M_x = \dot{H}_x + \omega_y H_z - \omega_z H_y \tag{3.40d}$$

$$M_y = \dot{H}_y + \omega_z H_x - \omega_x H_z \tag{3.40e}$$

$$M_z = \dot{H}_z + \omega_x H_y - \omega_y H_x \tag{3.40f}$$

where H_x, H_y, and H_z are given by Eq. (3.22). In the special case in which the rotating axes x, y, z coincide with the principal axes of moment of inertia, the components of the angular momentum are reduced to

$$
\begin{aligned}
H_x &= I_{xx}\omega_x \\
H_y &= I_{yy}\omega_y \\
H_z &= I_{zz}\omega_z
\end{aligned} \tag{3.41}
$$

Substituting Eq. (3.41) into Eqs. (3.40d) through (3.40f) the moment equations become

$$
\begin{aligned}
M_x &= I_{xx}\dot{\omega}_x + \omega_y\omega_z(I_{zz} - I_{yy}) \\
M_y &= I_{yy}\dot{\omega}_y + \omega_x\omega_z(I_{xx} - I_{zz}) \\
M_z &= I_{zz}\dot{\omega}_z + \omega_x\omega_y(I_{yy} - I_{xx})
\end{aligned} \tag{3.42}
$$

which are known as Euler's moment equations.

3.3 ATTITUDE STABILIZATION

Spacecraft attitude control systems can be classified into two broad categories: spin stabilization and three-axis stabilization. Spin stabilization is based on gyroscopic stiffness due to the rotation of all or part of the spacecraft body. There are two subsets of this class: single-spin stabilization and dual-spin stabilization. In single-spin stabilization, the whole body rotates about the axis of maximum or minimum principal moment of inertia. Early communications satellites, such as Syncom and Intelsat I and II were single-spin stabilized. The primary limitations of these satellites were that they could not use earth-oriented antennas, thus resulting in the requirement for an omnidirectional antenna with a consequential very low antenna gain toward the earth. This limitation is overcome in a dual-spin-stabilized spacecraft, which is divided into two parts: the platform and the bus, each part rotating at different rates. The platform, consisting of antennas and communications equipment, orients toward the earth by rotating at one revolution per day. The bus rotates at a higher spin rate, nominally at 60 rpm, to provide gyroscopic stiffness. In the case of three-axis stabilization, the entire spacecraft, except the solar arrays, which are sun oriented, is oriented toward the earth. The control torques along the three axes are provided by a combination of momentum wheels, reaction wheels, magnetic torque, and thrusters.

The coordinate systems used in attitude control are shown in Fig. 3.4. The inertially fixed coordinate system $X_0Y_0Z_0$ is used to determine the

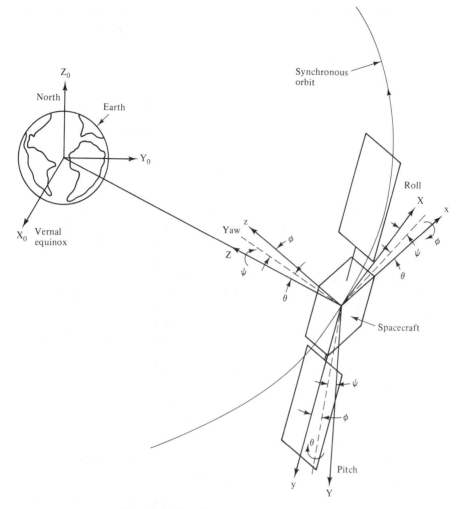

Figure 3.4 Coordinate systems in attitude control.

orbital position of the satellite. The attitude motion of a spacecraft is most commonly described in terms of an "airplane" three-axis coordinate system, namely, roll, pitch, and yaw.

The nominal roll axis, X, is along the orbit velocity vector; the nominal yaw axis, Z, is along the vector from the center of mass of the spacecraft to the center of mass of the earth; and the nominal pitch axis Y is normal to the orbit plane in such a way that the XYZ coordinate system is a right-handed mutually orthogonal frame. The coordinate system XYZ is also called the orbit coordinate system. The origin of the coordinate system is at the center of mass of the spacecraft. It is rotating with respect to the inertially fixed coordinate system $X_0Y_0Z_0$ at the angular rate of one revolution per day. The perturbed attitude of the spacecraft fixed coordinate system

xyz is obtained from the nominal attitude by the following rotations: ψ about the Z axis, θ about the once-displayed Y axis, and ϕ about the twice-displaced X axis. The angles ψ, θ, and ϕ are called yaw, pitch, and roll errors, respectively.

3.4 SPIN STABILIZATION OF A RIGID SPACECRAFT

In the absence of external moments, Euler's equations for a rigid body are

$$I_{xx}\dot{\omega}_x = (I_{yy} - I_{zz})\omega_y\omega_z \tag{3.43a}$$

$$I_{yy}\dot{\omega}_y = (I_{zz} - I_{xx})\omega_x\omega_z \tag{3.43b}$$

$$I_{zz}\dot{\omega}_z = (I_{xx} - I_{yy})\omega_x\omega_y \tag{3.43c}$$

where x, y, z are the axes of principal moments of inertia of the spacecraft. From Eqs. (3.43), it can be seen that pure rotation about any of the principal axes of inertia is possible. The question is which of the principal axes are the axes of stable spin. Let us assume that the spacecraft spins uniformly about the y-axis, the pitch axis, and allow a small perturbation from that motion to occur. The pitch axis is the preferred spin axis by spacecraft designers because it can be inertially fixed. The initial motion can be described by $\omega_y = \omega_s$, $\omega_x = \omega_z = 0$ and the perturbed motion by $\omega_y = \omega_s + \varepsilon$, $\omega_x \neq 0$, $\omega_z \neq 0$, where ω_x and ω_z are small. The resulting linearized equations are

$$I_{xx}\dot{\omega}_x = (I_{yy} - I_{zz})\omega_s\omega_z \tag{3.44a}$$

$$I_{yy}\,\dot{\varepsilon} = 0 \tag{3.44b}$$

$$I_{zz}\dot{\omega}_z = (I_{xx} - I_{yy})\omega_x\omega_s \tag{3.44c}$$

From Eq. (3.44b), we conclude that ε remains constant. Differentiating Eqs. (3.44a) and (3.44c), and substituting $\dot{\omega}_z$ and $\dot{\omega}_x$ from Eqs. (3.44c) and (3.44a), respectively, we obtain

$$\ddot{\omega}_x + \frac{(I_{yy} - I_{zz})(I_{yy} - I_{xx})}{I_{xx}I_{zz}}\,\omega_s^2\omega_x = 0 \tag{3.45a}$$

$$\ddot{\omega}_z + \frac{(I_{yy} - I_{xx})(I_{yy} - I_{zz})}{I_{xx}I_{zz}}\,\omega_s^2\omega_z = 0 \tag{3.45b}$$

For stability, the coefficient of ω_x and ω_z in Eq. (3.45) must be positive. Hence the stability conditions are (1) $I_{yy} > I_{xx}$ and $I_{yy} > I_{zz}$, in which case y is the axis of the maximum moment of inertia, or (2) $I_{yy} < I_{zz}$ and $I_{yy} < I_{xx}$, in which case y is the axis of minimum moment of inertia. Hence, for a perfectly rigid spacecraft, stable spin can take place about the axis of the maximum moment of inertia or about the axis of the minimum moment of inertia. The axis of the intermediate moment of inertia is unstable.

Let us consider a symmetric spacecraft, $I_{yy} = I_s$ and $I_{xx} = I_{zz} = I_T$. From Eq. (3.45a), we have

$$\omega_x = A \sin (\lambda t - \phi) \tag{3.46}$$

where

$$\lambda = \frac{I_s - I_T}{I_T} \omega_s = (\sigma - 1)\omega_s \tag{3.47}$$

in which A is the amplitude, ϕ is the phase angle, and σ is the inertia ratio I_s/I_T. Inserting Eq. (3.46) into Eq. (3.44a), we obtain

$$\omega_z = \frac{\dot{\omega}_x}{\lambda} = A \cos (\lambda t - \phi) \tag{3.48}$$

The resultant transverse angular velocity, ω_T, is the sum of velocity ω_x and ω_z and is given by

$$\omega_T = \omega_z + i\omega_x$$

$$= Ae^{i(\lambda t - \phi)}$$

The resultant transverse angular velocity ω_T is a vector in the xz plane, rotating with the angular velocity λ.

In the equations above, the spacecraft motion is given in the spacecraft fixed coordinate axes x, y, z. The spacecraft motion in space can best be explained by examining the angular momentum of the spacecraft for a symmetrical spacecraft. The angular momentum is

$$\mathbf{H} = I_s \boldsymbol{\omega}_s + I_T \boldsymbol{\omega}_T \tag{3.49}$$

Because the external moment is zero, the angular momentum vector is constant in magnitude and its direction is fixed in space. However, the transverse component of the angular momentum vector is rotating in the xz plane with angular velocity λ. In inertial space, the transverse component is rotating with angular velocity $\omega_n = \lambda + \omega_s$ because the body-fixed coordinate axes x and z are rotating with angular velocity ω_s (neglecting the effects of ω_x and ω_y). Hence ω_n is given by

$$\omega_n = \lambda + \omega_s = \frac{I_s}{I_T} \omega_s \tag{3.50}$$

Because the total angular momentum vector is fixed in space, the angular momentum component $I_s\omega_s$ is also rotating with the angular velocity ω_n. The spacecraft motion consists of the spacecraft rotation about its spin axis and the spin axis rotating about the angular momentum vector with angular velocity ω_n. This latter motion is called "nutational motion."

The nutational motion can be represented in terms of the body cone and the space cone, as shown in Figs. 3.5 and 3.6. The body cone is fixed in the body and its axis coincides with the spin axis. The space cone is fixed in space and its axis is along the direction of the angular momentum, $\mathbf{H}$. The total angular velocity is along the line of contact between the two cones. The plane containing angular momentum, $\mathbf{H}$, total angular velocity, $\boldsymbol{\omega}$, and spin axis rotates about $\mathbf{H}$ at angular velocity $\omega_n = I_s/I_T \omega_s$. For a disk-shaped body, $I_s/I_T > 1$, the inside surface of the body cone rolls on the outside surface of the space cone, as shown in Fig. 3.5. For a rod-shaped body, $I_s/I_T < 1$, the outside surface of the body cone rolls on the

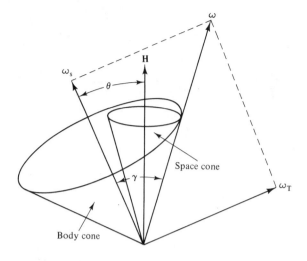

Figure 3.5 Nutational motion for a disk-shaped body, $I_s/I_T > 1$.

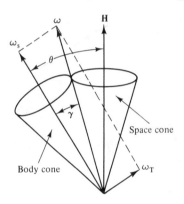

Figure 3.6 Nutational motion for a rod-shaped body, $I_s/I_T < 1$.

outside surface of the space cone as shown in Fig. 3.6. The nutation angle, θ, and the angle γ are given by

$$\sin \theta = \frac{I_T \omega_T}{H}$$

and
(3.51)

$$\tan \gamma = \frac{\omega_T}{\omega_s}$$

The acceleration of a point $p(x, y, z)$ in the spacecraft fixed coordinate system with its origin at the center of mass of the spacecraft is

$$a_x = -x\omega_s^2 + \omega_n yA \sin \lambda t$$
$$a_y = (\omega_s - \lambda) A (z \cos \lambda t + x \sin \lambda t) \qquad (3.52)$$
$$a_z = -z\omega_s^2 + \omega_n yA \cos \lambda t$$

3.5 SPIN STABILIZATION OF AN ENERGY-DISSIPATING SPACECRAFT

Single-Spin Stabilization

In single-spin stabilization, the entire spacecraft is spinning about a principal-moment-of-interia axis. As shown earlier, for a rigid spacecraft, stable spin can take place about the axis of either the maximum or the minimum moment of inertia. However, a communications satellite is likely to have several components experiencing relative motion, so that the satellite is not rigid. These components will be excited by oscillatory accelerations, as given by Eq. (3.52), resulting in energy dissipation. In the absence of external torques, the angular momentum of the spacecraft is constant, but the internal energy dissipation will cause the spacecraft to move toward a minimum-energy state. Assuming that the potential energy is constant, the spacecraft will tend to reach a minimum-kinetic-energy state. The kinetic energy, T, of the spacecraft is related to the magnitude of angular momentum, H, and to the moment of inertia, I_s, about the spin axis by

$$T = \frac{H^2}{2\,I_s} \qquad (3.53)$$

For a constant angular momentum, H, the kinetic energy is minimum when the moment of inertia, I_s, is maximum. Hence for a nonrigid spacecraft there is only one axis of stable spin, namely, the axis of the maximum moment of inertia.

In the case of energy dissipation, the nutation angle θ is no longer constant. The rate of change of the angle θ can be related to energy dissipation by assuming that the moments of inertia do not vary significantly as a result of the relative motion within the spacecraft and that the angular momentum of the relative motion is negligible compared to the rigid angular momentum of the motion. The kinetic energy, T, and the angular momentum magnitude, H, are given by

$$T = \frac{1}{2}[I_T(\omega_x^2 + \omega_z^2) + I_s\omega_s^2] \qquad (3.54)$$

$$H^2 = I_T^2(\omega_x^2 + \omega_z^2) + I_s^2\omega_s^2 \qquad (3.55)$$

The angle θ is still given by Eq. (3.51) and is no longer constant. Inserting Eqs. (3.54) and (3.55) into Eq. (3.51), we obtain

$$\sin^2\theta = \frac{I_T}{(I_s - I_T)\,H^2}(2\,I_sT - H^2) \qquad (3.56)$$

Differentiating Eq. (3.56) with respect to time, we have

$$\dot{\theta} = \frac{1}{\sin 2\theta}\frac{2\,I_TI_s}{(I_s - I_T)\,H^2}\frac{\dot{T}}{} = \frac{2\,I_T\omega_n\dot{T}}{\sin 2\theta\,\lambda\,H^2} \qquad (3.57)$$

Equation (3.57) gives the rate of change of the nutation angle in terms of the energy dissipation. Because $\dot{T}$ is negative, the nutation angle decreases only if $I_s > I_T$. This confirms that for a nonrigid spacecraft, stable spin takes place only about the axis of maximum moment of inertia. The above-mentioned analysis is sometimes known as the energy-sink approach.

The rate of nutation decay, given by Eq. (3.57), is proportional to the rate of energy dissipation. Nutation dampers are used to increase the energy dissipation rate, and hence the rate of nutation decay. The driving oscillatory forces for the dampers are proportional to the accelerations given by Eq. (3.52). Hence a damper on the spin axis should be placed as far as possible from the center of mass and be oriented to allow it to move laterally. For a damper in the xz plane, the damper should be placed at a maximum distance from the y axis and oriented so as to move parallel to the spin axis.

Dual-Spin Stabilization

The limitation of single-spin stabilization (i.e., it cannot have oriented antennas) is overcome in a dual-spin-stabilized spacecraft, which consists of a rotor providing gyroscopic stabilization and a platform pointing toward the earth. Iorillo (Ref. 5) presented a stability criterion for dual-spin-stabilized spacecraft in 1965. A similar stability criterion was discovered independently by V. D. Landon. In a dual-spin-stabilized spacecraft, the stable spin axis can be the axis of minimum moment of inertia if the rate of energy dissipation in the platform is higher than that in the rotor by a certain factor. The spin axis for Intelsat IV is the axis of minimum moment of inertia. Likins (Ref. 6), Pringle, and other investigators have proved the stability criterion of Iorillo in a rigorous manner. In the following equations, the stability criteria are obtained by using energy-sink arguments.

Figure 3.7 shows an axisymmetric dual-spin-stabilized spacecraft. The expressions for kinetic energy, T, the magnitude of the angular momentum,

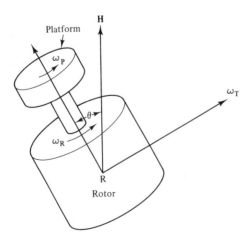

Figure 3.7 Motion of a dual-spin-stabilized spacecraft.

H, and the nutation frequency, ω_n are

$$T = \frac{1}{2}[I_T\omega_T^2 + I_{sP}\omega_P^2 + I_{sR}\omega_R^2] \tag{3.58}$$

$$H^2 = (I_{sP}\omega_P + I_{sR}\omega_R)^2 + (I_T\omega_T)^2 \tag{3.59}$$

$$\omega_n = \frac{I_{sP}\omega_P + I_{sR}\omega_R}{I_T} \tag{3.60}$$

where I_T is the transverse moment of inertia of the spacecraft, I_{sR} the spin moment of inertia of the rotor, I_{sP} the spin moment of inertia of the platform, ω_T the transverse angular velocity of the spacecraft, ω_P the spin angular velocity of the platform, and ω_R the spin angular velocity of the rotor.

Let us assume that there is no external torque and that the shaft is frictionless. In real systems, a servo-loop-controlled motor acts to counterbalance any shaft friction exactly. From the conservation of angular momentum principle, $\dot{H}$ is zero. Differentiating Eq. (3.59) with respect to time and using $\dot{H} = 0$, we obtain

$$
\begin{aligned}
I_T\omega_T\dot{\omega}_T &= -\frac{I_{sP}\omega_P + I_{sR}\omega_R}{I_T}(I_{sP}\dot{\omega}_P + I_{sR}\dot{\omega}_R) \\
&= -\omega_n(I_{sP}\dot{\omega}_P + I_{sR}\dot{\omega}_R)
\end{aligned} \tag{3.61}
$$

Differentiating Eq. (3.58) with respect to time, we have

$$\dot{T} = \dot{T}_P + \dot{T}_R = I_T\omega_T\dot{\omega}_T + I_{sP}\omega_P\dot{\omega}_P + I_{sR}\omega_R\dot{\omega}_R \tag{3.62}$$

where $\dot{T}_p$ and $\dot{T}_R$ are the rate of energy dissipation of the platform and of the rotor, respectively. Combining Eqs. (3.61) and (3.62), we obtain

$$
\begin{aligned}
\dot{T} &= \dot{T}_p + \dot{T}_R \\
&= -(\omega_n - \omega_P)I_{sP}\dot{\omega}_P - (\omega_n - \omega_R)I_{sR}\dot{\omega}_R \\
&= -\lambda_P I_{sP}\dot{\omega}_P - \lambda_R I_{sR}\dot{\omega}_R
\end{aligned} \tag{3.63}
$$

where $\lambda_P = \omega_n - \omega_P$, $\lambda_R = \omega_n - \omega_R$, in which λ_P and λ_R are the frequencies of oscillating accelerations in the platform and in the rotor, respectively. Because the rotor and the platform are assumed to be uncoupled about the spin axis, the reaction torques which tend to change the angular rates can be written from Eq. (3.63) as

$$I_{sP}\dot{\omega}_P = -\frac{\dot{T}_P}{\lambda_P} \tag{3.64a}$$

$$I_{sR}\dot{\omega}_R = -\frac{\dot{T}_R}{\lambda_R} \tag{3.64b}$$

Combining Eqs. (3.61) and (3.64), we obtain

$$I_T\omega_T\dot{\omega}_T = \omega_n\left(\frac{\dot{T}_P}{\lambda_P} + \frac{\dot{T}_R}{\lambda_R}\right) \tag{3.65}$$

Recalling that θ is no longer constant, differentiating Eq. (3.51) with respect to time, and using Eq. (3.65) we obtain

$$\dot{\theta} = \frac{2 I_T}{\sin 2\theta} \frac{\omega_n}{H^2} \left(\frac{\dot{T}_P}{\lambda_P} + \frac{\dot{T}_R}{\lambda_R} \right) \tag{3.66}$$

It should be noted that Eq. (3.57) for single-spin stabilization can be obtained as a special case of Eq. (3.66) by letting $\lambda_P = \lambda_R = \lambda$, $\dot{T}_P + \dot{T}_R = \dot{T}$ in the latter.

For a stable spacecraft, the nutation angle must decay with time (i.e., $\dot{\theta} < 0$). Hence the stability condition is

$$\frac{\dot{T}_P}{\lambda_P} + \frac{\dot{T}_R}{\lambda_R} < 0 \tag{3.67}$$

Assuming that the rotor angular momentum is much larger than the platform angular momentum (i.e., $I_{sR}\omega_R \gg I_{sP}\omega_P$), which is a reasonable assumption in view of the fact that ω_P is approximately equal to the orbit rate, we can write

$$\lambda_P = \omega_n - \omega_P = \frac{I_{sP}\omega_P + I_{sR}\omega_R}{I_T} - \omega_P \simeq \frac{I_{sR}\omega_R}{I_T} \tag{3.68a}$$

$$\lambda_R = \omega_n - \omega_R = \frac{I_{sP}\omega_P + I_{sR}\omega_R}{I_T} - \omega_R \simeq \left(\frac{I_{sR}}{I_T} - 1 \right) \omega_R \tag{3.68b}$$

Substituting Eq. (3.68) into inequality (3.67), the stability condition becomes

$$\frac{\dot{T}_P}{I_{sR}/I_T} + \frac{\dot{T}_R}{(I_{sR}/I_T) - 1} < 0 \tag{3.69}$$

There are two cases:

Case 1: $I_{sR} > I_T$. The spacecraft is stable if the energy dissipations occur in either the platform or the rotor. This implies that a nutation damper can be placed on either the platform or the rotor.

Case 2: $I_{sR} < I_T$. For this case the first term in Eq. (3.69), $\dot{T}_P/(I_{sR}/I_T)$, is negative and the second term, $\dot{T}_R/[(I_{sR}/I_T) - 1]$, is positive. The stability condition for this case becomes

$$|\dot{T}_P| > \left| \dot{T}_R \frac{I_{sR}/I_T}{(I_{sR}/I_T) - 1} \right| \tag{3.70}$$

As an example, for a dual-spin-stabilized spacecraft with $I_{sR}/I_T = \frac{2}{3}$, the magnitude of the energy dissipation rate of the platform should be at least twice the dissipation rate in the rotor. Hence the damper must be placed on the platform.

3.6 LIQUID-MOTION EFFECTS

Spacecraft carry liquid propellant for orbit maneuvers and attitude control. In some spacecraft where a liquid apogee motor is used, liquid constitutes approximately half of the spacecraft mass in the transfer orbit. For these spacecraft, liquid motion in the fuel tanks will have a significant influence on attitude dynamics. The influence will be mainly due to two effects: energy dissipation and change in inertia properties of the spacecraft. The rate of change of the nutation angle, given by Eq. (3.57), depends on the energy dissipation rate. The major source of energy dissipation is from liquid motion relative to the tank, known as liquid slosh. Therefore, a significant effort is normally spent on a spin-stabilized spacecraft program to estimate the energy dissipation rate due to liquid slosh. Some of the spacecraft attitude stability conditions derived in the previous sections are also modified because the spacecraft inertia properties change due to liquid motion.

Energy Dissipation Due to Liquid Slosh

The hydrodynamical equations of motion for liquids are very complex and, for complex geometries, are beyond the range of present analytical methods. In early works on the subject, liquid-slosh motion was assumed to be modeled by a spherical segment free to move as a spherical pendulum inside the tank. Energy dissipation was calculated at the boundary layer formed by the tank and the liquid. Ground slosh tests, however, indicated substantial discrepancies in the resultant analytical predictions. The current approach is to perform ground slosh tests and extrapolate the in-orbit energy dissipation by using scaling techniques from dimensional analysis.

The nutation angle is assumed to have exponential growth or decay, that is,

$$\theta = \theta_0 e^{t/\tau} \tag{3.71}$$

where τ is a time constant. The energy dissipation rate, $\dot{E}$, is assumed to be the function of the spacecraft parameters as follows:

$$\frac{\dot{E}}{\theta^2} = n f(\omega, \lambda, m, \rho, \mu, R, d, z, s, g) \tag{3.72}$$

where n = number of tanks
   ω = spin rate
   λ = body nutation frequency
   m = mass of fluid
   ρ = density of fluid
   μ = viscosity of fluid
   R = radial distance of tank center from spin axis
   z = axial offset of tank center from vehicle center of mass
   s = shape factor of the tank
   g = gravitational acceleration
   d = diameter of the tank

By applying dimensional analysis, the energy dissipation rate is expressed as a function of a dimensional group of variables.

$$\frac{\dot{E}}{n\rho d^5 \omega^3 \theta^2} = f(R_e, F_r, FF, s, \sigma, d/R, z/R) \qquad (3.73)$$

where $R_e = \rho \omega d^2 / \mu$ = Reynolds number

$\quad F_r = \omega^2 R / g$ = Froude number

$\quad FF$ = fraction fill of the tank

$\quad \sigma$ = inertia ratio, I_s / I_T

The time constant group, τ_{CG}, is given as

$$\tau_{CG} = \tau \frac{n d^5 \omega \rho}{I_T} = f_1(R_e, F_r, FF, s, \sigma, d/R, z/R) \qquad (3.74)$$

The ground slosh tests are normally performed on an air-bearing spin table, such as the one at the NASA Goddard Space Flight Center. The test procedure consists of spinning a scaled dynamic model of the spacecraft on the air-bearing spin table and then letting its nutation grow. The nutation angle is measured by an accelerometer on the model. To reduce air-drag effects, the table is operated in a vacuum chamber. The ground tests simulate nondimensional parameters, Reynolds number, fraction of fill, tank location (d/R and z/R), and inertia ratio. The Froude number is assumed to be at least equal to 8 in order to keep the fluid surface parallel to the spin axis, the zero-g condition. Since the time constant group, τ_{CG}, for the ground test and in-orbit conditions will be the same, the in-orbit time constant is extrapolated from the time constant determined by ground test by the relationship

$$\tau_0 = \frac{\tau_g (I_{To}/I_{Tg})}{(n_o/n_g)(\rho_o/\rho_g)(d_o/d_g)^5(\omega_o/\omega_g)} \qquad (3.75)$$

where the subscript o corresponds to the in-orbit condition and the subscript g corresponds to the ground tests.

Static Stability

The nominal spin axis of a single-spinning body is aligned along the axis of maximum or minimum moment of inertia. In the absence of nutation, the spin axis will be coincident with the angular momentum and total angular velocity vectors, resulting in every point on the rotating body experiencing a pure spin about the spin axis. The body is said to be statically stable if a small dynamic imbalance introduced into the body causes a bounded tilt of the spin axis.

Let the inertia matrix of the body be

$$I = \begin{bmatrix} I_{xx} & -I_{xy} & 0 \\ -I_{xy} & I_{yy} & -I_{yz} \\ 0 & -I_{yz} & I_{zz} \end{bmatrix} \qquad (3.76)$$

where I_{xx}, I_{yy}, and I_{zz} are the moments of inertia of the body about axes x, y, z, respectively, and I_{xy} and I_{yz} are products of inertia as defined in Eqs. (3.20) and (3.21). In the absence of dynamic imbalance, the products of inertia are zero, and x, y, and z are the principal moments of inertia axes and the body will be spinning about the y axis. When the products of inertia are nonzero, the spin axis will be tilted by an angle θ from the y axis whose components θ_x and θ_z along axes x and z will be

$$\theta_x = -\frac{I_{yz}}{I_{yy} - I_{zz}} \tag{3.77}$$

$$\theta_z = \frac{I_{xy}}{I_{yy} - I_{xx}} \tag{3.78}$$

Let us consider a single spinner in geosynchronous orbit. Let I_{yz} be zero and the product of inertia, I_{xy}, is represented by a dumbbell as shown in Fig. 3.8. Let us assume that the spacecraft contains two propellant tanks with their centers in the xy plane. In the absence of propellant motion due to dynamic imbalance I_{xy}, the spin axis will be tilted from the y axis by an angle, θ_z, given by Eq. (3.78) and shown in Fig. 3.8 for inertia ratio greater than unity. If we now allow the propellant to deform, shown by dashed lines in Fig. 3.8, it will have two types of motion. The first, the free surface will tilt in each tank parallel to the spin axis. The second, the propellant will migrate between interconnected tanks such that all points on the surface in both tanks are equidistant from the spin axis. The propellant deformation will cause an added imbalance and affect the product of inertia, as shown in Fig. 3.8. This imbalance, however, will be proportional to the tilt angle.

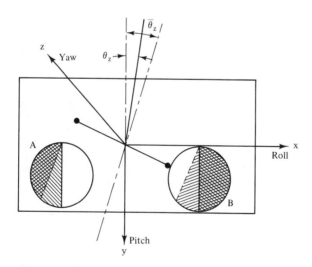

Figure 3.8 Dynamic imbalance due to liquid motion.

The final tilt angle with propellant motion, $\bar{\theta}_z$, is given by

$$\bar{\theta}_z = \frac{I_{xy} + K_p \bar{\theta}_z}{I_{yy} - I_{xx}} \tag{3.79}$$

or

$$\overline{\theta}_z = \frac{I_{xy}}{I_{yy} - I_{xx} - K_p}$$

$$= \frac{I_{xy}}{I_{yy} - I_{xx}} \frac{1}{1 - K_p/(I_{yy} - I_{xx})} \qquad (3.80)$$

$$= \theta_z \alpha$$

where θ_z is the rigid-body tilt angle and

$$\alpha = \frac{1}{1 - K_p/(I_{xx}(\sigma - 1))} \equiv \text{amplification factor}$$

$$\sigma = \frac{I_{yy}}{I_{xx}} \equiv \text{inertia ratio}$$

$$K_p = \text{positive parameter}$$

The equations above show that the propellant motion amplifies the rigid-body tilt for $\sigma > 1$. The static instability corresponds to infinite amplification or $\sigma = 1 + K_p/I_{xx}$. Hence, for static stability,

$$\sigma > 1 + \frac{K_p}{I_{xx}} \qquad (3.81)$$

For $\sigma < 1$, the propellant motion attenuates the initial rigid body tilt. For this case, static instability is not possible.

3.7 ACTIVE NUTATION CONTROL

As discussed in Section 3.5, for a single-spin-stabilized spacecraft, attitude stability is achieved if the spin axis is the axis of maximum moment of inertia. For a dual-spin-stabilized spacecraft, attitude stability is achieved even if the spin axis is the axis of minimum moment of inertia provided that the energy dissipation in the platform is greater than that in the rotor by a factor given by Eq. (3.70). For some spacecraft, it is not possible to satisfy attitude stability conditions by passive energy dissipation means due to launch vehicle and design constraints. For these spacecraft, the nutation angle is controlled actively by thrusters or other means.

Active Nutation Control (ANC) by Thrusters

A common technique to control the nutation angle of a passively unstable spacecraft is by using fixed jets on the rotor, as shown in Fig. 3.9. The basic idea is to provide jet torque in such a way that it reduces transverse angular velocity, ω_T, which in turn reduces the nutation angle. The torque could be generated by an axial jet which thrusts along an axis parallel to the spin axis but offset from it. For a passively unstable body, the transverse angular velocity, ω_T, rotates clockwise at the rotor

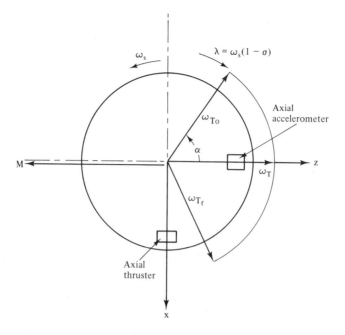

Figure 3.9 Active nutation control by thrusters.

nutation frequency, $\lambda = \omega_s (1 - \sigma)$, with respect to the rotor fixed zx plane. The axial jet torque is, however, fixed in the rotor body. Hence we need a sensor to phase the firing of the jets properly. The axial accelerometer, shown in Fig. 3.9, provides the turn-on and turn-off signals to the jets. The accelerometer during nutation senses a sinusoidal acceleration whose amplitude is proportional to the nutation angle and whose frequency is the rotor fixed nutation frequency, λ. The peak positive acceleration occurs when ω_T points toward the accelerometer. Therefore, the axial accelerometer is placed at approximately 90° from the axial jet axis in the zx plane. The axial jet is commanded on when the signal from the accelerometers is positive, resulting in a reduction in ω_T and nutation angle, θ. If another axial jet is placed at a diametrically opposite position, that jet should conversely be commanded on when the accelerometer signal is negative.

The expressions for the ANC time constant are derived in the following manner. Let us assume a symmetric spacecraft. Using Euler's equation (3.42), we get

$$M_z = I_T \dot{\omega}_z + \omega_s \omega_x (I_T - I_s) \tag{3.82a}$$

$$M_x = I_T \dot{\omega}_x + \omega_s \omega_z (I_s - I_T) \tag{3.82b}$$

where $I_T = I_{xx} = I_{zz}$

$I_s = I_{yy}$

$\omega_y = -\omega_s$

Multiplying Eq. (3.82b) by i and adding to Eq. (3.82a) and using a complex variable, we get

$$\frac{M}{I_T} = \dot{\omega}_T - i\lambda\omega_T \tag{3.83}$$

where

$$M = M_z + iM_x$$
$$\omega_T = \omega_z + i\omega_x \tag{3.84}$$
$$\lambda = \omega_s\left(1 - \frac{I_s}{I_T}\right)$$

At the start of the jet firing, $t = 0$, the transverse angular velocity, as shown in Fig. 3.9, is given by

$$\omega_{To} = |\omega_{To}|\, e^{-i\alpha} \tag{3.85}$$

From Eq. (3.83), by using the initial condition from Eq. (3.85), the transverse angular velocity during the jet firing is given by

$$\omega_T(t) = |\omega_{To}|\, e^{i(\lambda t - \alpha)} + \frac{iM}{\lambda I_T}(1 - e^{i\lambda t}) \tag{3.86}$$

The first term in Eq. (3.86) is the component of ω_T in the absence of the external moment and the second term is due to the external moment from the jet firing. Let the jet be fired for time T and the external moment be given by

$$M = |M|\, e^{i\beta} \tag{3.87}$$

The magnitude of the transverse angular velocity at the end of the jet firing is given by

$$|\omega_{Tf}| = \left| |\omega_{To}|\, e^{i(\lambda T - \alpha)} + \frac{i|M|\, e^{i\beta}}{\lambda I_T}(1 - e^{i\lambda T}) \right| \tag{3.88}$$

The jet torque, from Fig. 3.9, is in the $-z$ direction. The active nutation control is optimally phased to achieve minimum damping time constant when α is $\pi/2$ at the initiation of the control torque and when the torque exists for one-half of the body nutation period (i.e., $T = \pi/\lambda$). Substituting $\alpha = \pi/2$, $\beta = \pi$, and $T = \pi/\lambda$ in Eq. (3.88), we get

$$|\omega_{Tf}| = \left(|\omega_{To}| - \frac{2|M|}{I_T\lambda}\right) \tag{3.89}$$

The ANC time constant is

$$\tau_{\text{ANC}} = \frac{|\omega_{To}|}{|\omega_{Tf}| - |\omega_{To}|}\frac{2\pi}{\lambda} \tag{3.90}$$

where

$$|\omega_{To}| = \frac{I_s}{I_T}\omega_s \tan\theta$$
$$\theta = \text{nutation angle}$$

Let τ_d be the destabilizing time constant due to liquid slosh. The stability margin, M_s, is defined by

$$M_s = \frac{|\tau_d| - |\tau_{ANC}|}{|\tau_{ANC}|} \qquad (3.91)$$

The spacecraft overall time constant is

$$\tau_0 = \frac{1}{1/\tau_{ANC} + 1/\tau_d} \qquad (3.92)$$

Despin-Active Nutation Damping

In Intelsat IV, a dual-spin-stabilized spacecraft with an inertia ratio of less than unity, nutation damping was provided by eddy-current dampers on the platform. For recent spacecraft, SBS and Intelsat VI, the dynamic imbalance of the platform provides nutation damping. The basic concept is shown in Fig. 3.10.

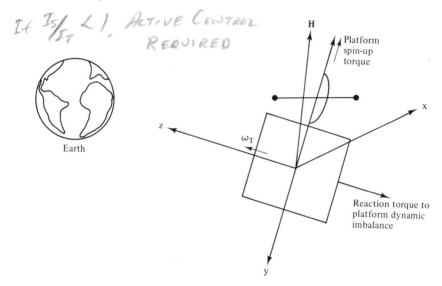

Figure 3.10 Despin-active nutation damping.

The despun platform has dynamic imbalance in the yz plane. In the presence of nutation, the spin axis cones about angular momentum vector **H** at inertial nutation frequency ω_n. When the platform is leaning backward from the earth, the transverse angular velocity, ω_T, will be along the z axis. If a platform spin-up torque is applied, the reaction torque will diametrically oppose the transverse angular vector, resulting in the reduction of the nutation angle. By similar reasoning, when the platform is leaning forward and ω_T is along the $-z$ axis, spin-down torque will create a reaction torque which will oppose ω_T and result in a lower nutation angle. A sinusoidal spin torque at the inertial nutation frequency is phased in such a manner that the peak spin-up torque occurs when ω_T is along the z, yaw, axis. Since the nutation frequency of the despun platform is the same as the inertial nutation frequency, the reaction torque will synchronize with the transverse angular velocity and will result in continuous reduction of the nutation angle. This is the basic principle of despin active nutation damping (DAND).

Example 3.1

A spacecraft with a liquid apogee motor has an inertia ratio of less than unity in transfer orbit, thus requiring an active control system. The spacecraft characteristics are: nominal spin rate = 30 rpm; inertia ratio, spin to transverse = 0.726; the thruster fired for half of the body nutation frequency period, $\alpha = \pi/2$ and $T = \pi/\lambda$; thruster torque = 26.64 $N \cdot m$ and $\beta = \pi$; transverse inertia = 10,876 kg $\cdot$ m²; initial nutation = 5°; number of tanks = four fuel and four oxidizer; tank diameter = 0.838 m; fuel density = 865.8 kg/m³; oxidizer density = 1443 kg/m³; time group constant = 1000. Determine the body nutation frequency, initial transverse angular velocity, time constant of nutation control, time constant due to liquid slosh in fuel tanks, time constant due to liquid slosh in oxidizer tanks, combined time constant due to liquid slosh, and overall nutation time constant.

Solution   The body-fixed nutation frequency, λ, is given by Eq. (3.84) as

$$\lambda = \omega_s(1 - \sigma)$$

where λ = body fixed nutation frequency, rad/s

   ω_s = spacecraft spin rate, rad/s = $(30 \times 2\,\pi)/60$ = 3.1415 rad/s

   σ = inertia ratio = 0.726

Substituting the values in the equation above yields

$$\lambda = 3.1415\,(1 - 0.726) = 0.8608 \text{ rad/s}$$

The transverse angular velocity, ω_T, is given by

$$\omega_T = \sigma\omega_s \tan \theta$$

where θ = nutation angle = 5°. Substituting the numerical values gives us

$$\omega_T = 0.726 \times 3.14 \times \tan 5°$$

$$= 0.1995 \text{ rad/s}$$

The time constant for active nutation control is given by

$$\tau_{ANC} = \frac{|\omega_{To}|}{|\omega_{Tf}| - |\omega_{To}|} \frac{2\pi}{\lambda}$$

where τ_{ANC} = nutation time constant,

   ω_{To} = initial transverse angular velocity

      = 0.1995 rad/s

   ω_{Tf} = transverse angular velocity after one time

      period of λ, rad/s

For this example with $\alpha = \pi/2$, $\beta = \pi$ and $T = \pi/\lambda$, $|\omega_{Tf}|$ is given by Eq. (3.89) as

$$|\omega_{Tf}| = \left(|\omega_{To}| - 2\frac{|M|}{I_T\lambda}\right)$$

where $|M|$ = magnitude of the thruster torque

$\quad\quad = 26.64 \text{ N} \cdot \text{m}$

Substituting the numerical values gives us

$$|\omega_{Tf}| = 0.1995 - \frac{2 \times 26.64}{10{,}876 \times 0.8608}$$

$$= 0.1938 \text{ rad/s}$$

The nutation time constant, τ_{ANC}, is given by

$$\tau_{ANC} = \frac{0.1995}{0.1938 - 0.1995} \frac{2\pi}{0.8608}$$

$$= -255.5 \text{ s}$$

The time constant due to liquid slosh, from Eq. (3.74), is given by

$$\tau = \frac{I_T \tau_{CG}}{n\rho\omega_s d^5}$$

where τ = time constant

$\quad \tau_{CG}$ = time group constant

$\quad\quad n$ = number of tanks

$\quad\quad \rho$ = liquid density, kg/m^3

$\quad\quad d$ = diameter of tanks, m

Substituting the numerical values yields:

Time constant for the fuel is

$$\tau_f = \frac{10{,}876 \times 1000}{4 \times 865.8 \times 3.1415 \times (0.838)^5}$$

$$= 2419 \text{ s}$$

Time constant for the oxidizer is

$$\tau_o = \frac{10{,}876 \times 1000}{4 \times 1443 \times 3.1415 \times (0.838)^5}$$

$$= 1452 \text{ s}$$

The combined dedamping time constant due to liquid slosh is

$$\tau_d = \frac{1}{1/\tau_f + 1/\tau_o} = \frac{1}{1/2419 + 1/1452}$$

$$= 907 \text{ s}$$

The overall nutation time constant is

$$\tau_o = \frac{1}{1/\tau_{ANC} + 1/\tau_d} = \frac{1}{-1/255.5 + 1/907}$$

$$= -355 \text{ s}$$

The stability margin is

$$M_s = \frac{|\tau_d| - |\tau_{ANC}|}{|\tau_{ANC}|} = \frac{907 - 255.5}{255.5}$$

$$= 2.54$$

3.8 THREE-AXIS STABILIZATION

The control torques along the axes of three-axis-stabilization systems are provided by various combinations of momentum wheels, reaction wheels, and thrusters. Broadly, however, there are two types of three-axis-stabilization systems: a momentum biased system with a momentum wheel along the pitch axis, and a zero-momentum system with a reaction wheel along each axis.

In momentum wheel systems, the angular momentum along the pitch axis provides gyroscopic stiffness. In these systems, the pitch and roll axes are controlled directly and the yaw axis is controlled indirectly due to gyroscopic coupling of yaw and roll errors, thus eliminating the need for a yaw sensor. In reaction wheel systems, all three axes are controlled independently, thus requiring a yaw sensor also. The momentum wheel can be fixed or gimbaled about one or two axes. The control torque along the pitch axis is provided by the change in the speed of the momentum wheel. The torque along the roll axis is provided by thrusters, a reaction wheel, or by changing the gimbal angles in the gimbaled momentum wheel.

A linearized analysis of the three-axis-stabilization system is given in this section. The linearized equations of motion of a three-axis-stabilized spacecraft are derived first. The angular velocities of the spacecraft can be expressed in terms of the orbital rate, ω_0, and the attitude error angles, ψ, θ, and ϕ, which are known as yaw, pitch, and roll errors, respectively, as follows:

$$
\boldsymbol{\omega} = \begin{Bmatrix} \omega_x \\ \omega_y \\ \omega_z \end{Bmatrix} = \begin{bmatrix} 1 & 0 & -\sin\theta \\ 0 & \cos\phi & \cos\theta\sin\phi \\ 0 & -\sin\phi & \cos\theta\cos\phi \end{bmatrix} \begin{Bmatrix} \dot\phi \\ \dot\theta \\ \dot\psi \end{Bmatrix}
$$

$$
- \omega_0 \begin{Bmatrix} \cos\theta\sin\psi \\ \cos\phi\cos\psi + \sin\phi\sin\theta\sin\psi \\ -\sin\phi\cos\psi + \cos\phi\sin\theta\sin\psi \end{Bmatrix} \qquad (3.93)
$$

$$
\simeq \begin{Bmatrix} \dot\phi - \omega_o\psi \\ \dot\theta - \omega_o \\ \dot\psi + \omega_o\phi \end{Bmatrix}
$$

The angular momentum of the system can be written in the form

$$
\mathbf{H} = \mathbf{H}_w + \mathbf{H}_b \qquad (3.94)
$$

where

$$
\mathbf{H}_w = [h_x \quad h_y \quad h_z]^T \qquad (3.95a)
$$

$$
\mathbf{H}_b = [I_{xx}\omega_x \quad I_{yy}\omega_y \quad I_{zz}\omega_z]^T \qquad (3.95b)
$$

are the angular momentum of the wheels and of the spacecraft, respectively, where I_{xx}, I_{yy}, and I_{zz} are the principal moments of inertia of the spacecraft and ω_x, ω_y, and ω_z are the angular velocity components of axes x, y, z as given by Eq. (3.93). Introducing Eq. (3.94), in conjunction with Eqs. (3.95) and (3.93) into Eq. (3.39) and ignoring nonlinear terms, we obtain the moment equation in the form

$$\mathbf{M} \cong \begin{cases} \dot{h}_x + I_{xx}(\ddot{\phi} - \omega_0\dot{\psi}) + \dot{h}_z(\dot{\theta} - \omega_0) - \dot{h}_y(\dot{\psi} + \omega_0\phi) \\ \quad - (I_{zz} - I_{yy}) \, \omega_0 \, (\dot{\psi} + \omega_0\phi) \\ \dot{h}_y + I_{yy} \, \ddot{\theta} + \dot{h}_x(\dot{\psi} + \omega_0\phi) - \dot{h}_z \, (\dot{\phi} - \omega_0\psi) \\ \dot{h}_z + I_{zz}(\ddot{\psi} + \omega_0\dot{\phi}) + \dot{h}_y(\dot{\phi} - \omega_0\psi) \\ \quad - \dot{h}_x(\dot{\theta} - \omega_0) - (I_{yy} - I_{xx}) \, \omega_0 \, (\dot{\phi} - \omega_0\psi) \end{cases} \quad (3.96)$$

The external moments arise from three major sources: the gravitational gradient, the solar radiation pressure, and the control moments from actuators. Denoting the individual torques by $\mathbf{M}_G$, $\mathbf{M}_s$ and $\mathbf{M}_c$, respectively, we can write

$$\mathbf{M} = \mathbf{M}_G + \mathbf{M}_s + \mathbf{M}_c \quad (3.97)$$

The expressions for gravity-gradient and solar pressure torques are derived in the following section.

3.9 DISTURBANCE TORQUES

Gravity-Gradient Torque

A spacecraft body experiences a gravity-gradient torque due to the variation of distances between the spacecraft mass points and the center of the mass of the earth. The gravity gradient has been used on early low-earth-orbit satellites to maintain the earth pointing of antennas or other instruments. In this section, forces and moments in a spacecraft due to gravity gradient are calculated.

The gravitational force corresponding to a differential element of mass, dm, shown as P in Fig. 3.11, is

$$\begin{aligned} \mathbf{F} &= \frac{\mu_e \mathbf{R} \, dm}{|\mathbf{R}|^3} = \mu_e \frac{\mathbf{R}_0 - \mathbf{r}}{|\mathbf{R}_0 - \mathbf{r}|^3} \, dm \\ &= \frac{\mu_e \, (\mathbf{R}_0 - \mathbf{r}) \, dm}{R_0^3} \left[1 + 3 \frac{\mathbf{r} \cdot \mathbf{R}_0}{R_0^2} + O\left(\frac{r^2}{R_0^2}\right) \right] \\ &\cong \frac{\mu_e \, (\mathbf{R}_0 - \mathbf{r}) \, dm}{R_0^3} \left(1 + 3 \frac{\mathbf{r} \cdot \mathbf{R}_0}{R_0^2} \right) \end{aligned} \quad (3.98)$$

where it is assumed that $r/R_0 \ll 1$. The gravitational moment on the spacecraft is obtained by using Eq. (3.98) and noting that $\int \mathbf{r} \, dm = 0$ because $\mathbf{r}$ is a vector measured from the center of the spacecraft mass. The result is

$$\begin{aligned} \mathbf{M}_G &= \int \mathbf{r} \times \mathbf{F} \, dm \\ &= \frac{3\mu_e}{R_0^5} \int (\mathbf{r} \times \mathbf{R}_0)(\mathbf{r} \cdot \mathbf{R}_0) \, dm \end{aligned} \quad (3.99)$$

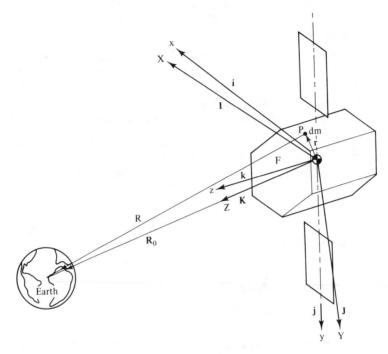

Figure 3.11 Gravity-gradient moment on a spacecraft.

The unit vectors **i**, **j**, **k** along the spacecraft body axes, also principal moments of inertia axes, are related to the unit vectors **I**, **J**, **K** along the orbital axes as follows:

$$\begin{Bmatrix} \mathbf{i} \\ \mathbf{j} \\ \mathbf{k} \end{Bmatrix} = \begin{bmatrix} 1 & 0 & 0 \\ 0 & \cos\phi & \sin\phi \\ 0 & -\sin\phi & \cos\phi \end{bmatrix} \begin{bmatrix} \cos\theta & 0 & -\sin\theta \\ 0 & 1 & 0 \\ \sin\theta & 0 & \cos\theta \end{bmatrix} \begin{bmatrix} \cos\psi & \sin\psi & 0 \\ -\sin\psi & \cos\psi & 0 \\ 0 & 0 & 1 \end{bmatrix} \begin{Bmatrix} \mathbf{I} \\ \mathbf{J} \\ \mathbf{K} \end{Bmatrix}$$

$$= \begin{bmatrix} \cos\theta\cos\psi & \cos\theta\sin\psi \\ -\cos\phi\sin\psi + \sin\phi\sin\theta\cos\psi & \cos\phi\cos\psi + \sin\phi\sin\theta\sin\psi \\ \sin\phi\sin\psi + \cos\phi\sin\theta\cos\psi & -\sin\phi\cos\psi + \cos\phi\sin\theta\sin\psi \end{bmatrix}$$

$$\begin{bmatrix} -\sin\theta \\ \sin\phi\cos\theta \\ \cos\phi\cos\theta \end{bmatrix} \begin{Bmatrix} \mathbf{I} \\ \mathbf{J} \\ \mathbf{K} \end{Bmatrix} \tag{3.100}$$

Using Eq. (3.100), the vector $\mathbf{R}_0 = R_0\mathbf{K}$ can be written in the spacecraft-fixed coordinates as

$$\mathbf{R}_0 = R_0(-\sin\theta\mathbf{i} + \sin\phi\cos\theta\mathbf{j} + \cos\phi\cos\theta\mathbf{k}) \tag{3.101}$$

Substituting Eq. (3.101) into Eq. (3.99), the gravity gradient moment becomes

$$\mathbf{M}_G = \frac{3\mu_e}{R_0^3} \begin{Bmatrix} (I_{zz} - I_{yy})\sin\phi\cos\phi\cos^2\theta \\ -(I_{xx} - I_{zz})\sin\theta\cos\theta\cos\phi \\ -(I_{yy} - I_{xx})\sin\theta\cos\theta\sin\phi \end{Bmatrix} \tag{3.102}$$

From Eq. (2.31),

$$\omega_0^2 = \frac{\mu_e}{R_0^3}$$

(3.103)

where ω_0 is the orbital angular velocity. Inserting Eq. (3.103) into Eq. (3.102) and assuming that the angles θ and ϕ are small, the gravity-gradient moment is reduced to

$$\mathbf{M}_G \cong 3\omega_0^2 \begin{Bmatrix} \phi \, (I_{zz} - I_{yy}) \\ \theta \, (I_{zz} - I_{xx}) \\ 0 \end{Bmatrix}$$

(3.104)

Solar Radiation Pressure Torque

The solar pressure torque is the major long-term disturbance torque for a geosynchronous spacecraft. The solar radiation forces are due to photons impinging on the spacecraft surfaces. In general a fraction, ρ_s, of the impinging photons will be specularly reflected, a fraction, ρ_d, will be diffusely reflected, and a fraction, ρ_a, will be absorbed by the surface.

The surface, A, intercepts a beam of radiation with cross section $A \cos \psi$, where ψ, as shown in Fig. 3.12, is the angle between the unit vector $\mathbf{n}$ along the surface normal and the unit vector $\mathbf{S}$ along the direction of the incoming photons. If the entire beam is absorbed (i.e., if $\rho_a = 1$), the force developed is $PA \, (\mathbf{n} \cdot \mathbf{S})\mathbf{S}$, where P is the solar radiation pressure. Because the momentum of the photons intercepted by the surface is changed in various ways, as shown in Fig. 3.12, the radiation force, $\mathbf{F}_a$, due to the absorbed photons is

$$\mathbf{F}_a = \rho_a PA(\mathbf{n} \cdot \mathbf{S})\mathbf{S}$$

(3.105)

If the fraction of the incoming photons reflected specularly, as from a mirror, is ρ_s, the force, $\mathbf{F}_s$, due to specularly reflected photons is given by

$$\mathbf{F}_s = 2\rho_s PA(\mathbf{n} \cdot \mathbf{S})\mathbf{n}$$

(3.106)

A fraction, ρ_d, of the incoming photons is assumed to be diffusely reflected. The incoming photons's momentum may be considered as stopped at the

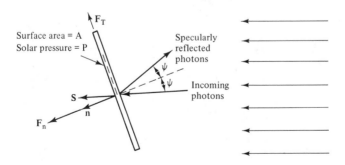

Figure 3.12 Solar radiation force on a surface.

P - solar pressure (constant)

surface and subsequently reradiated uniformly into the hemisphere. Integrating the momentum corresponding to the hemisphere, it can be seen that the tangential component of the outgoing momentum is canceled due to symmetry, leaving only the component normal to the surface. The total force, $\mathbf{F}_d$, caused by diffusely reflected photons is (Ref. 9)

A - area force is acting on

$$\mathbf{F}_d = \rho_d PA(\mathbf{n} \cdot \mathbf{S})\mathbf{S} + \frac{2}{3}\rho_d PA(\mathbf{n} \cdot \mathbf{S})\mathbf{n} \qquad (3.107)$$

Combining Eqs. (3.105) through (3.107), the total force due to solar radiation pressure is

$$\mathbf{F} = PA(\mathbf{n} \cdot \mathbf{S})\left\{ (\rho_a + \rho_d)\mathbf{S} + \left(2\rho_s + \frac{2}{3}\rho_d\right)\mathbf{n} \right\} \qquad (3.108)$$

Noting that $\rho_a + \rho_d + \rho_s = 1$, the solar radiation pressure moment $\mathbf{M}_s$ is

$$\mathbf{M}_s = PA(\mathbf{n} \cdot \mathbf{S})\mathbf{r} \times \left\{ (1 - \rho_s)\mathbf{S} + 2\left(\rho_s + \frac{1}{3}\rho_d\right)\mathbf{n} \right\} \qquad (3.109)$$

where $\mathbf{r}$ is the vector from the center of mass of the spacecraft to the center of pressure of a given area, A. The solar radiation pressure, P, is generally assumed to be constant and to have the value $4 \cdot 644 \times 10^{-6}$ N/m^2. From Fig. 3.13, the unit vector $\mathbf{S}$ is *4.644×10^{-6} N/m^2*

$$\mathbf{S} = \sin \delta\, \mathbf{J}_0 + \cos \delta\, \mathbf{K}_0$$

$$= \sin \alpha \cos \delta\, \mathbf{I} + \sin \delta\, \mathbf{J} + \cos \alpha \cos \delta\, \mathbf{K} \qquad (3.110)$$

where δ is the declination of the sun, positive from the vernal equinox to the autumnal equinox; α is the orbit angle measured from spacecraft local noon; $\mathbf{I}_0$, $\mathbf{J}_0$, $\mathbf{K}_0$ are the vectors along the orbital axes at spacecraft local noon; and $\mathbf{I}$, $\mathbf{J}$, $\mathbf{K}$ are the unit vectors along orbital axes at α orbit angle.

For a three-axis-stabilized spacecraft, the solar array contributes the

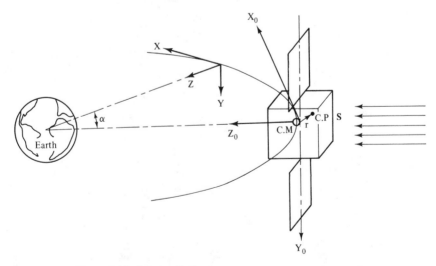

Figure 3.13 Solar radiation pressure moment on a spacecraft.

major portion of the solar pressure moment. Because the solar array is suntracking about the pitch axis, as shown in Fig. 3.13, the vector **n** is given by

$$\mathbf{n} = \sin \alpha \, \mathbf{I} + \cos \alpha \, \mathbf{K} \qquad (3.111)$$

The vector **n**, normal to the solar array, coincides with the vector **S** at equinox, where $\delta = 0$. Substituting Eqs. (3.110) and (3.111) into Eq. (3.109), we obtain

$$\mathbf{M}_s = \begin{Bmatrix} PA(yK_1 \cos \alpha - zK_2)\mathbf{I} \\ PA(zK_1 \sin \alpha - xK_1 \cos \alpha)\mathbf{J} \\ PA(xK_2 - yK_1 \sin \alpha)\mathbf{K} \end{Bmatrix} \qquad (3.112)$$

where we introduced the notation

$$K_1 = \left[(1 - \rho_s) \cos \delta + 2 \left(\rho_s + \frac{1}{3} \rho_d \right) \right] \cos \delta \qquad (3.113a)$$

$$K_2 = (1 - \rho_s) \cos \delta \sin \delta \qquad (3.113b)$$

Moreover x, y and z are the components of the vector **r**. Transforming the solar pressure moment $\mathbf{M}_s$ into the coordinate system X_0, Y_0, Z_0, which is rotating at the rate of 1°/day, we obtain

$$\mathbf{M}_s = PA \begin{Bmatrix} (yK_1 - zK_2 \cos \alpha - xK_2 \sin \alpha) \, \mathbf{I}_0 \\ (zK_1 \sin \alpha - xK_1 \cos \alpha) \, \mathbf{J}_0 \\ (-zK_2 \sin \alpha + xK_2 \cos \alpha) \, \mathbf{K}_0 \end{Bmatrix} \qquad (3.114)$$

From Eqs. (3.112) and (3.114), it should be noted that the solar radiation torque component along the pitch axis (Y and Y_0) is periodic, with period equal to the orbit period. Hence the net effect of this torque component over a full orbit period is zero. On the other hand, the components along the other axes have secular (non-zero unidirectional) components, resulting in a net change in the spacecraft angular momentum over a full orbit. It should be further noted that the secular torque in one coordinate system becomes periodic in the other coordinate system.

3.10 FIXED MOMENTUM WHEEL WITH THRUSTERS

Most of the recent three-axes-stabilized communications satellites, such as RCA Satcom, Intelsat V, and OTS/Marecs, have used an attitude control system consisting of a momentum wheel and thrusters (magnetic torquers on RCA Satcom). The momentum wheel provides the gyroscopic stiffness. The attitude dynamics of the fixed-momentum-wheel system is similar to that of dual-spin-stabilized spacecraft. The major difference lies in the fact that the angular momentum of a dual-spin-stabilized spacecraft is in general significantly higher than that of a three-axis-stabilized spacecraft with a fixed momentum wheel. Hence the rate of attitude error buildup due to disturbing moments is higher for a fixed-momentum-wheel system. For a dual-spin-stabilized spacecraft, the attitude errors are corrected periodically by ground commands or autonomously as in Intelsat VI. In a fixed-momentum-wheel system, however, the thrusters are fired by on-board control electronics to correct the attitude errors automatically. The system is shown in Fig. 3.14. The pitch and roll errors are detected by earth sensors.

The momentum wheel is nominally aligned with the pitch axis, with the angular momentum vector along the negative pitch axis. Hence the components of the angular momentum of the wheel in a spacecraft-fixed coordinate system are

$$h_x = 0 \qquad h_y = -h \qquad h_z = 0 \qquad (3.115)$$

where $h = I_w\Omega$ is the magnitude of the angular momentum of the wheel, in which I_w is its moment of inertia and Ω the angular velocity. Inserting Eqs. (3.97), (3.104), and (3.115) into Eq. (3.96), we obtain

$$\begin{Bmatrix} M_{cx} + M_{sx} \\ M_{cy} + M_{sy} \\ M_{cz} + M_{sz} \end{Bmatrix} = \begin{Bmatrix} I_{xx}\ddot{\phi} + [4\omega_0^2(I_{yy} - I_{zz}) + \omega_0 h]\phi \\ \quad + [h - \omega_0(I_{xx} - I_{yy} + I_{zz})]\dot{\psi} \\ I_{yy}\ddot{\theta} + 3\omega_0^2\theta\,(I_{xx} - I_{zz}) - \dot{h} \\ I_{zz}\ddot{\psi} + [\omega_0^2(I_{yy} - I_{xx}) + \omega_0 h]\psi \\ \quad - [h - \omega_0(I_{xx} + I_{zz} - I_{yy})]\dot{\phi} \end{Bmatrix} \qquad (3.116)$$

Assuming that

$$h \gg \max[4\omega_0(I_{yy} - I_{zz}), \omega_0(I_{xx} - I_{yy} + I_{zz}), \omega_0(I_{yy} - I_{xx})] \quad (3.117)$$

the equations of motion become

$$\begin{Bmatrix} M_{cx} + M_{sx} \\ M_{cy} + M_{sy} \\ M_{cz} + M_{sz} \end{Bmatrix} = \begin{Bmatrix} I_{xx}\ddot{\phi} + \omega_0 h\phi + h\dot{\psi} \\ I_{yy}\ddot{\theta} + 3\omega_0^2\theta\,(I_{xx} - I_{zz}) - \dot{h} \\ I_{zz}\ddot{\psi} + \omega_0 h\psi - h\dot{\phi} \end{Bmatrix} \qquad (3.118)$$

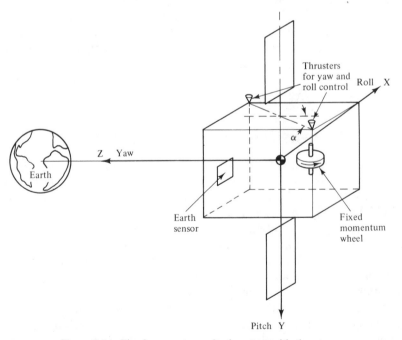

Figure 3.14 Fixed-momentum-wheel system with thrusters.

From Eq. (3.118), it is clear that for small attitude errors, the equation about the pitch axis is uncoupled from the equations about the roll and yaw axes. The equations about the roll and yaw axes are coupled by the angular momentum, h. Because of this, we can discuss the pitch axis and roll–yaw axes controls separately.

Pitch Axis Control

From Eq. (3.118), the linearized pitch axis equation is

$$M_{cy} + M_{sy} = I_{yy}\ddot{\theta} + 3\omega_0^2(I_{xx} - I_{zz})\theta - \dot{h} \tag{3.119}$$

In the case of $I_{xx} = I_{zz}$, Eq. (3.119) reduces to

$$M_{cy} + M_{sy} = I_{yy}\ddot{\theta} - \dot{h} \tag{3.120}$$

The solar pressure moment, M_{sy}, along the pitch axis is periodic, as shown in Eq. (3.114). The pitch error can be controlled by applying a torque proportional to the pitch error and its rate. The control torque is provided by changing the angular momentum of the wheel according to

$$\dot{h} = -K_\theta(\tau_\theta\dot{\theta} + \theta) \tag{3.121}$$

Substituting Eq. (3.121) into Eq. (3.120) and noting that $M_{cy} = 0$, the pitch equation becomes

$$M_{sy} = I_{yy}\ddot{\theta} + K_\theta\tau_\theta\dot{\theta} + K_\theta\theta \tag{3.122}$$

For K_θ, $\tau_\theta > 0$, Eq. (3.122) is the equation of the motion of a damped second-order system. The natural frequency, ω_θ, and damping ratio, ζ_θ, are given by

$$\omega_\theta = \sqrt{\frac{K_\theta}{I_{yy}}} \qquad \zeta_\theta = \frac{\tau_\theta}{2}\sqrt{\frac{K_\theta}{I_{yy}}} \tag{3.123}$$

The control block diagram is shown in Fig. 3.15. The momentum wheel is intended to apply a moment to the spacecraft to nullify the external moment. The transfer functions of the pitch control are

$$G(s) = \frac{1}{I_{yy}s^2} \tag{3.124}$$

and

$$H(s) = K_\theta(\tau_\theta s + 1) \tag{3.125}$$

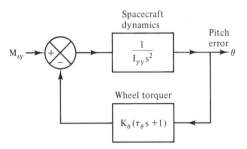

Figure 3.15 Block diagram for pitch axis control.

yielding

$$G(s)H(s) = \frac{K_\theta(\tau_\theta s + 1)}{I_{yy}s^2} \qquad (3.126)$$

Hence the open-loop transfer function, $G(s)H(s)$, has two poles at $s = 0$ and one zero at $s = -1/\tau_\theta$. The root locus for the pitch control is shown in Fig. 3.16.

Point A, which is generally the design point, corresponds to a critically damped system, $\zeta_\theta = 1$. Using Eq. (3.123), we conclude that for a critically damped system,

$$\tau_\theta = 2\sqrt{\frac{I_{yy}}{K_\theta}} \qquad (3.127)$$

and the closed-loop transfer function is

$$\frac{\theta(s)}{M_{sy}(s)} = \frac{1}{I_{yy}(s + \sqrt{K_\theta/I_{yy}})^2} \qquad (3.128)$$

The pitch error is introduced from several sources, such as initial pitch error, impulse moments from a thruster during desaturation period (to keep the wheel speed within normal allowable range) and during station keeping because of misalignment of the thrusters, and cyclic moment due to solar pressure. The disturbance torque during station keeping is generally large enough so that control by wheel is abandoned in favor of thruster control. In this case, wheel speed is kept constant while the thrusters are used to provide attitude control torque. The parameters of the control system, such as K_θ and τ_θ, are selected such that the attitude errors due to these disturbances are within allowable limits.

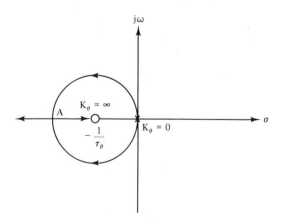

Figure 3.16 Root-locus plot for pitch axis control.

For the critically damped system, the response due to initial pitch error is

$$\theta(t) = \theta(0)\left(1 + \frac{t}{\tau}\right)e^{-t/\tau}$$

and   (3.129)

$$\tau = \frac{1}{\omega_\theta} = \sqrt{I_y/K_\theta}$$

where $\theta(0)$ is the initial pitch error. If M_y is an impulse moment such that $M_y = M_0\,\delta(t)$, use of Eq. (3.128) yields

$$\theta(t) = \frac{M_0}{I_{yy}}te^{-t/\tau}$$ (3.130)

The maximum pitch error occurs at $t = \tau$ and has the value

$$\theta_{max} = \frac{M_0\tau}{I_{yy}}\frac{1}{e}$$ (3.131)

For a sinusoidal disturbing moment, of the type

$$M_y = M_0\cos\omega_0 t$$ (3.132)

such as in the case of solar radiation pressure moment, the steady-state response is

$$\theta = \theta_{max}\cos(\omega_0 t - \psi_0)$$ (3.133)

where the amplitude is given by

$$\theta_{max} = \frac{M_0\tau^2}{I_{yy}}\frac{1}{[(1 - \tau^2\omega_0^2)^2 + (2\tau\omega_0)^2]^{1/2}}$$ (3.134)

and the phase angle by

$$\psi_0 = \tan^{-1}\frac{2\tau\omega_0}{1 - \tau^2\omega_0^2}$$ (3.135)

For $\tau\omega_0 \ll 1$, Eq. (3.133) can be reduced to

$$\theta = \frac{M_0\tau^2}{I_{yy}}\cos\omega_0 t$$ (3.136)

The gain, K_θ, is selected on the basis of the steady error and the time constant of the system.

Noting that the control torque on the spacecraft is provided by the momentum wheel, we can use Eq. (3.121) and obtain

$$I_W\,\dot\Omega = -K_\theta(\tau_\theta\dot\theta + \theta)$$ (3.137)

The Laplace transformation of Eq. (3.137) for zero initial angle, $\theta(0) = 0$, is

$$\Omega(s) = -\frac{K_\theta}{I_W s}(\tau_\theta s + 1)\theta(s) + \frac{\Omega_n}{S}$$ (3.138)

where Ω_n denotes the initial angular velocity of the wheel. Considering a critically damped system where $\zeta_\theta = 1$, $\tau_\theta = 2\tau$, we can use Eqs. (3.127) and (3.128) and write

$$\Omega(s) = -\frac{2\tau s + 1}{I_W s(\tau s + 1)^2} M_{ys}(s) + \frac{\Omega_n}{s} \quad (3.139)$$

For an impulsive disturbance of magnitude $M_y = M_0 \delta(t)$, $M_{ys}(s) = M_0$, the momentum wheel speed response is

$$\Omega(t) = -\frac{M_0}{I_W}\left[1 + \left(\frac{t}{\tau} - 1\right)e^{-t/\tau}\right] + \Omega_n \quad (3.140)$$

The steady-state response, obtained by letting $t \to \infty$, is simply

$$\Omega_{ss} = -\frac{M_0}{I_W} + \Omega_n \quad (3.141)$$

It is clear from Eqs. (3.130) and (3.141) that the long-term effects of an impulsive disturbance moment is that the spacecraft attitude does not change but the speed of the momentum wheel changes as if the disturbance moment acted on the wheel directly. The negative sign in Eq. (3.141) is due to the fact that the angular momentum is along the negative pitch axis.

For a cyclic disturbing torque, such as that given by Eq. (3.132), the Laplace transform of the wheel speed response is

$$\Omega(s) = -\frac{(2\tau s + 1)s M_0}{I_W s(\tau s + 1)^2(s^2 + \omega_0^2)} + \frac{\Omega_n}{s} \quad (3.142)$$

Noting that for $\tau \omega_0 \ll 1$, the steady-state response can be reduced to

$$\Omega_{ss} = -\frac{M_0}{I_W \omega_0} \sin \omega_0 t + \Omega_n \quad (3.143)$$

Similarly, a secular (unidirectional nonzero) disturbing torque will lead to an infinite wheel speed unless an external moment along the pitch axis is applied, such as by control jets, to reduce the change in wheel speed. Thus the secular torques will require periodic control jets firing to keep the wheel speed within the allowable limits. This procedure is called the momentum wheel desaturation. Cyclic disturbance moments result in a cyclic wheel speed with an average value Ω_n. The momentum wheel is generally designed in such a way that the wheel speed variation due to cyclic disturbance moments is within allowable limits.

As shown in Fig. 3.15, the control torque is proportional to the pitch error, θ, and the pitch error rate, $\dot{\theta}$. The pitch error can be determined by attitude sensors, such as earth sensors. However, to determine the error rate, rate gyros may be required. Rate gyros cannot detect very low rates while satisfying long-life requirements. The common practice is to use lead compensation.

Roll–Yaw Axes Control

Using Eq. (3.118), the Laplace transform of the equation of motion in the roll–yaw axes is

$$\begin{bmatrix} M_{cx} + M_{sx} \\ M_{cz} + M_{sz} \end{bmatrix} = \begin{bmatrix} I_{xx}s^2 + \omega_0 h & hs \\ -hs & I_{zz}s^2 + \omega_0 h \end{bmatrix} \begin{Bmatrix} \phi(s) \\ \psi(s) \end{Bmatrix} \quad (3.144)$$

         Attitude Dynamics and Control   Chap. 3

For an uncontrolled spacecraft, $M_{cx} = M_{cz} = 0$, the motion is the same as that of a dual-spin-stabilized spacecraft. In the symmetric case, $I_{xx} = I_{zz}$, the transfer function between the yaw angle and the yaw torque is

$$\frac{\psi(s)}{M_z(s)} = \frac{I_{xx}s^2 + \omega_0 h}{(I_{xx}s^2 + \omega_0 h)(I_{xx}s^2 + \omega_0 h) + h^2 s^2} \tag{3.145}$$

so that the characteristic equation can be written in the form

$$s^4 + (2\omega_0\omega_n + \omega_n^2)s^2 + \omega_0^2\omega_n^2 = 0 \tag{3.146}$$

where $\omega_n = h/I_{xx}$ is the nutation frequency. The roots of the characteristic equation are

$$s^2 = -\frac{\omega_n^2}{2}\left[\left(1 + 2\frac{\omega_0}{\omega_n}\right) \pm \sqrt{1 + 4\frac{\omega_0}{\omega_n}}\right]$$

$$= -\frac{\omega_n^2}{2}\left[\left(1 + 2\frac{\omega_0}{\omega_n}\right) \pm \left(1 + 2\frac{\omega_0}{\omega_n} - 2\left(\frac{\omega_0}{\omega_n}\right)^2 + \cdots\right)\right] \tag{3.147}$$

Note that when $\omega_0/\omega_n \ll 1$, the poles are $s = \pm j\omega_n$ and $s = \pm j\omega_0$. Hence the force-free motion consists of two periodic motions. The first motion is the nutational motion with frequency equal to ω_n and the second

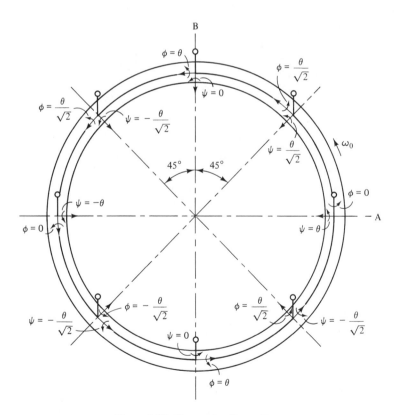

Figure 3.17 Yaw and roll coupling.

motion consists of coupled yaw and roll motions at the orbital rate ω_0. Figure 3.17 shows the yaw–roll coupling with zero nutation. The angle θ between the orbit normal and the angular momentum vector appears as yaw error θ at point A, with zero roll error. After a quarter period, at point B, the angle θ appears as roll error alone. Hence yaw and roll errors interchange every quarter of the orbit period. It is necessary only to sense and control one error directly, either yaw or roll, as the other error is controlled indirectly due to the coupling effect. It is shown later that for communications satellites, the roll error has a greater impact on the antenna pointing accuracy and is easier to sense than the yaw error. Therefore, for a fixed momentum wheel system, the roll error is sensed and controlled directly and the yaw error is controlled indirectly. The following methods can be used to control roll–yaw errors.

WHECON control system (Ref. 12). In this method the control torque is applied along the roll axis by control jets and it is proportional to the roll error, ϕ, and the roll error rate, $\dot{\phi}$. Control jets are commanded only on or off. So a modulator is necessary to transform the control signal to the on-off signal for control jets. The common practice is to use a pseudorate modulator, as shown in Fig. 3.18. A Schmitt trigger with hysteresis and real pole feedback produces a train of pulses whose average is proportional to the error and the error rate. This process is sometimes called a "derived-rate increment system" because it offers a method of synthesizing angular rate when direct measurement is not possible. The control jets are offset so as to provide control torque about the yaw axis also. The control torques along the roll and yaw axes are

$$M_{cx} = -K \cos \alpha \, (\tau s + 1) \, \phi(s) \tag{3.148}$$

$$M_{cz} = K \sin \alpha \, (\tau s + 1) \, \phi(s) \tag{3.149}$$

where K is the gain, determined by the thrust force and moment arm, α is the offset angle, and τ is the lead time constant of the pseudorate circuit. Substituting Eqs. (3.148) and (3.149) into Eq. (3.144), we obtain

$$\begin{Bmatrix} M_{sx} \\ M_{sz} \end{Bmatrix} = \begin{bmatrix} I_{xx}s^2 + \omega_0 h + K \cos \alpha \, (\tau s + 1) & hs \\ -K \sin \alpha \, (\tau s + 1) - hs & I_{zz}s^2 + \omega_0 h \end{bmatrix} \begin{Bmatrix} \phi(s) \\ \psi(s) \end{Bmatrix} \tag{3.150}$$

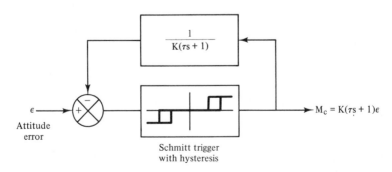

Figure 3.18 Block diagram of pseudorate modulation.

The characteristic equation corresponding to Eq. (3.150) is

$$s^4(I_{xx}I_{zz}) + s^3(K\tau I_{zz} \cos \alpha) + s^2[h(\omega_0 I_{xx} + \omega_0 I_{zz}$$
$$+ h + K\tau \sin \alpha) + K \cos \alpha I_{zz}] + s[h(K\tau\omega_0 \cos \alpha \quad (3.151)$$
$$+ K \sin \alpha)] + (\omega_0^2 h^2 + \omega_0 hK \cos \alpha) = 0$$

According to Routh's criterion, a necessary condition for stability is that all coefficients of the characteristic equation be positive. The coefficients are positive stable for positive values of K, τ, α, and h. Therefore, in a fixed-momentum-wheel system, the angular momentum of the wheel is kept along the negative pitch axis.

The angular momentum of the wheel is determined mainly by the required yaw accuracy. In the absence of notation, and disturbance and control torques, the orbital roll–yaw cycle will be a circle, as shown in Fig. 3.17. However, in the presence of a roll deadband and a secular roll–yaw disturbance torque the orbital roll–yaw cycle will be as shown in Fig. 3.19. When the roll error crosses the roll deadband, the offset thruster fires a series of pulses to keep the roll error within the roll deadband. This continues until, due to orbital motion, the roll error leaves the boundary and crosses the deadband zone. The yaw error ψ_1 and time t_1 of leaving the roll deadband limit ϕ_D, and the yaw error ψ_2 and time t_2 of reaching the opposite limit $-\phi_D$ are derived in Ref. 13, and are given as follows:

$$\cos \nu_1 = \frac{\omega_0 \phi_D h}{2 T_s} \tag{3.152}$$

$$\psi_1 = -\frac{T_s \sin \nu_1}{\omega_0 h} \tag{3.153}$$

$$\phi_D (1 + \cos (\nu_2 - \nu_1)) + \psi_1 \sin (\nu_2 - \nu_1) + (\nu_2 - \nu_1) \frac{\sin \nu_2 T_s}{\omega_0 h} = 0 \tag{3.154}$$

$$\psi_2 = \psi_1 \cos (\nu_2 - \nu_1) - \phi_D \sin (\nu_2 - \nu_1) + (\nu_2 - \nu_1) \cos \nu_2 \frac{T_s}{\omega_0 h} \tag{3.155}$$

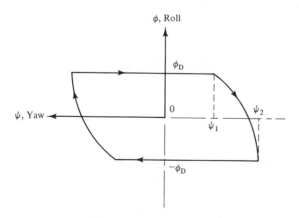

Figure 3.19 Orbital roll–yaw cycle.

where
$$\nu_1 = \omega_0 t_1 \quad \text{and} \quad \nu_2 = \omega_0 t_2 \qquad (3.156)$$
In Eqs. (3.152)–(3.155), T_s is the secular roll/yaw disturbance torque. The yaw angle ψ_2 at the deadband boundary $-\phi_D$ arrival represents the maximum yaw error and is of interest in angular momentum sizing. To obtain ψ_1, one may use iterative linearization.

For the disturbing torques, $T_{bx}(s) = T_{bx}/s$ and $T_{bz}(s) = T_{bz}/s$, which are constant in the body-fixed coordinates and cyclic in inertial coordinates, the steady-state errors are from Eq. (3.150),

$$\phi_{ss} = \frac{T_{bx}}{\omega_0 h + K \cos \alpha}$$
$$\cong \frac{T_{bx}}{K \cos \alpha} \quad \text{for } K \cos \alpha \gg \omega_0 h \qquad (3.157)$$

$$\psi_{ss} = \frac{1}{\omega_0 h}\left(T_{bz} + \frac{T_{bx} K \sin \alpha}{\omega_0 h + K \cos \alpha}\right)$$
$$\cong \frac{T_{bz} + T_{bx} \tan \alpha}{\omega_0 h} \quad \text{for } K \cos \alpha \gg \omega_0 h \qquad (3.158)$$

From Eq. (3.158), the yaw steady-state error is inversely proportional to the magnitude of the angular momentum. Due to this yaw offset, the orbital roll–yaw cycle, as shown in Fig. 3.19, will no longer be symmetric. The combined maximum yaw error will be the sum of the yaw error ψ_2 in the orbital roll–yaw cycle and the yaw error offset due to the body-fixed constant disturbance torques. The angular momentum of the wheel is selected such that the yaw error is within the specified yaw accuracy limits.

The gain, K, is selected to be large enough to ensure a roll capture capability and rapid roll transient response. To select the offset angle, α, and time constant, τ, it is necessary to analyze the roots of the characteristic equation (3.151). It should be noted that the linearized analysis applies only when the roll error is merely greater than ϕ_D. Parameters selected on this basis may not be the best parameters for normal on-orbit operation when the roll motion is within the deadband.

Assuming that $\omega_0 \ll \omega_n$ and $h \tan^2 \alpha \gg \omega_0 I_{zz}$, the open-loop transfer function of the roll axis is

$$G(s)H(s) = \frac{K I_{zz} \cos \alpha \, (s + \omega_0/\tan \alpha)(s + h \tan \alpha/I_{zz})(\tau s + 1)}{I_{zz}I_{xx}\,(s^2 + h^2/I_{xx}I_{zz})(s^2 + \omega_0^2)} \qquad (3.159)$$

The root locus for the closed-loop roll–yaw control is shown in Fig. 3.20. The zero from the derived-rate compensator provides the damping of the nutation frequency pole. The damping of the orbit frequency pole is due to the offset angle, α. Assuming that

$$h \gg \frac{\tau \omega_0}{\tan \alpha} \qquad (3.160)$$

and

$$K \cos \alpha \gg \frac{4 \omega_0 h I_{xx}}{I_{zz}} \qquad (3.161)$$

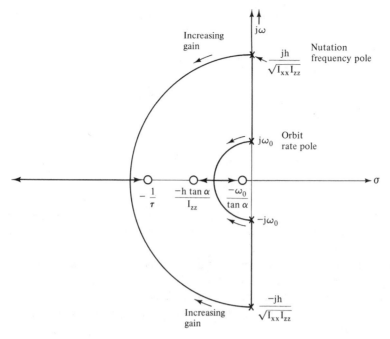

Figure 3.20 Root-locus for roll–yaw coupling.

the closed-loop natural frequencies and associated damping ratios are given by

$$\omega_1 = \sqrt{\frac{K \cos \alpha}{N I_{xx}}} \qquad \zeta_1 = \frac{\tau}{2} \sqrt{\frac{N K \cos \alpha}{I_{xx}}} \qquad (3.162)$$

$$\omega_2 = \sqrt{\frac{N \omega_0 h}{I_{zz}}} \qquad \zeta_2 = \frac{\tan \alpha}{2} \sqrt{\frac{N h}{I_{zz} \omega_0}} \qquad (3.163)$$

where

$$N = \frac{1}{1 + h^2/(I_{zz} K \cos \alpha)} \qquad (3.164)$$

The parameter N from Eq. (3.164) can be considered a correction factor with a nominal value of unity.

To provide high damping and rapid response to disturbance, values of τ and α are nominally chosen to critically damp the system. The thruster offset angle, α, is selected to critically damp the orbital frequency mode ω_2. Setting $\zeta_2 = 1$ in Eq. (3.163) gives the design equation for α,

$$\alpha = \tan^{-1} 2 \sqrt{\frac{I_{zz} \omega_0}{N h}} \qquad (3.165)$$

The lead-time constant, τ, is selected to critically damp the nutation frequency motion. Setting $\zeta_1 = 1$ in Eq. (3.162), the design equation for τ is given by

$$\tau = 2 \sqrt{\frac{I_{xx}}{N K \cos \alpha}} \qquad (3.166)$$

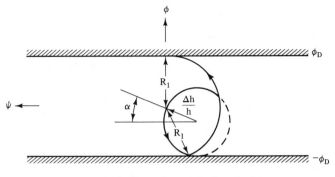

Figure 3.21 Upper bound for impulse bit.

Impulse bit (Ref. 13). One of the most critical design parameters for a fixed momentum wheel system is the minimum impulse bit of the offset thrusters. The upper bound is that which drives the roll error completely across the dead zone by means of a single pulse, as shown in Fig. 3.21.

It is assumed that the contact with the $-\phi_D$ deadband boundary is made with an initial nutation of magnitude $\Delta h/h$ and a single pulse drives the roll error to the other boundary, ϕ_D. From the geometry, the upper bound for Δh is

$$\Delta h \leq \frac{2\phi_D h}{1 + \sin \alpha + 2 \sin (45° + \alpha/2)} \tag{3.167}$$

It is also desirable to make the minimum impulse bit large enough so that a two-sided limit cycle, if it occurs, leads to a firing of only one pulse per boundary contact. If the roll error is not reduced quickly enough, the presence of the sensor noise will cause multiple thruster firings. From Fig. 3.22 it follows that the limiting case is when AB (or CD) is normal to the deadband boundary. From this, we conclude that the lower bound on Δh must satisfy the inequality

$$\Delta h \geq \frac{2 h \phi_D \tan \alpha}{\cos \alpha} \tag{3.168}$$

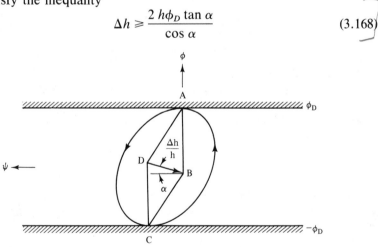

Figure 3.22 Lower bound for impulse bit.

Attitude Dynamics and Control   Chap. 3

Two-pulse modulation (Ref. 13). The principal feature of this method, which also uses offset control jets that provide control torques along both the roll and yaw axes, is that when the roll error stays within the deadband zone, the thrusters are fired in a two-pulse modulation sequence. Figure 3.23 shows the block diagram of the control sequence of the system. As the spacecraft roll attitude reaches the positive deadband boundary at point A, shown in Fig. 3.24, the X/Z thruster is fired to change the angular momentum by Δh. After a time ΔT, which is normally between $\frac{7}{12}$ and $\frac{5}{8}$ of the nutation period and is equivalent to a δ between 30 and 45°, the second pulse of the same thruster is fired. The second pulse not only changes the angular momentum by Δh but also reduces the nutation angle. The two-pulse modulation technique dampens the nutation and reduces the roll error.

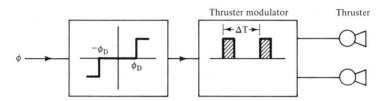

Figure 3.23 Block diagram of two-pulse modulation control.

Magnetic torquer. The control torque for the roll–yaw axes can be provided by using the earth's magnetic field. Such a system consists of a simple coil(s) which produces a magnetic dipole when current flows around the loop. The magnetic dipole, M, is proportional to the ampere-turns and the area enclosed by the coil, and the direction is normal to the plane of the coil. The torque, $\mathbf{T}_c$, acting on the spacecraft is

$$\mathbf{T}_c = \mathbf{M} \times \mathbf{B} \tag{3.169}$$

where $\mathbf{M}$ is the magnetic dipole of the coil and $\mathbf{B}$ is the earth's magnetic field. As shown in Fig. 3.25, the earth's magnetic field, $\mathbf{B}$, is modeled as a tilted dipole located at 69° west longitude and 78.3° north latitude. The

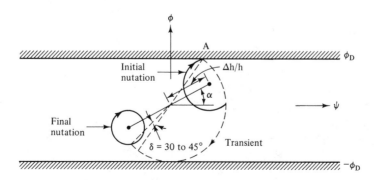

Figure 3.24 Diagram of roll–yaw control.

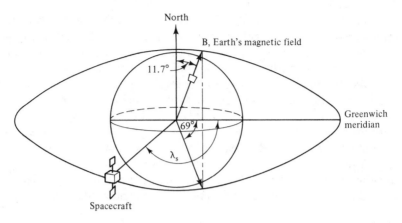

Figure 3.25 Earth's magnetic field.

earth's magnetic field in the orbit coordinate is

$$B_x = B_e \sin(11.7°) \sin(\lambda_s - 69°)$$
$$B_y = -B_e \cos(11.7°) \tag{3.170}$$
$$B_z = -B_e \sin(11.7°) \cos(\lambda_s - 69°)$$

where λ_s is the west longitude of the synchronous spacecraft. Let us consider a coil whose magnetic dipole is

$$M_x = -M \sin \alpha$$
$$M_y = 0 \tag{3.171}$$
$$M_z = -M \cos \alpha$$

Using Eqs. (3.169) through (3.171), magnetic torque on the spacecraft becomes

$$T_x = M \cos \alpha \, B_y = -MB_e \cos(11.7) \cos \alpha$$
$$T_y = M \sin \alpha \, B_z - M \cos \alpha \, B_x \tag{3.172}$$
$$= -MB_e \sin(11.7) \sin(\alpha + \lambda_s - 69)$$
$$T_z = -M \sin \alpha \, B_y = MB_e \cos(11.7) \sin \alpha$$

From Eqs. (3.148), (3.149) and (3.172), we conclude that the control torques along the yaw and roll axes can be made equivalent to those of a control jet with the offset angle by setting

$$K(\tau\dot{\phi} + \phi) = MB_e \cos(11.7) \tag{3.173}$$

The control dynamics for a magnetic torquer are similar to that of WHECON, except that the gain, K, for the magnetic torquer is much smaller. The magnetic dipole can be produced by two coils, a yaw coil which produces a magnetic dipole along the yaw axis, M_z, and a roll coil which produces a magnetic dipole along the roll axis, M_x. The direction of the control torque can be reversed by changing the direction of the current in the coil(s). It should be noted that the spacecraft can have magnetic dipoles which interact with the earth's magnetic field and produce magnetic torque on the spacecraft, which can be considered as a part of the disturbing torque. Magnetic torquers have also been used to compensate for solar disturbance torques, as in Intelsat V.

3.11 THREE-AXIS REACTION WHEEL SYSTEM

A three-axis reaction wheel system can be considered as a combination of three independent pitch, roll, and yaw control systems (Fig. 3.26). Each axis is controlled by varying the speed of the reaction wheel in response to the attitude error. The system requires a reaction wheel with zero nominal angular momentum and an attitude sensor for each axis (i.e., pitch, roll, and yaw).

Introducing Eqs. (3.96) and (3.104) into Eq. (3.97) and neglecting the transverse inertia of the wheel, the equations of motion for a three-axis reaction wheel system are

$$M_{cx} + M_{sx} = I_{xx}\ddot{\phi} + [4\omega_0^2(I_{yy} - I_{zz}) - \omega_0 h_y]\phi$$
$$+ [-h_y - \omega_0(I_{xx} - I_{yy} + I_{zz})]\dot{\psi} + \dot{\theta}h_z - \omega_0 h_z + \dot{h}_x$$

$$M_{cy} + M_{sy} = I_{yy}\ddot{\theta} + 3\omega_0^2\theta(I_{xx} - I_{zz}) + \omega_0 h_x\phi$$
$$- h_z\dot{\phi} + \omega_0 h_z\psi + h_x\dot{\psi}_x + \dot{h}_y \qquad (3.174)$$

$$M_{cz} + M_{sz} = I_{zz}\ddot{\psi} + [\omega_0^2(I_{yy} - I_{xx}) - \omega_0 h_y]\psi$$
$$+ [h_y + \omega_0(I_{xx} + I_{zz} - I_{yy})]\dot{\phi} - h_x\dot{\theta} + \omega_0 h_x + \dot{h}_z$$

Although Eqs. (3.174) are coupled, because h_x, h_y, h_z and ω_0 are small, the coupling terms are small. If the coupling terms are neglected, the equations of motion about the pitch, roll, and yaw axes become independent, and

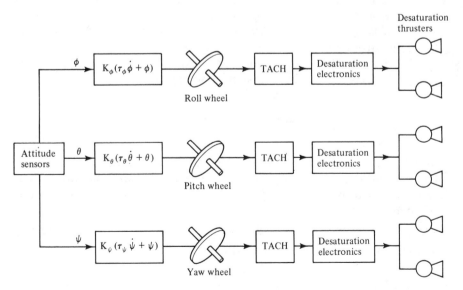

Figure 3.26 Block diagram of the three-axis reaction control system.

hence they can be controlled independently. The control torques are applied by letting the rates of change of the angular moment of the reaction wheels have

$$\dot{h}_x = K_\phi(\tau_\phi\dot{\phi} + \phi)$$

$$\dot{h}_y = K_\theta(\tau_\theta\dot{\theta} + \theta) \tag{3.175}$$

$$\dot{h}_z = K_\psi(\tau_\psi\dot{\psi} + \psi)$$

The control dynamics of each axis is very similar to the pitch axis control dynamics of the fixed-momentum-wheel system. The secular disturbing moments (in the orbital coordinate system) will lead to unacceptably high speed unless control jets or other devices are used to apply an external moment along the wheel axis to reduce the wheel speed. Therefore, the secular torques will require desaturation of the reaction wheels. Cyclic disturbing moments result in cyclic wheel speed and do not require desaturation, if the variation in the speed is within allowable wheel speed limits.

3.12 EFFECTS OF STRUCTURAL FLEXIBILITY

In previous sections, the spacecraft was assumed to be a rigid body. In a spacecraft, however, flexibility is introduced from many sources, such as solar arrays, antennas, and antenna support structure. In future, spacecraft are expected to become much more mechanically flexible, due to large deployable antennas and solar arrays. These spacecraft will require advanced attitude control concepts. In the last decade, extensive research has been performed to study the interaction between the structural flexibility and attitude control systems. Several advanced concepts have been proposed for attitude control of such spacecraft. It is beyond the scope of this book to provide a detailed discussion of these concepts. However, a simple example will be discussed to introduce the fundamentals of flexibility and attitude control interactions.

Let us consider that the torsional flexibility of a solar array in the pitch axis is represented by a torsional spring and a disk, as shown in Fig. 3.27. The equations of motion of the system are

$$I_b\ddot{\theta}_b + K(\theta_b - \theta_s) = T \tag{3.176}$$

$$I_s\ddot{\theta}_s + K(\theta_s - \theta_b) = 0 \tag{3.177}$$

where I_b and I_s are the pitch axis inertias of the central body and the solar array, θ_b and θ_s are the pitch angles of the body and the solar array, K is the torsional stiffness of the solar array, and T is the external torque along the pitch axis. Let

$$\theta_r = \theta_s - \theta_b \tag{3.178}$$

Adding Eqs. (3.176) and (3.177) and using Eq. (3.178), we get

$$I_T\ddot{\theta}_b = T - I_s\ddot{\theta}_r \tag{3.179}$$

$$\ddot{\theta}_r + \omega_n^2\theta_r = -\ddot{\theta}_b \tag{3.180}$$

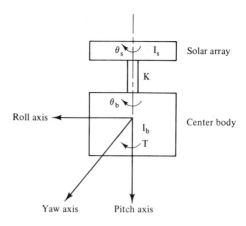

Figure 3.27 Model of solar array flexibility.

Roll axis

Yaw axis   Pitch axis

Solar array

Center body

where $I_T = I_b + I_s$ = total pitch inertia

$\omega_n = \sqrt{K/I_s}$ = fixed base natural frequency of the solar array

Taking the Laplace transform of Eqs. (3.179) and (3.180), θ_b can be expressed as

$$I_T s^2 \theta_b(s) = T(s) + \frac{I_s s^4 \theta_b(s)}{s^2 + \omega_n^2} \tag{3.181}$$

The transfer function in this form is shown in Fig. 3.28(a). Here the flexibility effects are introduced as a feedback torque. By rearranging the terms in Eq. (3.181), we get

$$\frac{\theta_b(s)}{T} = \frac{s^2 + \omega_n^2}{s^2 I_b (s^2 + \Omega_n^2)} \tag{3.182}$$

where Ω_n is the free–free natural frequency of the spacecraft and is given by

$$\Omega_n = \omega_n \sqrt{1 + \frac{I_s}{I_b}} \tag{3.183}$$

The transfer function, by factoring the denominator terms, is given by

$$\theta_b = T \left[\frac{1}{I_T s^2} + \frac{I_s}{I_b I_T (s^2 + \Omega_n^2)} \right] \tag{3.184}$$

The transfer function in the form of Eq. (3.184) is represented in Fig. 3.28(b). Here the pitch angle of the body consists of two components. The first component, θ_R, is obtained by assuming the entire spacecraft to be a rigid body and the second component, θ_f, is due to the flexibility of the solar array. By using the transfer function from Eq. (3.182), the pitch axis control is modified as shown in Fig 3.28(c). Here the base fixed natural frequency of the solar array, ω_n, is the zero of the open-loop transfer function and the free–free natural frequency of the spacecraft, Ω_n, is the pole. The root locus of the pitch control is shown in Fig. 3.29. The system is stable, although the flexible modes are lightly damped. To avoid interaction

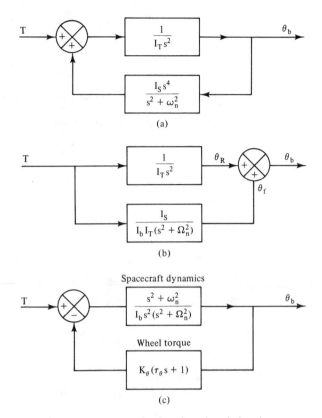

Figure 3.28 Transfer functions for pitch axis.

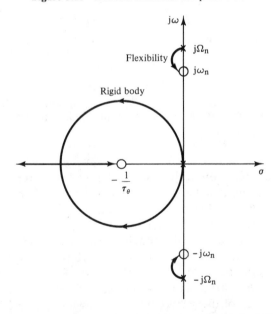

Figure 3.29 Root locus for pitch axis control with flexibility.

Attitude Dynamics and Control   Chap. 3

of the flexibility with the attitude control, it is desirable to have high structural damping and wide separation between the control frequencies and structural frequencies.

Example 3.2

Determine the parameters for the attitude control system consisting of a fixed momentum wheel with thrusters for a spacecraft whose mass properties are: $I_{xx} = 2700$ kg $\cdot$ m^2, $I_{yy} = 1360$ kg $\cdot$ m^2, and $I_{zz} = 2200$ kg $\cdot$ m^2. The solar array area is 16.3 m^2. The coefficient of specular reflection $\rho_s = 0.2$ and coefficient of diffuse reflection $\rho_d = 0$. The components of CM-CP offset are $x = 0.03$ m, $y = 0.03$ m, and $z = -0.33$ m. The roll thruster torque is 0.615 N $\cdot$ m. Wheel desaturation thruster torque $= 0.422$ N $\cdot$ m with a pulse time of 0.2 s. The permissible attitude control errors are 0.05° in roll and pitch and 0.2° in yaw. Determine the solar disturbance torques, angular momentum of the wheel, offset angle for the roll thruster, and lead time constant for the pitch and roll–yaw control.

Solution The solar disturbance torques are calculated by using Eqs. (3.112), (3.113), and (3.114). At equinox, $\delta = 0$, so

$$K_1 = 1 + \rho_s = 1.2 \qquad K_2 = 0 \qquad \omega_0 = 2\pi \text{ rad/day}$$

In spacecraft body coordinates, the solar torques are determined by using Eq. (3.112) in N $\cdot$ m as

$$M_s = 4.644 \times 10^{-6} \times 16.3 \left\{ \begin{array}{l} 0.03 \times 1.2 \cos \alpha \, \mathbf{I} \\ (-0.33 \times 1.2 \sin \alpha - 0.03 \times 1.2 \cos \alpha)\mathbf{J} \\ -0.03 \times 1.2 \sin \alpha \, \mathbf{K} \end{array} \right\}$$

$$= 2.72 \times 10^{-6} \cos \alpha \, \mathbf{I} + (3.0 \times 10^{-5} \sin \alpha - 2.72 \times 10^{-6} \cos \alpha)\mathbf{J}$$
$$- 2.72 \times 10^{-6} \sin \alpha \, \mathbf{K} \quad \sim 3.0$$

Similarly, at summer solstice, with $\delta = 23.5°$, the solar disturbance torques are

$$M_s = (2.36 \times 10^{-6} \cos \alpha + 7.30 \times 10^{-6})\mathbf{I} - (2.60 \times 10^{-5} \sin \alpha$$
$$+ 2.36 \times 10^{-6} \cos \alpha)\mathbf{J} + (0.664 \times 10^{-6} - 2.36 \times 10^{-6} \sin \alpha)\mathbf{K}$$

In inertial coordinates, the solar torques are determined by using Eq. (3.114) as

Equinox:

$$M_s = 2.72 \times 10^{-6}I_0 + (3.0 \times 10^{-5} \sin \alpha - 2.72 \times 10^{-6} \cos \alpha)J_0$$

Summer solstice:

$$M_s = (2.36 \times 10^{-6} + 7.30 \times 10^{-6} \cos \alpha - 0.664 \times 10^{-6} \sin \alpha)I_0$$
$$- (2.60 \times 10^{-5} \sin \alpha + 2.36 \times 10^{-6} \cos \alpha)J_0$$
$$+ (7.30 \times 10^{-6} \sin \alpha + 0.664 \times 10^{-6} \cos \alpha)K_0$$

Roll–yaw control. From the calculations above, constant roll–yaw solar torque in inertial coordinates, T_s, is 2.72×10^{-6} N $\cdot$ m at equinox. Let us assume that roll deadband is 0.03° or $\phi_D = 0.03°$. For a given angular momentum value, the yaw error, ψ_2, can be calculated from Eqs. (3.152) through (3.155). For $h = 35$ N $\cdot$ m $\cdot$ s, ψ_2 is 0.15°, as calculated in Ref. 13. The constant body fixed roll–yaw torques at equinox are zero. At summer solstice, these torques are: $T_{bx} = 7.30 \times 10^{-6}$ N $\cdot$ m and $T_{bz} = 0.634 \times 10^{-6}$ N $\cdot$ m. Substituting the numerical values in Eq. (3.158), and assuming that $\alpha = 8°$,

$$\psi_{ss} = \frac{0.664 \times 10^{-6} + 7.30 \times 10^{-6} \tan 8}{2\pi/(24 \times 60 \times 60) \times 35} \frac{180}{\pi}$$

$$= 0.038°$$

Hence the total yaw error due to inertially fixed roll–yaw solar torque, ψ_2, and body-fixed solar torques will be less than 0.2°. Therefore, the angular momentum of 35 N · m · s is selected for the wheel.

The autopilot gain, K, is the maximum control torque available divided by the linear range of the sensor. Assuming that the linear range is $\pm 3°$, the gain, K, is

$$K = \frac{0.615 \times 180}{3\pi} = 11.74 \text{ N} \cdot \text{m/rad}$$

Assuming that $\cos \alpha \simeq 1$, the correction factor N from Eq. (3.164) is

$$N = \frac{1}{1 + h^2/I_{zz}K} = \frac{1}{1 + 35^2/(2200 \times 11.74)} = 0.955$$

The offset angle, α, is selected by using Eq. (3.165) as

$$\alpha = \tan^{-1} 2 \sqrt{\frac{I_{zz}\omega_0}{Nh}}$$

$$= \tan^{-1} 2 \sqrt{\frac{2200 \times 2\pi}{24 \times 60 \times 60 \times 0.955 \times 35}} = 7.8°$$

The lead time constant, τ, is selected by using Eq. (3.166) as

$$\tau = 2 \sqrt{\frac{I_{xx}}{NK \cos \alpha}} = 2 \sqrt{\frac{2700}{0.955 \times 11.74 \times 0.99}}$$

$$= 31.2 \text{ s}$$

The upper bound for the roll thruster impulse bit, Δh, from Eq. (3.167) is

$$\Delta h \leq \frac{2\phi_D h}{1 + \sin \alpha + 2 \sin (45 + \alpha/2)}$$

$$\leq \frac{2 \times 0.03 \times \pi \times 35}{180(1 + \sin 7.8 + 2 \sin 48.9)}$$

$$\leq 0.01387 \text{ N} \cdot \text{m} \cdot \text{s}$$

The lower bound for the roll thruster impulse bit from Eq. (3.168) is

$$\Delta h \geq \frac{2h\phi_D \tan \alpha}{\cos \alpha}$$

$$\geq \frac{2 \times 35 \times 0.03 \times \pi \tan 7.8}{180 \times \cos 7.8}$$

$$> 0.00507 \text{ N} \cdot \text{m} \cdot \text{s}$$

These limits on the impulse bit imply that the impulse time should be less than 0.022 s and greater than 0.0082 s. The impulse time selected is 0.015 s. (average)

Pitch control. The pitch control design can be based on the maximum allowable pitch error during wheel desaturation. The desaturation torque impulse, M_y, is

$$M_y = 0.422 \times 0.2 = 0.0844 \text{ N} \cdot \text{m} \cdot \text{s}$$

The pitch error due to desaturation impulse is given by Eq. (3.130). From Eq. (3.131), the time constant of the system is given by

$$\tau = \frac{\theta_{max}I_{yy}e}{M_y}$$

Assuming that $\theta_{max} = 0.04°$, keeping $0.01°$ for other error sources,

$$\tau = \frac{0.04 \times \pi \times 1360 \times 2.718}{180 \times 0.0844}$$

$$= 31 \text{ s}$$

The time constant in Eq. (3.129) is defined as

$$\tau = \sqrt{\frac{I_{yy}}{K}}$$

or

$$K = \frac{I_{yy}}{\tau^2}$$

Substituting the numerical values in the equation above yields

$$K = \frac{1360}{(31)^2} = 1.41 \text{ N} \cdot \text{m/rad}$$

The lead-time constant τ_θ from Eq. (3.127) is given by

$$\tau_\theta = 2\sqrt{\frac{I_{yy}}{K_\theta}} = 2\tau = 2 \times 31 = 62 \text{ s}$$

3.13 ATTITUDE DETERMINATION

The selection of sensors for attitude determination depends on several factors, such as the type of spacecraft stabilization, orbital parameters, operational procedures, and required accuracy. The requirements for attitude determination can be divided into four functions: determination of spin axis attitude of the spacecraft in the transfer orbit, the acquisition or reacquisition of the operational mode, the attitude errors in the operational mode, and station keeping. The sensors used for these functions may not be the same.

Transfer Orbit

The attitude of the spin axis of spin-stabilized spacecraft in transfer orbit is normally determined by the cone intercept method. The sun sensor, shown in Fig. 3.30, determines the angle between the spin axis and the sun vector. The locus of the spin axis is a cone about the sun vector with half-angle θ_s. The earth sensor determines the angle between the spin axis and the earth radius vector from the mass center of the spacecraft to the center of the earth. The locus of the spin axis is a cone about the radius vector with half angle θ_e. Since one spin axis is common to both cones, it represents the intersection of these two cones. As shown in Fig. 3.30, there are in general two intersections, the real and imaginary spin axes. The ambiguity in the solution is resolved by taking two additional measurements at a later time. The true solution remains unchanged but the imaginary value changes. Figure 3.31 shows the commonly used sensors for transfer orbit.

Sun sensors. Sun sensors are normally used to determine θ_s. For a typical sun sensor, the beam shape is a wide, thin fan with a beamwidth

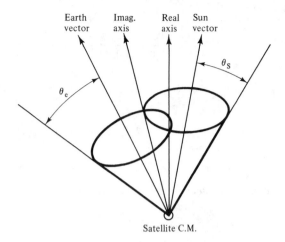

Figure 3.30 Cone intercept method.

of 160° in one plane and 0.7° normal to that plane. A sun sensor assembly consists of two sensors, which are often designated as ψ and ψ_2. The sensor ψ is in the plane XZ and produces an output pulse when the sun line is in the XZ plane. The ψ_2 sensor is rotated about the X axis relative to the ψ sensor through an angle i. The pair of sensors forms a "V beam" system and the relative time between their output pulses is used to measure the angle between the sun line and the spin axis.

Let ϕ be the angle through which the spacecraft rotates between the

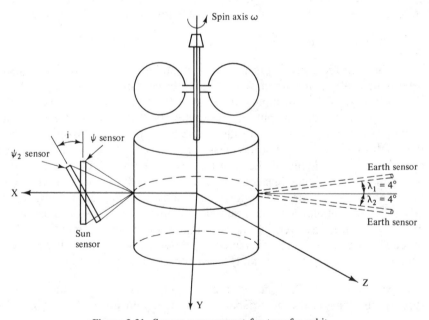

Figure 3.31 Sensor arrangement for transfer orbit.

time the sun is in the ψ sensor field of view (FOV) and the time it is in the ψ_2 sensor FOV. To determine the angle θ_s in terms of ϕ, let us look at Fig. 3.32. The center of the unit sphere is located at the satellite mass center with the spin axis along the north pole of the sphere. It is visualized that the satellite is fixed and that the sun is rotating relative to the satellite. The sun follows the path shown in Fig. 3.32. At points B and D, the sun is in the field of view of the sun sensor ψ and ψ_2, respectively. The angle between the sun positions B and D is ϕ. Using spherical trigonometry for triangles $ABCD$ and ADE, we can write

$$\frac{\sin \theta_s}{\sin i} = \frac{\sin AD}{\sin \phi} \tag{3.185}$$

and

$$\frac{\sin (90 - \theta_s)}{\sin (90 - i)} = \frac{\sin AD}{\sin 90°} \tag{3.186}$$

Equation (3.186) can be replaced by

$$\frac{\cos \theta_s}{\cos i} = \sin AD \tag{3.187}$$

Eliminating $\sin AD$ from Eqs. (3.185) and (3.187), we obtain

$$\cot \theta_s = \sin \phi \cot i \tag{3.188}$$

$$\theta_s = \cot^{-1}(\sin \phi \cot i) \tag{3.189}$$

The angle θ_s between the spin axis and the sun vector is determined from Eq. (3.189) by knowing the angle of rotation of the spaecraft between the time the sun is in the ψ sensor FOV and the time it is in the ψ_2 sensor FOV.

Earth sensors. Earth sensors commonly operate in the 14 to 16-μm CO_2 absorption band. An earth sensor assembly normally consists of two earth sensors each having a 1.1° FOV, and in some cases two sun-guard

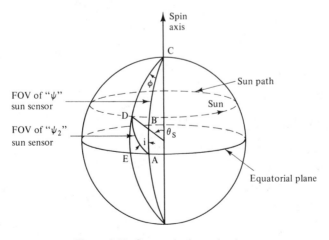

Figure 3.32 Sun angle determination.

sensors each having a 7.2° × 18° FOV. The sun-guard sensor's field view masks the FOV of the associated earth sensor and blocks the output of the earth sensor when it sees the sun, thus eliminating spurious data (the sun will saturate the earth sensor). The optical axes of the sensors are normally canted from the axis normal to the spin axis.

To calculate the angle θ_e, let us examine Fig. 3.33. The center of the unit sphere is located at the spacecraft mass center with the spin axis along the north pole of the sphere. The earth is now considered as a projection on this sphere. The angular distance between the projections of the earth's center and the earth's surface is β and is equal to $\sin^{-1}(R_e/R)$, where R_e is the earth radius and R is the satellite radius from the earth mass center. The cant angle of the earth sensor's to the satellite equator is λ. The earth sensor measures the angle through which the satellite rotates during the time the earth is in the sensor's field of view. Let this angle be μ_1 for earth sensor 1 and μ_2 for earth sensor 2. Using spherical trigonometry in conjunction with triangles 1 and 2 in Fig. 3.33, we obtain

$$\cos \beta = \cos \theta_e \sin \lambda + \sin \theta_e \cos \lambda \cos \frac{\mu_1}{2} \tag{3.190}$$

and

$$\cos \beta = \cos (180 - \theta_e) \sin \lambda + \sin (180 - \theta_e) \cos \lambda \cos \frac{\mu_2}{2}$$

$$= -\cos \theta_e \sin \lambda + \sin \theta_e \cos \lambda \cos \frac{\mu_2}{2} \tag{3.191}$$

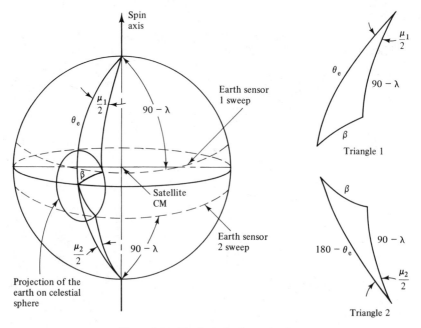

Figure 3.33 Earth angle determination.

Equations (3.190) and (3.191) yield

$$\theta_e = \tan^{-1} \frac{2 \tan \lambda}{\cos(\mu_2/2) - \cos(\mu_1/2)} \qquad (3.192)$$

Assuming that the angles θ_s and θ_e are known, the right ascension, α_w, and declination, δ_w, of the spin axis in inertial coordinates can be obtained by solving the equations

$$\cos \theta_s = \sin \delta_w \sin \delta_s + \cos \delta_w \cos \delta_s \cos (\alpha_w - \alpha_s) \qquad (3.193)$$

and

$$\cos \theta_e = -\sin \delta_w \sin \delta_R - \cos \delta_w \cos \delta_R \cos (\alpha_w - \alpha_R) \qquad (3.194)$$

where α_s and δ_s are the right ascension and declination of the sun, respectively, and α_R and δ_R are the right ascension and declination of the satellite position, respectively.

Geosynchronous Orbit

For a spin-stabilized spacecraft, the earth and sun sensors used in the transfer orbit can also be used for attitude determination in geosynchronous orbits. However, a three-axis-stabilized spacecraft, whose main body is not spinning, requires different sensors for the synchronous orbit from those used for the transfer orbit. The pitch and roll attitudes are easily determined by earth sensors. The yaw attitude, however, is difficult to determine.

There are several types of earth sensors, such as scanning, balanced radiation, and edge tracking types. The scanning sensor scans the earth's 14- to 16-μm infrared radiance profile with a resonant torsion-bar suspended mirror. The scanning movement is provided by a closed-loop feedback system. The sensor provides two scanning beams separated by an angle which is less than that subtended by the earth. The scanning is in the east-west direction and for a nominal spacecraft orientation, the sensor is centered about the earth's center with one beam scanning across the northern half of the earth disk and the other across the southern half. Attitude is determined by comparing the angle between the first horizon crossing and a center reference point to the angle from the reference to the second horizon crossing. Figure 3.34 shows the radiance image of the earth. H_1 is the angle of the scan at the entrance to the radiance image and H_2 is the angle of scan at the departure from the earth radiance image. H_0 is the reference angular position, normally at mirror center position. The pitch attitude is proportional to the difference between the average of angles H_1 and H_2 and reference angle H_0 (i.e., $(H_1 + H_2)/2 - H_0$). Either of the scanning beams, A or B, can be used for this computation. The roll attitude error is proportional to the difference between the scan angles of the earth's radiance image for scanning beam A and B [i.e., $(H_1 - H_2)_A - (H_1 - H_2)_B$].

One way to determine these angles is to have an optical encoder mounted with a mirror so as to generate a pulse for each incremental angle of the scan motion, normally 0.01°, and to generate a center reference pulse

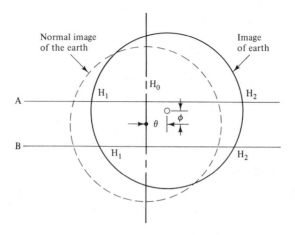

Figure 3.34 Scanning earth sensor.

at the mirror center position. For pitch attitude sensing, a count-up of the encoder pulses is started at H_1 until the center reference pulse is reached and a count-down is then started until H_2 is reached. The output count is proportional to the pitch error and will be zero for zero pitch error. For roll attitude sensing, a count-up is started at H_1 and terminated at H_2. The difference in the count for the north and south scan is proportional to the roll error.

The yaw attitude can be determined by a sun sensor which determines the angle between the sun line and the y axis (pitch). Assuming small attitude errors, Eq. (3.100) yields

$$\mathbf{j} = -\psi\mathbf{I} + \mathbf{J} + \phi\mathbf{K} \tag{3.195}$$

Moreover, according to Eq. (3.110),

$$\mathbf{S} = \sin\alpha\cos\delta\mathbf{I} + \sin\delta\mathbf{J} + \cos\alpha\cos\delta\mathbf{K} \tag{3.196}$$

where $\mathbf{S}$ is the sun line unit vector, δ the declination of the sun, α the orbit angle of the spacecraft measured from spacecraft local noon, and $\mathbf{I}$, $\mathbf{J}$, $\mathbf{K}$ are unit vectors along orbit coordinates axes. The angle between the vector $\mathbf{S}$ and the y axis is given by

$$\cos\theta_s = \mathbf{S}\cdot\mathbf{j} = -\psi\sin\alpha\cos\delta + \sin\delta + \phi\cos\alpha\cos\delta \tag{3.197}$$

From Eq. (3.197) it can be seen that the angle θ_s changes with the yaw error, ψ. Differentiating Eq. (3.197) with respect to ψ, we obtain

$$\frac{\partial\cos\theta_s}{\partial\psi} = -\sin\alpha\cos\delta \tag{3.198}$$

The yaw sensitivity of the sun sensor is a function of the spacecraft position. Referenced to local spacecraft time, it is maximum at 6 P.M. and 6 A.M. ($\alpha = 90°, 270°$) and zero at noon or midnight. Close to local midnight, the spacecraft is eclipsed during certain times of the year. Hence a sun sensor gives poor yaw attitude information close to noon/midnight. It should be noted from Eq. (3.197) that the sun sensor can also provide roll attitude information and that the roll sensitivity has a maximum at noon/midnight and is zero at 6 P.M./6 A.M. Other yaw sensors which have been used include a polarization sensor on Syncom, an interferometer/monopulse, and a Polaris star tracker on ATS-6.

3.14 ANTENNA BEAM POINTING ACCURACY

Attitude control errors, as determined in the previous sections by analyses, are only a part of the total antenna beam pointing error. The total antenna beam pointing error, as given in Table 3.1, can be divided into four categories: attitude control, thermal distortion, spacecraft alignments, and orbit errors. All errors are classified as either random or deterministic. Random errors refer to rapidly varying errors whose limits are defined, but whose values

TABLE 3.1 ANTENNA POINTING ERRORS (DEGREES) DURING OPERATION

Error Source	Pitch		Roll		Yaw	
	Random	Deter-ministic	Random	Deter-ministic	Random	Deter-ministic
Attitude control errors						
Earth sensor	± 0.035	± 0.036	—	± 0.018	—	—
Wheel speed/ torque error	—	± 0.020	—	—	—	—
Electronics effects	—	± 0.030	—	± 0.030	—	—
Deadbands and mo-mentum deviation	—	—	—	± 0.050	—	± 0.485
Wheel wobble effects/solar array pertur-bation	± 0.010	—	± 0.010	—	± 0.010	—
Subtotal/attitude control (RSS)	± 0.036	± 0.051	± 0.010	± 0.061	± 0.010	± 0.485
Thermal distortion		+ 0.024 − 0.018		+ 0.061 − 0.093		+ 0.068 − 0.028
Spacecraft alignment		± 0.020		± 0.020		± 0.043
Total	± 0.036	+ 0.095 − 0.089	± 0.010	+ 0.142 − 0.174	± 0.01	+ 0.596 − 0.556
Total spacecraft error		+ 0.131 − 0.125		+ 0.152 − 0.184		+ 0.606 − 0.566
Orbit error (worst case)		± 0.018		± 0.018		± 0.10
Antenna pointing error		+ 0.149 − 0.143		+ 0.170 − 0.202		+ 0.706 − 0.666

and likelihood of occurrence are uncertain. Deterministic errors are those whose values are predictable.

Attitude errors are primarily the result of inherent inaccuracies in the performances of attitude control equipment and control subsystem performance as established by analysis and simulation and test. Sensors are the major source of errors in equipment. The thermal distortion errors consist of errors due to reflector distortion, feed translation, and reflector translation and rotation due to reflector support structure distortion. The alignment errors are due to a lack of exact reflector deployment repeatability and alignment errors in the sensors and actuators.

The orbit errors, north-south and east-west, introduce corresponding spacecraft pointing errors in the pitch, roll, and yaw. The effect of these errors will depend on the specific earth station location. The worst-case errors (at the subsatellite point) are

$$\text{Roll error: } \phi_{max} \leq i \frac{R_e}{R'}$$

$$\text{Yaw error: } \quad \psi = i$$

$$\text{Pitch error: } \quad \theta_{max} \leq \Delta\lambda \frac{R_e}{R'}$$

where i is the inclination error, R_e is the radius of the earth, R' is the orbit altitude, and $\Delta\lambda$ is the east-west error.

In Table 3.1, the antenna beam pointing errors are expressed in terms of roll, pitch, and yaw. Figure 3.35 shows a crossed-pole antenna whose ground target is the sub-satellite point. Figure 3.35 shows the impact of each error individually. The pitch and roll errors transform directly into antenna beam pointing errors in azimuth (east-west) and elevation (north-south). The contribution of the yaw errors depends on the location of the ground target. The combined contribution of roll, pitch, and yaw error is expressed as

$$\Delta AZ = \theta + \frac{EL}{57.3} \psi$$

$$\Delta EL = \phi - \frac{AZ}{57.3} \psi$$

(3.199)

where $\Delta AZ, \Delta EL$ = azimuth and elevation errors, deg
ϕ, θ, ψ = roll, pitch and yaw errors, deg
EL, AZ = elevation and azimuth of the ground target in degrees with respect to spacecraft antenna; at the subsatellite point these values are zero; the maximum values are 8.7° at the earth edge

From Eq. (3.199) it is clear that only a fraction of yaw error is contributed to antenna beam errors. Therefore, for a geosynchronous communications satellite, the typical attitude errors are $\phi_{max} = \theta_{max} = \frac{1}{5}\psi_{max}$.

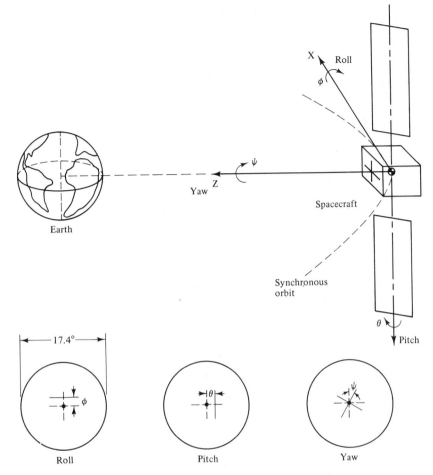

Figure 3.35 Antenna beam pointing error.

3.15 PROPULSION SYSTEMS

The functions of a spacecraft propulsion system normally start after the separation of the spacecraft from the launch vehicle/transfer stage. The main functions are reorientation of the spacecraft for apogee injection, apogee injection, correction of orbit error, attitude acquisition, normal-mode attitude control, north-south and east-west station keeping, repositioning, and so on. Although every spacecraft may not require all these functions, it is apparent that the propulsion system is an important subsystem of a spacecraft.

Spacecraft propulsion systems can be divided into three categories: solid propellant, liquid propellant, and electric propulsion. In a solid-propellant motor, the propellant to be burned is contained within the combustion chamber. Once ignited, it usually burns smoothly at a nearly constant rate on the exposed surface of the charge. The solid rockets are usually simple in construction. The apogee motors for Intelsat V were solid motors. Liquid-propellant thrusters use liquid propellants that are fed under pressure from tanks into a thruster chamber. Liquid thrusters can be subdivided into two groups: monopropellant and bipropellant. In monopropellant systems, hydrazine has been the most extensively used propellant for attitude control and station keeping. Bipropellant systems, consisting of monomethylhydrazine (MMH) as fuel and nitrogen tetroxide (N_2O_4) as oxidizer, have been used on several spacecraft for both apogee injection and on-orbit control. Electric propulsion thrusters use electric energy for ejecting propellant mass. The basic principles of these thrusters are discussed in this section.

Basic Equations

The propulsive force is obtained by ejecting the propellant at high velocities. The basic equation is

$$m \frac{dV}{dt} = F + F_a - mg \qquad (3.200)$$

where m is the mass of the vehicle, V the velocity of the vehicle, F the thrust of the jet, F_a the aerodynamic force, and g the gravitational acceleration. The thrust of the jet is given by

$$F = -u\dot{m} \qquad (3.201)$$

where $\dot{m}$ is the propellant flow rate and u is the propellant exhaust velocity. Spacecraft thrusters are used in either the transfer orbit or geosynchronous orbit where the aerodynamic forces are zero and the spacecraft is under zero gravitational effect. Introducing Eq. (3.201) into Eq. (3.200) and letting F_a and g be zero, we obtain

$$m \, dV = -u \, dm \qquad (3.202)$$

For chemical propellants, the ejection velocity, u, relative to the rocket's nozzle depends on heat energy released per unit weight of propellant. The principal measure of the performance of a thruster is specific impulse, I, which is the force impulse of a unit weight of a propellant and is defined as the thrust of a unit weight of propellant multiplied by the time required to burn it. The relationship of specific impulse to u is

$$I = \int_0^t F \, dt = -\int_0^t u \frac{dm}{dt} \, dt = -\int_{1/g}^0 u \, dm = \frac{u}{g} \qquad (3.203)$$

Introducing Eq. (3.203) into Eq. (3.202), we obtain

$$m \, dV = -Ig \, dm \qquad (3.204)$$

Equation (3.204) is the basic equation for a propulsion system. From it, the following relationships can be derived:

$$\Delta V = -Ig \log \frac{m_f}{m_i} \qquad (3.205)$$

$$m_f = m_i e^{-\Delta V/Ig} \qquad (3.206)$$

$$m_p = m_i(1 - e^{-\Delta V/Ig}) = m_f(e^{\Delta V/Ig} - 1) \qquad (3.207)$$

where ΔV is the change in the velocity imparted to the vehicle, m_i the initial mass of the vehicle before burnout, m_f the final mass of the vehicle after burnout, and $m_p = m_i - m_f$ is the propellant mass expended.

The impulse from a thruster mounted radially on a spinning platform is delivered over some fraction of the revolution of the body. This results in only a portion of the total impulse being delivered along the desired correction direction.

As shown in Fig. 3.36, the thruster starts firing at point A and stops firing at point B. Hence the thruster is fired over an angle ϕ_0. The angular speed of the platform is ω. The total force impulse provided by the thruster is

$$I_t = \int F \, dt = \int_0^{\phi_0} F \frac{d\phi}{\omega} = \frac{F\phi_0}{\omega} \tag{3.208}$$

The total effective force impulse, however, is

$$I_{\text{eff}} = \sqrt{\left(\int_0^{\phi_0} F \cos \phi \frac{d\phi}{\omega}\right)^2 + \left(\int_0^{\phi_0} F \sin \phi \frac{d\phi}{\omega}\right)^2}$$

$$= \frac{2F}{\omega} \sin \left(\frac{\phi_0}{2}\right) \tag{3.209}$$

The effective impulse, which is less than I_t, is at an angle, α, from the angular position at point A. This angle, called the lead angle, is given by

$$\alpha = \tan^{-1} \frac{\displaystyle\int_0^{\phi_0} F \sin \phi \, (d\phi/\omega)}{\displaystyle\int_0^{\phi_0} F \cos \phi \, (d\phi/\omega)} = \frac{\phi_0}{2} \tag{3.210}$$

The rotational efficiency η_{CR} is defined as *eff. of thrust maneuver*

$$\eta_{\text{CR}} = \frac{I_{\text{eff}}}{I_t} = \frac{\sin (\phi_0/2)}{\phi_0/2} \tag{3.211}$$

From Eq. (3.211), it is clear that it is desirable to fire thrusters in the spinning mode for only a small angular duration. This, however, will require high-level thrusters. For $\phi = 2\pi$, the efficiency is zero. The specific impulses for various propulsion systems are given in Table 3.2.

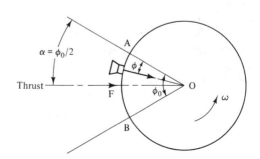

Figure 3.36 Thruster firing.

TABLE 3.2 SPECIFIC IMPULSE (SECONDS) OF VARIOUS PROPULSION SYSTEMS

	Function		
Propulsion System	High-Thrust Steady State (>450 N) Apogee Injection	Low-Thrust Steady State (0.05–22 N) Station Keeping	Low-Thrust Pulse (0.05–22 N) Attitude Control
Monopropellant hydrazine (N_2H_4)	235	220	135
Electrothermal hydrazine	—	290	—
Bipropellant (N_2O_4–MMH)	300	285	175
Ion thruster		3000	
Solid propellant	285		

3.16 LIQUID-PROPELLANT THRUSTERS

Monopropellant Thrusters

Since the development of the Shell 405 catalyst for the spontaneous decomposition of hydrazine, hydrazine thrusters have been used almost exclusively for the reaction control of spacecraft. Before the development of the Shell 405 catalyst, most spacecraft used catalytic decomposition of hydrogen peroxide with a number using cold gas jets. There are several disadvantages in using hydrogen peroxide. It has a relatively low specific impulse, is shock sensitive and requires tank relief valves to relieve the gaseous products of decomposition.

Figure 3.37 shows a catalytic thruster. The monopropellant injected at the flow rate, $\dot{m}$, is converted to a larger volume of gas in the catalyst bed that flows at sonic velocity through the throat and supersonic velocity from the nozzle. Thrust is generated by reaction to the gas mass being exhausted at high velocity.

Anhydrous hydrazine is a clear, colorless liquid with a distinct ammonialike odor. Hydrazine is a relatively stable chemical which can be stored for long periods of time without decomposition. Anhydrous hydrazine has a boiling point of 113.5°C and it freezes at 2°C. Thrust chambers contain Shell 405 catalyst, which decomposes hydrazine according to the following chemical reaction:

$$3N_2H_4 \xrightarrow{\text{cat}} 4NH_3 + N_2 + 36,360,714 \text{ cal} \qquad (3.212)$$

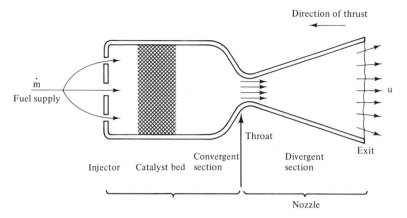

Figure 3.37 Fundamental thruster design.

A portion of the ammonia is then further decomposed according to

$$4NH_3 \rightarrow 2N_2 + 6H_2 - 19,956,816 \text{ cal} \qquad (3.213)$$

The fraction of the ammonia that is decomposed is called the ammonia decomposition fraction and is usually equal to about 0.4. The decomposition fraction is dependent on flow variables and the geometry of the reaction control chamber. The ammonia, nitrogen, and hydrogen are then exhausted through a converging–diverging nozzle to produce thrust. Shell 405, a spontaneous catalyst for hydrazine was developed by Shell Development Company under a NASA contract from JPL. The catalyst consists primarily of an alumina substrate with large surface area onto which the active catalytic material (iridium) is deposited in a finely dispersed state.

The life-limiting factor of a hydrazine thruster is the degradation of the Shell 405 catalyst bed. There are at least two types of catalyst degradation: physical loss of the catalyst (void formation) and the loss of catalytic activity. It has generally been found that the starting temperature of the catalyst bed is the single most important parameter for the catalyst attrition. The number of cold starts (ambient temperature ignitions) has been correlated with the useful catalyst bed life in a number of applications. The use of electric heaters to maintain moderate bed temperature is necessary for long-life applications. The minimum temperature maintained in the catalyst bed depends on how severe the requirements are. For example, a 2.2-N thruster with life requirements of 1 million cycles is maintained at 121°C. The 0.45-N thrusters of Fleetsatcom are maintained at 315°C. As shown in Fig. 3.38, the capability of the currently developed hydrazine thrusters to survive a large number of starts is inversely proportional to their thrust levels. One of the main mechanisms of catalyst deactivation is poisoning from minute quantities of impurities in the hydrazine.

The thruster performance changes over the mission life of the spacecraft because the pressure in the system decreases as the propellant is expelled.

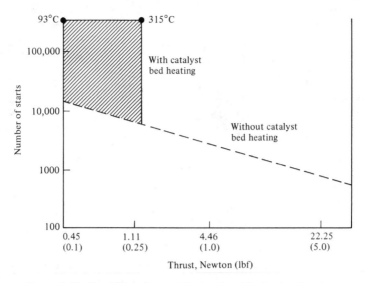

Figure 3.38 Capability of currently developed hydrazine thrusters.

Figure 3.39 shows a plot of thrust versus tank pressure for an Intelsat IV thruster in steady state thrusting mode used for both spin-up maneuvers and north-south station keeping.

A major propulsion advantage in spin-stabilized attitude control is that the thruster duty cycle is well defined and relatively benign. The total number of pulses required is typically 20,000 to 100,000, with only a few hundred cold starts being required. In a three-axis-stabilized spacecraft, the duty-cycle requirements are usually more severe. In many cases, the exact duty cycles are known only approximately. Typically, the attitude

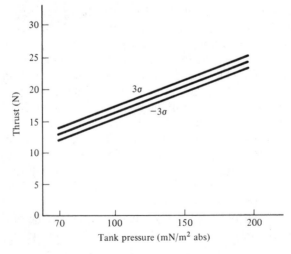

Figure 3.39 Thrust versus tank pressure.

control thrusters for three-axis-stabilized spacecraft are required to deliver between 200,000 and 1 million pulses at very low duty cycles.

The functions and locations of hydrazine thrusters in dual-spin-stabilized spacecraft such as Intelsat IV are shown in Fig. 3.40. The reaction control system consists of two axial thrusters, two radial thrusters, and two spin-up thrusters. The maneuvers performed by these thrusters are: spin axis reorientation and attitude change by one axial thruster in a pulsing mode, east-west station keeping and repositioning by one radial thruster in a pulsing mode, north-south station keeping by one or two axial thrusters in a continuous mode, and spin-up control by one spin-up thruster in a pulsing mode. It should be noted that there is a redundant thruster for each maneuver. The axial thrusters provide a major portion of the thruster requirements. A small misalignment of the axial thruster may result in a spin-up of the spacecraft and the spacecraft does not have spin-down thrusters. To provide a positive means of spin-speed control, the axial jets are biased 0.5° in the spin-down direction.

Figure 3.41 shows a schematic diagram of the propulsion system of Intelsat V, a three-axis-stabilized spacecraft. The system consists of two surface-tension propellant/pressurant tanks that are connected to two redundant sets of thrusters. Latching isolation valves separate the tanks into two half-systems. By using isolation valves, either tank can feed one or both sets of thrusters. Two 22.2-N thrusters are used mainly for maneuvers

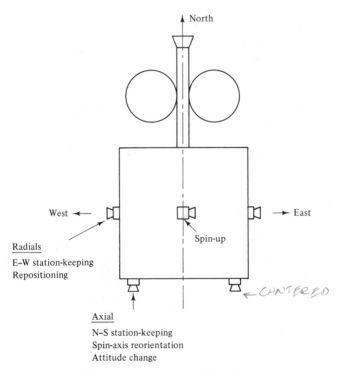

Figure 3.40 Thruster location for a dual-spin spacecraft.

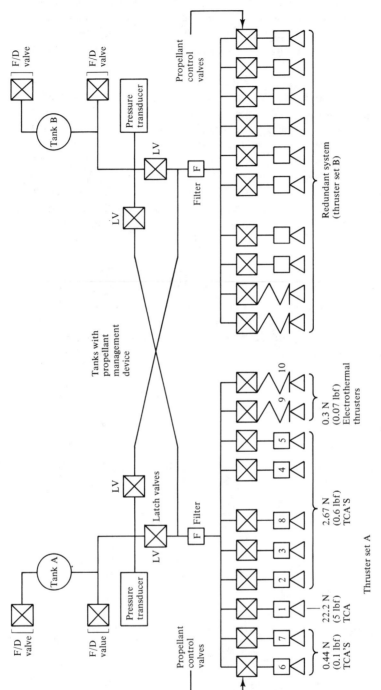

Figure 3.41 Schematic diagram of Intelsat V propulsion system. (Courtesy of INTELSAT and FACC)

during transfer orbit before apogee motor ignition. North-south station keeping is performed by 0.3-N electrothermal thrusters. Roll maneuvers are performed by 0.44-N thrusters. East-west station keeping, pitch and yaw control, and backup for N/S station keeping are performed by 2.67-N thrusters.

The thrust vector geometry of the thrusters is given in Fig. 3.42. The maneuvers performed by these thrusters are: spin-up by 4A/5B, spin-down by 4B/5A, active nutation control during transfer orbit by 1A/1B, ΔV correction by 1A and 1B, south station keeping by 9A and 10A, north station keeping by 9B and 10B, redundant south station keeping by 4A and 5A, redundant north station keeping by 4B and 5B, east station keeping by 2B and 3B, west station keeping by 2A and 3A, positive pitch by 2A/2B, negative pitch by 3A/3B, positive yaw by 4A/5B, negative yaw by 4B/5A, positive roll and negative yaw by 6A/6B, negative roll and positive yaw by 7A/7B, redundant east station keeping by 8A, and redundant west station keeping by 8B. It should be noted that Intelsat V uses 20 thrusters compared to six thrusters in Intelsat IV, a dual-spin-stabilized spacecraft.

Bipropellant Thrusters

Liquid bipropellant systems for spacecraft are a lower-thrust version of the relatively standard nitrogen tetroxide (N_2O_4)/monomethylhydrazine (MMH) propellant systems, commonly used for expendable launch vehicles as well as for the Space Shuttle Orbiter. These high-performance hypergolic

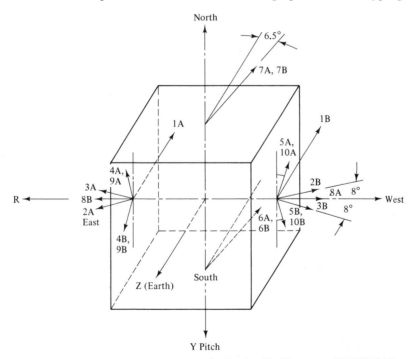

Figure 3.42 Thruster vector geometry for Intelsat V. (Courtesy of INTELSAT and FACC)

propellants, with N_2O_4 as an oxidizer and MMH as a fuel, eliminate the requirement for an ignition system usually associated with a bipropellant propulsion system. The major driver for the use of bipropellant systems is the higher specific impulse (300–310 s) and the ability to combine the apogee injection function with the on-orbit control.

A liquid bipropellant system is being used on Intelsat VI for apogee injection and on-orbit control. The subsystem consists of two functionally independent half-systems. Four thrusters, each with 22 N of thrust, are used for east-west station keeping and spin-up and spin-down control. Two similar axial thrusters provide north-south station keeping and attitude control. Two 490-N apogee thrusters perform the apogee injection and reorientation maneuvers.

3.17 ELECTRIC PROPULSION

Electric propulsion devices use electrical energy for ejecting propellant mass. The electrical propulsion devices can be grouped into the following three major classes:

1. *Electrothermal thrusters:* use a resistance or an electric arc to convert electricity to heat, thereby increasing the enthalpy of the liquid propellant and its specific impulse.

2. *Electrostatic thrusters:* produced by accelerating electrically charged particles through an electrostatic field.

3. *Electromagnetic thrusters:* use an electrical discharge to drive a large current through the working fluid, producing a plasma. The interaction of this current with a properly configured magnetic field produces volumetric body forces which directly accelerate the propellant.

Electrothermal Hydrazine Thrusters

The major driver for the electrothermal hydrazine thruster is the approximately 28% increase in specific impulse over standard catalytic thrusters (300 versus 235 s). The high performance is achieved by electrically augmenting the enthalpy of the decomposition products of hydrazine in a second-stage vortex chamber as shown in Fig. 3.43. These thrusters are used on Intelsat V for north-south station keeping. The power consumption is low, about 1W/mN of thrust. Specific impulse in the pulse mode depends on both the pulse width and duty cycle. It is lowest for short pulses and low duty cycles where the heat loss to the thrusters absorbs a large percentage of the heat decomposition. The supply pressure of the hydrazine is approximately the same for both catalytic and electrothermal hydrazine thrusters. Hence a common propellant feed system is used for these thrusters on Intelsat V. Catalytic thrusters are used as a backup to electrothermal thrusters for north-south station keeping on Intelsat V.

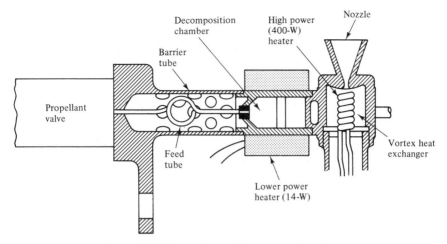

Figure 3.43 Electrothermal thruster configuration.

Ion Thrusters

Among the electrostatic and electromagnetic thrusters, ion thrusters have reached the most advanced stage of development. Ion thrusters have an order-of-magnitude higher specific impulse than that of chemical propulsion thrusters. The exhaust velocity of these thrusters is not limited, as with conventional thrusters, by the energy released in chemical reaction. The only physical limitation is the velocity of light, assuming that an adequate power source can be made available. Hence specific impulse is dependent upon the available power source.

An ion thruster subsystem consists of four parts: the electric power source (usually a solar array and batteries), the power conditioner, the discharge chamber, and the accelerator. Figure 3.44 is a schematic illustration of an ion thruster. The propellant atoms injected into the chamber through the rear walls are ionized by an electrical discharge generated between a central hollow cathode and a cylindrical anode. The efficiency of creating the ions by electron collisions is increased by an axially divergent magnetic field formed by pole pieces at the downstream end of the anode and around the hollow cathode. At the end of the discharge chamber a screen electrode is maintained at approximately the same positive potential as the cathode. A negatively biased accelerator grid is mounted less than 1 mm downstream of the screen grid. The ions passing through the grid screens are accelerated. Hollow cathodes, through which a small fraction of the propellant is fed, are very efficient as electron emitters. The baffle increases the voltage drop in the discharge in order to produce higher-energy electrons which have a larger ionization cross section. A second smaller hollow cathode, positioned immediately downstream of the accelerator grid and to the side of the exhaust beam, supplies electrons for neutralization for the high velocity positive beam.

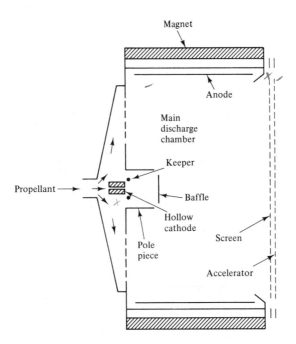

Figure 3.44 Schematic diagram of ion thruster (Ref. 14).

Ion thrusters require external power to ionize, accelerate, and neutralize the propellant. Of all these processes, only the acceleration step produces thrust. To maintain a low power supply mass, ion thrusters operate at very low thrust levels. For north-south station keeping, thrust levels of 5 to 10 mN will be adequate. However, due to a low thrust level, electric propulsion systems must operate over much longer thrust periods than do chemical propulsion systems.

It is more difficult to calculate propellant mass for electric propulsion in comparison with chemical propulsion since the specific impulse, related to the exhaust velocity, is not constant. The propellant mass decreases exponentially as the specific impulse increases, while the mass of power supply increases with the increase in the specific impulse. Hence a suitable exhaust velocity is nearly always chosen on the basis of a trade-off between power and propulsion system mass.

The variation of the power/thrust ratio with propellant exhaust velocity is shown in Fig. 3.45. The lowest curve represents the power/thrust ratio for an ideal electric thruster (i.e., one in which the input power is entirely converted into thrust). Real thrusters have additional power losses which do not produce thrust. The low shaded area represents the range of performance which has been demonstrated for large (>20-mN thrust) ion thrusters, and the upper shaded area represents the range of performance for small (2 to 10 mN thrust) ion thrusters.

The ion thrusters are potential candidates for north-south station keeping. To minimize both the size and power required, it is desirable to use low

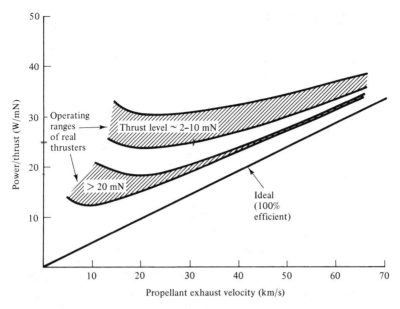

Figure 3.45 Power thruster ratios for ion thrusters (Ref. 15).

thrust over two long periods centered on the nodes. The thrust is most effective in removing orbit inclination at the orbit nodes. Between the nodes and antinodes, the effective thrust is equal to the normal component of the thrust, T_N, multiplied by the cosine of the angle between the line of nodes and the satellite position vector. The mean effective thrust is given by

$$
T_e = \frac{T_N \int_0^\theta \cos \theta \, d\theta}{\int_0^\theta d\theta} = T_N \frac{\sin \theta}{\theta} \tag{3.214}
$$

The thrusters are normally canted from the north-south axis to avoid any interaction between the exhaust plume and other spacecraft components. So the component of the thrust normal to the orbit plane, T_N, is $T \cos \alpha$, where T is the total thrust and α is the canting angle.

Example 3.3

Determine the propellant mass for north-south station keeping of a spacecraft whose mass at the beginning of life is 2250 kg. The velocity required for north-south station keeping is 460 m/s for 10 years.

Solution   The possible propulsion systems for north-south station keeping are monopropellant hydrazine, electrothermal hydrazine, bipropellant, and ion thrusters.

To calculate the propellant mass, Eq. (3.207) can be used as

$$
m_p = m_i (1 - e^{-\Delta V / \eta I g}) \tag{A}
$$

The initial spacecraft mass, m_i, is 2250 kg and ΔV is 460 m/s. The specific impulses for monopropellant, electrothermal, bipropellant, and ion thrusters, from Table 3.2, are assumed to be 220, 290, 285, and 3000 s, respectively. A typical thruster efficiency,

η, for the first three propulsion systems is 0.9. This accounts for the cant angle, the plume impingement, and the finite burn loss. The term g, gravitational acceleration, is equal to 9.8066 m/s^2.

By substituting the parameters above into the equations, the propellant masses are

474.6 kg for monopropellant hydrazine

370.1 kg for electrothermal hydrazine

376.0 kg for bipropellant

From these numbers it is clear that approximately a 20% propellant mass saving can be achieved by using either electrothermal hydrazine or bipropellant instead of monopropellant hydrazine. It should be noted that electrothermal thrusters will need additional electrical power.

For ion thrusters, the determination of spacecraft mass is more complex. Let us assume that ion thrusters are on every day for two periods centered around nodes, except during eclipses. The thrusters for the descending and ascending nodes will be different because of opposite thrust directions. It is assumed that the lifetime of the current ion thrusters is 10,000 h. The number of days allowed for a thruster to be on, excluding eclipse periods, is 275 days/yr. Therefore, the maximum period for each thrust on, T_{ON}, is

$$T_{ON} = \frac{10,000}{10 \times 275} = 3.63 \text{ h}$$

Let us assume that each thruster is on for 3 hours per day or 45° and that the cant angle is 30°. So the thruster efficiency for the ion thruster becomes

$$\eta = \cos 30° \frac{\sin 22.5°}{\pi/8} = 0.844$$

By using Eq. (A), the propellant mass for the ion thruster is

$$m_p = 2250[1 - e^{-460/(0.844 \times 3000 \times 9.8066)}]$$

$$= 41.3 \text{ kg}$$

The average ΔV removed by each thruster during one maneuver is

$$\Delta V = \frac{460}{10 \times 275 \times 2 \times 0.844} = 0.099 \text{ m/s}$$

The thrust required for the thruster, T_e, is ~thrusters

$$T_e = \frac{2250 \times 0.099}{3 \times 60 \times 60} = 0.0206 \text{ N}$$

Assuming approximately a 20% margin to take into account the extreme conditions, the selected thrust level is 25 mN. The power/thrust ratio for this thruster is 25 W/mN, from Fig. 3.45. Therefore, the power required for the ion thruster is 625 W. Assuming 20 W/kg as the power density for a solar array, the solar array mass for this ion thruster is 31.25 kg.

PROBLEMS

3.1. A spacecraft is spinning at 45 rpm. Its spin and transverse moments of inertia are 757 kg $\cdot$ m^2 and 1847 kg $\cdot$ m^2, respectively. The nutation angle is 5°.

Determine (a) the nutation frequency λ relative to a spacecraft fixed coordinate system, and (b) the magnitude of the transverse angular velocity.

3.2. Ground liquid-slosh tests were performed on a scale model of the spacecraft of Problem 3.1. The dimensional scale factor, ratio of model to spacecraft dimensions, was 0.595. The test spin rate was 127 rpm. The test fluid has the same density and Reynolds number as the propellant. Both the model and the spacecraft have two tanks. The ratio of the transverse moment of inertias of the spacecraft to the model is 14. The nutation time constant from the test was measured to be 18 minutes. Determine the spacecraft nutation time constant.

3.3. For the spacecraft in Problem 3.1, active nutation damping by a thruster is used. The thruster is fired for half the body nutation frequency period, $\alpha = \pi/2$, $T = \pi/\lambda$, and $\beta = \pi$. The thruster torque is 24.5 N · m. For a 15° nutation angle, determine the time constant for the active nutation control. Using the nutation time constant due to liquid slosh from Problem 3.2, calculate the spacecraft overall nutation time constant.

3.4. A spacecraft BOL mass in synchronous orbit is 1000 kg. The velocity required for north-south station keeping is 450 m/s for 10 years. Assume that the possible propulsion systems are monopropellant hydrazine, electrothermal hydrazine, bipropellant, and ion thrusters. Determine the propellant mass required for each system.

3.5. A three-axis-stabilized spacecraft has the following mass properties: $I_{xx} = 15{,}053$ kg · m², $I_{yy} = 6{,}510$ kg · m², and $I_{zz} = 11{,}122$ kg · m². The solar disturbance torque at summer solstice in body coordinates, M_s (N · m) $= [-7.3 \times 10^{-5} + (-1.6 \times 10^{-5} \cos \alpha + 1.4 \times 10^{-6} \sin \alpha)]\mathbf{I} + [3.0 \times 10^{-6} + (7.2 \times 10^{-5} \cos \alpha - 1.4 \times 10^{-4} \sin \alpha)]\mathbf{J} + [2.2 \times 10^{-5} + (2.4 \times 10^{-6} \cos \alpha + 3.5 \times 10^{-6} \sin \alpha)]\mathbf{K}$. The angle α is the orbit angle measured from spacecraft local noon. The permissible attitude control errors are 0.05° in roll and pitch and 0.2° in yaw. Assume the stabilization system to be a fixed momentum wheel with thrusters. The roll torque is 11.7 N · m. The wheel desaturation torque is 17.8 N · m with a pulse time of 0.01 s. Determine (a) the angular momentum of the wheel, (b) the offset angle for the roll thruster, and (c) the lead-time constants for the pitch and the roll–yaw control.

3.6. A spacecraft has a three-axis reaction wheel system. The moment of inertia along the pitch axis is 7,000 kg · m². The disturbance torque impulse along the pitch axis is 0.178 N · m · s. The maximum allowable pitch error is 0.05°. Determine (a) the time constant and (b) the gain of the control system. For an initial pitch error of 5°, how long will it take to reach the allowable error limit of 0.05°?

REFERENCES

1. W. T. Thomson, *Introduction to Space Dynamics,* Wiley, New York, 1961.

2. M. H. Kaplan, *Modern Spacecraft Dynamics and Control,* Wiley, New York, 1976.

3. L. Meirovitch, *Methods of Analytical Dynamics,* McGraw-Hill, New York, 1970.

4. J. J. D'Azzo and C. H. Houpis, *Feedback Control System Analysis and Synthesis,* McGraw-Hill, New York, 1966.

5. A. J. Iorillo, "Nutation Damping Dynamics of Axisymmetric Rotor Stabilized Satellites," ASME Winter Meeting, Chicago, Nov. 1965.

6. P. W. Likins, "Attitude Stability Criteria for Dual-Spin Spacecraft," *Journal of Spacecraft and Rockets,* Vol. 4, No. 12, pp. 1638–1643, Dec. 1967.

7. A. E. Sabroff, "Advanced Spacecraft Stabilization and Control Techniques," *Journal of Spacecraft and Rockets,* Vol. 5, No. 12, pp. 1377–1393, Dec. 1968.

8. R. H. Canon, Jr., "Some Basic Response Relations for Reaction-Wheel Attitude Control," *ARS Journal,* Jan. 1962.

9. J. D. Acord and J. C. Nicklas, "Theoretical and Practical Aspects of Solar Pressure Attitude Control for Interplanetary Spacecraft," in *Progress in Astronautics and Aeronautics,* Vol. 13; *Guidance and Control II,* R. C. Langford and C. J. Mundo, eds., Academic Press, New York, 1964.

10. M. E. Ellion, D. P. Frizell, and R. A. Meese, "Hydrazine Thrusters—Present Limitations and Possible Solutions," AIAA/SAE 9th Propulsion Conference, Las Vegas, Nev., Nov. 5–7, 1973, AIAA Paper No. 73-1265.

11. V. J. Sansevero, Jr., C. D. Arvidson, W. D. Boyce, and S. F. Archer, "Monopropellant Hydrazine Reaction Control Subsystem for Communications Technology Satellite," AIAA/SAE 9th Propulsion Conference, Las Vegas, Nev., Nov. 5–7, 1973, AIAA Paper No. 73-1268.

12. H. J. Dougherty, K. L. Lebsock, and J. J. Rodden, "Attitude Stabilization of Synchronous Communications Satellites Employing Narrow-Beam Antennas," Journal of Spacecraft and Rockets, Vol. 8, No. 8, August 1971, pp. 834–841.

13. P. P. Iwens, A. W. Fleming, and V. A. Spector, "Precision Attitude Control with a Single Body-Fixed Momentum Wheel," AIAA Mechanics and Control of Flight Conference, 1974, Paper No. 79-894.

14. K. E. Clark, "Survey of Electric Propulsion Capability," AIAA/SAE 10th Propulsion Conference, 1974.

15. B. A. Free, "Chemical and Electrical Propulsion Tradeoffs for Communications Satellites," *COMSAT Technical Review,* Volume 2, No. 1, Spring, 1972.

4

SPACECRAFT
STRUCTURES

4.1 INTRODUCTION

The primary function of a spacecraft structure is to provide mechanical support to all the subsystems within the framework of the spacecraft configuration. It also satisfies subsystem requirements, such as alignments of the sensors, actuators, and antennas; and system requirements for launch vehicle interfaces, integration, and tests. A spacecraft is subjected to major mechanical loads during the launch period. Therefore, the spacecraft structure is designed to survive the launch loads and to protect the other subsystems from excessive launch loads. On-orbit loads, in a zero-gravity environment, are significantly lower than the launch loads. The on-orbit requirements in a structural design are mainly high stiffness for the deployed appendages to avoid the interaction between their vibrations and the attitude control system, and low thermal distortion of the antenna structure to achieve high antenna pointing accuracies.

A typical spacecraft structural design process is shown in Fig. 4.1. Initially, system trade-offs are performed to determine the spacecraft configuration which would meet launch vehicle and subsystem constraints. Next, a preliminary structural design is performed on the basis of launch loads from the user's manual of the launch vehicle and other mechanical constraints. Based on this design, a finite element structural model is generated by using a general-purpose computer program, such as NASTRAN. Using this model, static, dynamic, and stress analyses are performed. The thermal distortions of the antenna structures are calculated by using the predicted temperature distributions from the thermal analysis. The spacecraft structure is redesigned if the stress margins are not adequate and/or the natural frequencies and thermal distortion requirements are not met. The preliminary

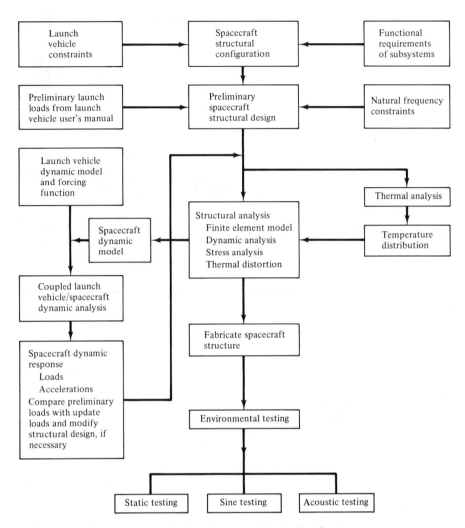

Figure 4.1 Spacecraft structure design/verification.

launch loads from the user's manual of the launch vehicle are updated by performing a coupled launch vehicle/spacecraft dynamic analysis. The coupled analysis is performed by synthesizing spacecraft and launch vehicle structural models and applying the launch forces. The dynamic response of the spacecraft determines launch loads and accelerations within the spacecraft. The spacecraft structural design is evaluated for these loads and modified if necessary. After the completion of the structural design, the spacecraft structure is fabricated and tested for the launch loads. A typical design verification program consists of the static testing of the primary structure and sinusoidal and acoustic testing of the complete structure. This chapter presents the basic principles in the design and testing of spacecraft structures.

4.2 SPACECRAFT STRUCTURAL CONFIGURATION

A spacecraft structural configuration is influenced by several factors, such as the launch vehicle, the stabilization system, and subsystem requirements. The launch vehicle will determine whether the spacecraft structure has to provide support to the apogee and perigee motors. For the Space Shuttle, which inserts the spacecraft into a low-altitude circular parking orbit, both perigee and apogee motors will be part of the spacecraft if geostationary orbit is planned. For Ariane, however, which inserts the spacecraft into a geotransfer orbit, the spacecraft will require only an apogee motor. The other factor that has a major impact on a spacecraft configuration is the selected attitude stabilization system. Based on the type of attitude stabilization system used, spacecraft structural configurations can be divided into two groups: dual-spin stabilization and three-axis stabilization.

Figure 4.2 shows the spacecraft structure of Intelsat IV, which was a dual-spin-stabilized spacecraft. The structure can be divided into two major parts: the spun section and the despun section. The two sections were joined by a rotary interface unit called the bearing and power transfer assembly (BAPTA).

The spun section can be subdivided into the thrust cone, the platform subassembly, and the solar substrates. The thrust cone consists of three monocoque magnesium shells joined together by magnesium rings and by a separation clamp to the booster adapter. The apogee motor and the equipment platform were supported by the thrust cone at the ring joints. The platform supported the positioning and orientation subsystem and the batteries. The substrates for the solar array were a one-piece aluminum honeycomb and a fiberglass face-skin cylinder. The two solar panels enclosed the periphery of the spinning section and were cantilevered from the brackets on the spinning section.

The despun compartment, which supported the communications subsystem, was composed of forward and aft aluminum honeycomb shelves. The antenna mast assembly supported all the antennas. The mast was basically an aluminum tube, concentric with the spin axis, with the tubular aluminum crossarm normal to the mast. The launch vehicle for Intelsat IV was an Atlas-Centaur.

Intelsat V is a three-axis-stabilized spacecraft. The structure, as shown in Fig. 4.3, can be divided into three modules: the communications module, the spacecraft bus module, and the antenna module. The communications module, which supports all communications subsystems except the antennas, consists of north and south equipment panels, the antenna deck, and the north and south structural webs. The spacecraft bus module houses the remaining subsystems. It consists of a central tube, a horizontal platform called the attitude dynamics control system (ADCS) deck, and north and south spacecraft bus module equipment panels. The antenna module structure is a truss structure which supports all the antennas. The truss structure with all the antennas installed is self-supporting and requires no additional

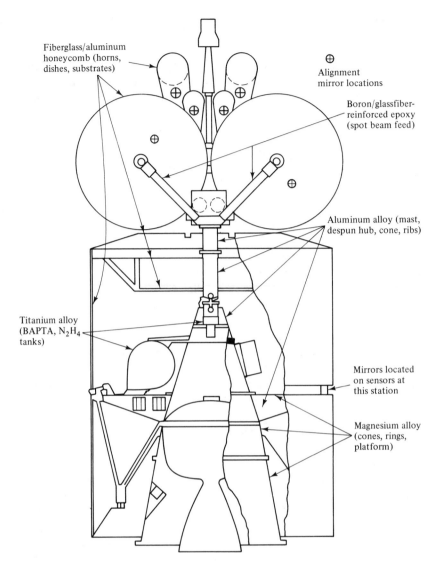

Fiberglass/aluminum
honeycomb (horns,
dishes, substrates)

Alignment
mirror locations

Boron/glassfiber-
reinforced epoxy
(spot beam feed)

Aluminum alloy (mast,
despun hub, cone, ribs)

Titanium alloy
(BAPTA, N_2H_4
tanks)

Mirrors located
on sensors at
this station

Magnesium alloy
(cones, rings,
platform)

Figure 4.2 Structural configuration of Intelsat IV dual-spin stabilized spacecraft (Ref. 21).

support for handling. It is attached to the top of the central tube with four bolts.

The central tube, which is a semi-monocoque aluminum construction, is the "backbone" of the spacecraft. The loads from all spacecraft equipment are carried by the tube to the launch vehicle. It extends from the base of the spacecraft, where it attaches to the launch vehicle, to the top of the main body, where it supports the antenna module. The apogee motor is supported inside the tube. The diameter of the upper cylinder of the tube is selected to meet the apogee flange diameter. The lower part of the tube

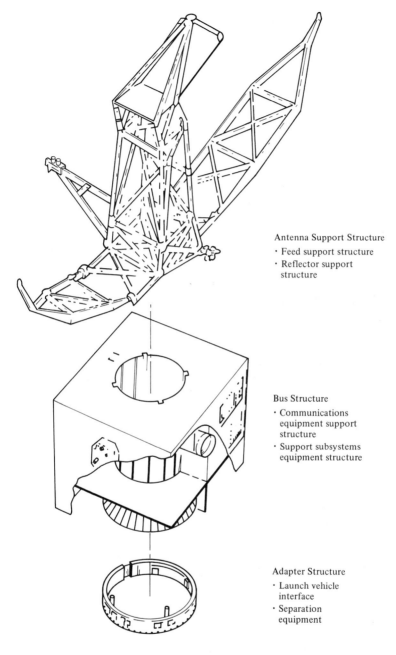

Antenna Support Structure
· Feed support structure
· Reflector support
  structure

Bus Structure
· Communications
  equipment support
  structure
· Support subsystems
  equipment structure

Adapter Structure
· Launch vehicle
  interface
· Separation
  equipment

Figure 4.3 Structural configuration of Intelsat V. (Courtesy of INTELSAT and FACC)

is conical to permit it to span the larger diameter of the launch vehicle interface. The two horizontal decks, antenna and ADCS, are supported by the center tube. The horizontal deck provides the load paths for transmission of lateral loads from north-south equipment panels to the center tube. The load paths for the longitudinal loads on the panels are provided by the vertical webs that extend between the north-south panels and the central tube. Aluminum sandwich construction is used for all the panels. Graphite/epoxy is used for the antenna module, for thermal dimensional stability considerations.

4.3 LAUNCH LOADS

A spacecraft structure is normally subjected to major mechanical loads during launch for less than half an hour. On-orbit mechanical loads under zero-gravity accelerations are so low that the lightweight structures, which would not be able to support their own weight on the earth, will function properly in space. Such structures are stowed during launch. A spacecraft is subjected to launch loads through two paths: accelerations transmitted through the launch vehicle interface and direct acoustic noise through the shroud. Acceleration through the interface consists of steady state, low-frequency transients and random vibrations. The acoustic noise also generates random vibrations in the spacecraft. At separation from the launch vehicle, separation shocks are generated due to the firing of the pyrotechnic bolts.

During launch, the dynamic accelerations are superimposed on the steady-state accelerations. The steady-state axial acceleration is developed due to engine thrust. The acceleration increases as the propellant is burned, reducing the vehicle mass, resulting in peak acceleration at burnout. The compressive loading develops stored strain energy which is released at burnout, resulting in a vibration. The lateral accelerations occur most significantly at lift-off, maximum αq (α = angle of attack and q = dynamic pressure), and engine ignition and cutoff. Maximum random and acoustic excitations occur during STS lift-off, with lower excitation levels occurring during the transonic flight period.

The interface acceleration at a given time will consist of a combination of steady-state, low-frequency transient (sinusoidal), and random accelerations in all the directions. To simplify the analysis and testing, launch loads are normally divided into three groups: quasi-static, sinusoidal, and random (or acoustic). Quasi-static accelerations represent an equivalent static acceleration for a combination of steady-state acceleration and low-frequency transients. Sinusoidal accelerations represent low-frequency transients. The primary structure of a spacecraft is generally designed for quasi-static loads. The secondary structures are designed for dynamic loads. The dynamic launch loads for a spacecraft depend on the dynamic coupling of the launch vehicle and the spacecraft natural vibration modes. To reduce the dynamic coupling, the spacecraft structures are designed such that their fundamental natural frequencies are above certain minimum values. In the following sections, the spacecraft launch loads for the Space Shuttle and Ariane on the basis of their user's manuals are given.

Space Shuttle

The Space Shuttle inserts the payload into a low-altitude circular parking orbit. Therefore, a spacecraft launched by the Space Shuttle will need both perigee and apogee motors. Several spacecraft have been designed for launch by either Space Shuttle or an expandable launch vehicle (ELV). This has resulted in the development of perigee stages for Delta-sized spacecraft, PAM-D, and Atlas-Centaur-sized spacecraft, PAM-A.

The cradle, the perigee motor, and the spacecraft payload envelope for PAM-D are shown in Fig. 4.4. It occupies approximately one-eighth of the STS cargo bay length. Space Shuttle–launched spacecraft are designed not only for launch loads but also for landing and emergency landing loads, because the Space Shuttle can bring the spacecraft back to earth in case of problems occurring in orbit prior to release.

The quasi-static limit loads (2σ levels) for PAM-D spacecraft are given in Table 4.1. These loads are used for initial design of the spacecraft structure. The loads should be considered as acting at the center of the mass of the spacecraft. The upper part of the spacecraft and secondary structures can experience considerably higher accelerations. To update these loads, a coupled PAM-D/spacecraft/STS dynamic analysis is performed. The dynamic loads in the spacecraft are a function not only of the launch vehicle characteristics, but also of the spacecraft/cradle mass and dynamic characteristics and of the particular location within the orbiter payload bay. The maximum lateral accelerations occur for the lift-off condition and result from the dynamic response to the thruster buildup transients. The maximum axial compression acceleration occurs for the normal landing conditions and results from the dynamic response to the landing impact. The maximum axial tension accelerations occur for the lift-off condition and result from the dynamic response to thrust buildup transients. The spacecraft is also designed to withstand loads occurring during an STS emergency landing condition in a manner so as not to jeopardize crew safety. To avoid dynamic coupling between the low-frequency vehicle and spacecraft modes, it is recommended that the spacecraft structure fundamental natural frequencies be above 35 Hz in the thrust axis and 15 Hz in the lateral axis for a spacecraft hard-mounted at the spacecraft separation plane without an attach fitting and separation clamp.

The sinusoidal flight levels (2σ) for the base excitation to the spacecraft are given in Table 4.2. The spacecraft loads due to this base excitation at the primary frequencies of the spacecraft, however, are usually much more severe than those experienced in flight. To avoid overtesting of the critical structural elements, the base excitation is reduced, commonly known as "notched," in such a manner that the loads in the critical structural elements

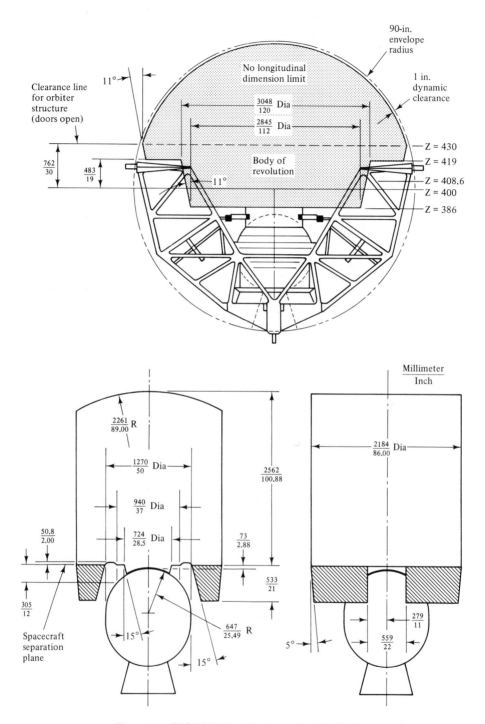

Figure 4.4 STS PAM-D payload envelope (Ref. 22).

TABLE 4.1 STS PAM-D SPACECRAFT LIMIT LOAD
FACTORS, SPACECRAFT MASS (907–1247 kg)

| | Limit Load Factors (g's) | |
Condition	Lateral	Axial
Maximum lateral	5.1	3.3
Maximum axial (compression)	3.5	5.0
Maximum axial (tension)	5.1	− 3.3

TABLE 4.2 STS PAM-D SINUSOIDAL VIBRATION
ACCEPTANCE

Axis	Frequency Range (Hz)	Acceptance Level, g (0–peak)
All three axes	5–35,	0.75

do not exceed the expected maximum flight loads. The high-frequency acoustic pressure excitations that occur during launch are given in Table 4.3. Acoustic testing is the best simulation of the high-frequency flight environment. However, random testing with inputs at the base of the attach fitting has been found to be sufficient in many cases in lieu of acoustic testing. The maximum random levels for the STS PAM-D vehicle are given in Table 4.4. These levels are meant to excite the main equipment area of the spacecraft to the levels generated in flight mainly by acoustics. For external spacecraft surfaces with a large area and low density, such as solar panels and antennas, either separate acoustic testing or hard-mounted random testing of these structures is required.

Ariane

For Ariane launch, quasi-static accelerations at the center of the mass of the spacecraft are given in Table 4.5. For Ariane, as is also true in general for all expendable launch vehicles, the axial compressive loads are significantly higher than the lateral and axial tension loads. For STS/PAM-D, as given in Table 4.1, the maximum compressive axial loads and the maximum lateral loads are of approximately the same magnitude. The axial tension loads are approximately 60% of the axial compression loads. Therefore, STS-launched spacecraft are subjected to higher lateral and axial tension loads and expendable launch vehicle launched spacecraft are subjected to higher axial compressive loads. For a spacecraft which is designed to be compatible for launch by both STS and an expendable launch vehicle, the worst lateral and axial load conditions are determined by STS and the expendable launch vehicle, respectively.

TABLE 4.3 MAXIMUM FLIGHT ACOUSTIC LEVELS FOR STS PAM-D

⅓ Octave Band Center Frequency (Hz)	Sound Pressure Level dB (ref. $2 \times 10^{-5} N/m^2$)	Time (s)
31.5	122.0	
40.0	124.0	
50.0	125.5	
63.0	127.0	
80.0	128.0	
100.0	128.5	
125.0	129.0	
160.0	129.0	
200.0	128.5	
250.0	127.0	5
315.0	126.0	
400.0	125.0	
500.0	123.0	
630.0	121.5	
800.0	120.0	
1000.0	117.5	
1250.0	116.0	
1600.0	114.0	
2000.0	112.0	
2500.0	110.0	
Overall	138.0	

[handwritten annotation: natural freq.]

TABLE 4.4 MAXIMUM RANDOM FLIGHT LEVELS FOR STS PAM-D

Frequency (Hz)	Power Spectral Density (g^2/Hz)	g_{rms}	Time (s)
10	0.0020		
10 – 80	+4 dB/octave		
80 – 170	0.033	3.7	5
170 – 2000	−4 dB/octave		
2000	0.0012		

TABLE 4.5 ARIANE LIMIT LOADS (g's) (Ref. 24)

	Acceleration			
	Axial Axis			Lateral Axis
Flight Event	Static	Dynamic	Total	
Maximum dynamic pressure	−1.9	±1.6	−3.5	±2
Second-stage Burnout	−4.7 ⎱ 0 ⎰	±3.2	−7.9 ⎱ +3.2 ⎰	±1

TABLE 4.6 ARIANE SINUSOIDAL VIBRATION
ACCEPTANCE LEVELS

Axis	Frequency Range (Hz)	Acceptance Level (0–peak)
Axial	5–10	3.8 mm
	10–100	1.5 g
Lateral	5–7	7.7 mm
	7–15	1.5 g
	15–100	1.0 g

The recommended sinusoidal base excitations for Ariane IV launched spacecraft are given in Table 4.6. The levels are reduced for a particular spacecraft testing to avoid overtesting of the critical structural elements at the primary resonances. To avoid dynamic coupling between the launch vehicle and the spacecraft, it is recommended that the spacecraft fundamental natural frequencies be greater than 40 Hz in the axial axis and greater than 10 Hz in the lateral axis for hard-mounted spacecraft.

Structural Design Criteria

Launch loads have statistical variations from launch to launch. Therefore statistical properties are used to define the launch loads. The flight limit loads, as defined by NASA/GSFC, are the estimated 97.7% (2σ) probability of occurrence with 50% confidence based on a one-sided tolerance limit. The duration of the loads are derived by considering their nominal durations during flight. The design loads, also called ultimate loads, are obtained by applying a factor of 1.5 to the flight limit loads. The acoustic design loads are computed by adding a factor of 4 dB to the flight loads. The U.S. Air Force, per MIL-STD-1540, adds a factor of 6 dB to flight loads to determine design/ultimate loads. For spacecraft structures, inelastic deformations are not allowed under design loads due to alignment considerations. In addition, the structural design criteria for several spacecraft, including Intelsat VI, require a minimum margin of safety for all primary structures to be 10% for yield stress and elastic buckling stress at ultimate design loads.

4.4 STRESS–STRAIN ANALYSIS

When a body is under the action of external forces, it undergoes distortion and the effect of the forces is transmitted throughout the body. "Stress" denotes internal force per unit area. The stress at any point across a small area, ΔA, is defined as

$$\text{stress} = \lim_{\Delta A \to 0} \frac{\Delta F}{\Delta A} \qquad (4.1)$$

where ΔF is the internal force on the area ΔA surrounding the given point.
If the internal force acts in a direction perpendicular to the area, the

stress is called normal stress and is normally denoted by σ. The normal stress is called tensile stress if it pulls on the area, and compressive stress if it pushes on the area. If the internal force acts in the plane of the area, the stress is called shear stress and is normally denoted by τ.

Strain in any direction is defined to be the deformation per unit of length in that direction. It is denoted by ε. For a continuous material, ε at any point in a segment of initially ΔL length may be defined as

$$\varepsilon = \lim_{\Delta L \to 0} \frac{\Delta \delta}{\Delta L} \tag{4.2}$$

in which $\Delta \delta$ is the change in length of the given segment. Figure 4.5 represents an elemental cube of the material at a point at which shearing stresses exist in the member. The original shape is shown by the dashed lines in Fig. 4.5 and the deformed shape by the solid lines. The shear strain, γ, is defined as

$$\gamma = \frac{e_s}{l} = \tan \phi \simeq \phi \tag{4.3}$$

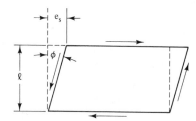

Figure 4.5 Shear strain at a point.

Thus a small shearing strain at any point in a member is measured by the change in angle of two lines in the member that pass through the point and that were originally at right angles and in the directions of the shearing stresses.

Stress–Strain Relationship   *In Elastic range*

Let a bar, as shown in Fig. 4.6, be subjected to an increasing load, P, in a testing machine such that the stress and strain are given by

$$\sigma = \frac{P}{a} \quad \text{and} \quad \varepsilon = \frac{e}{l} \tag{4.4}$$

The plot of a stress versus strain curve is known as a stress–strain diagram. For many materials, σ is proportional to ε up to a certain value of stress that is called the proportional limit. The ratio between stress and strain is a constant and it is called the modulus of elasticity, E, such that

$$E = \frac{\sigma}{\varepsilon} \qquad \text{*Young's modulus*}$$

Hooke's law also states that for small strains, stress is proportional to strain. Beyond the proportional limit, there is a stress value called the

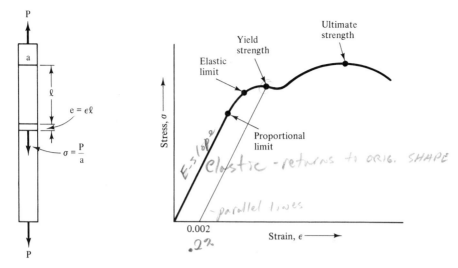

Figure 4.6 Longitudinal stress–strain.

elastic limit, such that the application of a stress of less than the elastic limit will not leave a permanent set. The yield strength of a material is defined to be the maximum stress that can be developed in a test specimen of the material without causing more than a specified permanent set. A permissible permanent set of 0.10 to 0.20% is frequently specified as the permissible set for metals. The ultimate strength, or tensile strength, is the maximum stress reached on the stress–strain diagram.

The ratio of shear stress, τ, to shear strain, γ, called the modulus of rigidity, is constant and is denoted by the symbol G such that

$$G = \frac{\tau}{\gamma} \qquad (4.5)$$

Experiments demonstrate that when a material is placed in tension, there exists not only an axial strain but also a lateral contraction strain. The ratio of the lateral strain to the axial strain is a constant, denoted by ν, and is called Poisson's ratio. The modulus of elasticity, modulus of rigidity and Poisson's ratio are related by

$$E = 2G\,(1 + \nu) \qquad (4.6)$$

Tangent Modulus and Secant Modulus

When the compressive stresses in a member fall within the inelastic range, the tangent modulus of elasticity, E_t, and the secant modulus of elasticity, E_s, measure the stiffness of the member. The tangent modulus, E_t, shown in Fig. 4.7, is determined by drawing a tangent to the stress–strain diagram at the point under consideration. The slope of the tangent gives the local rate of change of stress with strain. The secant modulus, E_s, is determined by drawing a straight line from the origin to the point in question. This modulus measures the ratio between stress and actual strain.

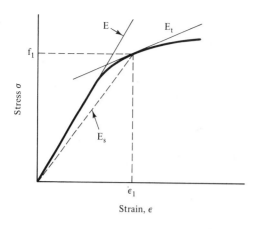

Figure 4.7 Tangent modulus and secant modulus.

In the elastic range, the tangent modulus will be the same as the modulus of elasticity, but it gets smaller in magnitude as the stress gets higher in the plastic range.

Bending Stress

A structure will normally be subjected not only to normal axial loads, as discussed earlier, but also to bending forces. Let us consider a beam, shown in Fig. 4.8, supported by reactions R_1 and R_2 and subjected to loads F_1, F_2, and F_3.

If the beam is imagined to be cut at $x = x_1$ and the left-hand portion is removed as a free body in order to maintain equilibrium, as shown in Fig. 4.8, it will be necessary to replace the action of the right-hand portion. This action is replaced by a vertical force, V, called the shearing force, and a couple, M, called the bending moment. The shear force is obtained by summing the forces to the left of the section under consideration. The bending moment is found as a summation of all the forces to the left of the section multiplied by their respective distances to the section, taking into account the signs of the quantities. The relation between the shearing force and the bending moment is

$$V = \frac{dM}{dx} \tag{4.7}$$

Let us consider a beam of a rectangular cross section under the action of bending moment, as shown in Fig. 4.8. The fibers in the top plane will be elongated while those in the bottom plane will be shortened, and those on plane A-A will remain at their original length. Plane A-A is called the neutral plane and represents the points of zero stress. The stress at a distance c from the neutral plane is given by

$$\sigma = \frac{Mc}{I} \tag{4.8}$$

Spacecraft Structures   Chap. 4

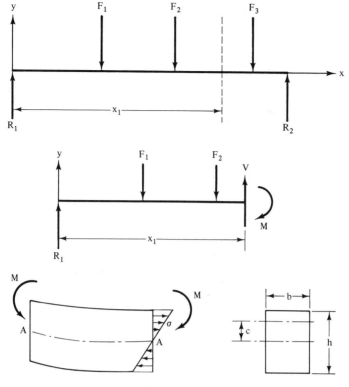

Figure 4.8 Beam under bending moment.

where σ = normal stress at the fiber
M = bending moment at the cross section
c = distance of the fiber from the neutral plane
I = cross-sectional moment of inertia
$= \frac{1}{12} bh^3$ for rectangular cross section

Torsional Stress

Let us consider a simple case of a solid circular bar subjected to a torque T along the axis of the bar as shown in Fig. 4.9. The shear stress

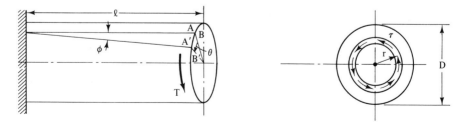

Figure 4.9 Circular bar under torque.

at any point in the cross section will be normal to the radius vector to that point and will be given by

$$\tau = \frac{T}{I_p} r \qquad (4.9)$$

where τ = shear stress
T = torque
r = radial distance of the point *in QUESTION*
I_p = polar moment of inertia of the cross section

The stress will be zero at the center and a maximum at the surface. For a solid round bar, $I_p = \pi D^4/32$ and Eq. (4.9) becomes

$$\tau = \frac{32Tr}{\pi D^4} \qquad (4.10)$$

and

$$\tau_{max} = \frac{16T}{\pi D^3} \qquad (4.11)$$

where D is the diameter of the bar. For a hollow circular shaft

$$I_p = \frac{\pi(D_o^4 - D_i^4)}{32} \qquad (4.12)$$

$$\tau_{max} = \frac{16TD_o}{\pi(D_o^4 - D_i^4)} \qquad (4.13)$$

where D_o and D_i are the outside and inside diameters, respectively. For a thin-walled cylindrical bar of radius, R, and thickness, t,

$$I_p = 2\pi R^3 t \qquad (4.14)$$

$$\tau = \frac{T}{2\pi R^2 t} \qquad (4.15)$$

For a rectangularly shaped cross section, the analysis is complicated due to warping of the cross section. Reference 2 gives the stress formulas for these cases.

The shear strain at radius r is given, as shown in Fig. 4.9, by

$$\gamma_c = \frac{BB'}{l} = \frac{r\theta}{l} \qquad (4.16)$$

The shear strain is related to shear stress by

$$\gamma_c = \frac{\tau}{G} = \frac{r\theta}{l} \qquad (4.17)$$

Hence, from Eqs. (4.9) and (4.17), we can write

$$\frac{\tau}{r} = \frac{T}{I_p} = \frac{G\theta}{l} \qquad (4.18)$$

Three-Dimensional Analysis

Stresses. For a complete description of a stress, we have to specify not only its magnitude, direction, and sense, but also the surface on which it acts. Stress at a point can be specified as shown in Fig. 4.10.

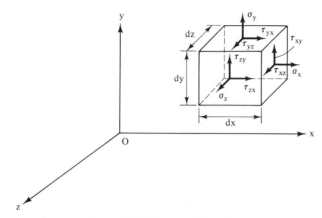

Figure 4.10 Stresses at a point.

The stress at a point on a surface may be resolved into two components: a normal stress perpendicular to the surface and a shearing stress acting in the plane of the surface. The normal stress component is denoted by σ and the shear stress component by τ. For a normal stress component, the direction of the stress and the normal to the surface are the same, so we identify them by one subscript. For example, σ_x is normal stress on the surface whose normal is along the x axis. In the case of a shear stress component, it can again be resolved into two components in the directions of the coordinate axes. So the shear stress is denoted by two subscripts. The first subscript denotes the normal to the plane on which it acts and the second subscript denotes its direction.

A normal stress is defined as positive if it is a tensile stress and negative if it is a compressive stress. For shear stress, the sign convention is as follows: on any surface where the tensile stress is in the positive direction of the coordinate axis, the shearing stresses are positive if they are in the positive directions of the other two coordinate axes, and vice versa. It can be shown for a general three-dimensional case that shear stress components are symmetrical, that is,

$$\tau_{xy} = \tau_{yx} \qquad \tau_{xz} = \tau_{zx} \qquad \tau_{yz} = \tau_{zy} \qquad (4.19)$$

So we need three normal stress components and three shear stress components to define the stress at a point. Let us consider a simpler case of plane stress, where the stresses in the plane normal to z the axis are zero. Let us also consider a plane whose normal makes an angle α with

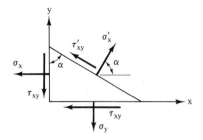

Figure 4.11 Plane stress.

the x axis as shown in Fig. 4.11. The normal stress, σ_x', and shear stress, τ_{xy}', on the plane may be calculated by considering the equilibrium of the element as

$$\sigma_x' = \sigma_x \cos^2\alpha + \sigma_y \sin^2\alpha + 2\tau_{xy} \sin\alpha \cos\alpha \qquad (4.20)$$

$$\tau_{xy}' = \tau_{xy} \cos 2\alpha + \tfrac{1}{2}(\sigma_y - \sigma_x) \sin 2\alpha \qquad (4.21)$$

The normal stress and shear stress are dependent on the angle α. The shear stress will be zero for the plane for which

$$\tan 2\alpha = \frac{2\tau_{xy}}{\sigma_x - \sigma_y} \qquad (4.22)$$

Since $\tan 2\alpha = \tan(\pi + 2\alpha)$, we find two perpendicular planes where the shear stresses are zero. It can be shown that along one plane the normal stress will be a maximum and along the other plane a minimum. These planes are called normal planes and the corresponding stresses are called principal stresses and given by

$$\sigma_{\max} = \frac{\sigma_x + \sigma_y}{2} + \sqrt{\left(\frac{\sigma_x - \sigma_y}{2}\right)^2 + \tau_{xy}^2} \qquad (4.23)$$

$$\sigma_{\min} = \frac{\sigma_x + \sigma_y}{2} - \sqrt{\left(\frac{\sigma_x - \sigma_y}{2}\right)^2 + \tau_{xy}^2} \qquad (4.24)$$

Similarly, we can determine principal stresses for a general case.

Strains. A body is said to be strained whenever the relative positions of the points in the body are altered. Let the coordinates of the particle before strain be x, y, z. After strain, the particle will undergo displacements u, v, w in the x, y, z directions, respectively, and will now have the coordinates $x + u$, $y + v$, $z + w$. In general, the displacements u, v, w vary from point to point in the body and are therefore functions of x, y, z. The strain condition in a two-dimensional case is shown in Fig. 4.12.

The strains can be classified into two types: longitudinal strains and shear strains. The ratio of the change in length to the original length of a strain line element is defined as the longitudinal strain and is denoted by ε. ε_x is defined as the longitudinal strain of an element which is in the x direction before strain. The change in angle from an initial right angle in the unstrained state is defined as shear strain and is denoted by γ. The strains, in terms of the displacement u, v, w, are defined as follows:

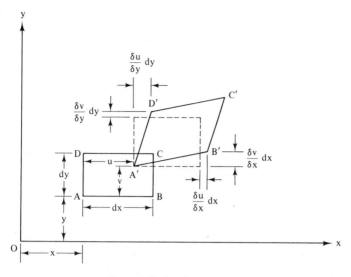

Figure 4.12 Strain at a point.

$$\varepsilon_x = \frac{\partial u}{\partial x} \qquad \varepsilon_y = \frac{\partial v}{\partial y} \qquad \varepsilon_z = \frac{\partial w}{\partial z}$$

$$\gamma_{xy} = \frac{\partial u}{\partial y} + \frac{\partial v}{\partial x} \qquad \gamma_{yz} = \frac{\partial v}{\partial z} + \frac{\partial w}{\partial y} \qquad (4.25)$$

$$\gamma_{zx} = \frac{\partial w}{\partial x} + \frac{\partial u}{\partial z}$$

The modulus of elasticity, modulus of rigidity, and Poisson's ratio are related by Eq. (4.6). Therefore, isotropic materials, whose properties are the same in all directions, have only two independent property constants.

In terms of strains, the stress equations become

$$\sigma_x = \frac{\nu E}{(1 + \nu)(1 - 2\nu)}e + \frac{E}{1 + \nu}\varepsilon_x \qquad \tau_{xy} = G\gamma_{xy}$$

$$\sigma_y = \frac{\nu E}{(1 + \nu)(1 - 2\nu)}e + \frac{E}{1 + \nu}\varepsilon_y \qquad \tau_{yz} = G\gamma_{yz} \qquad (4.26)$$

$$\sigma_z = \frac{\nu E}{(1 + \nu)(1 - 2\nu)}e + \frac{E}{1 + \nu}\varepsilon_z \qquad \tau_{zx} = G\gamma_{zx}$$

where $e = \varepsilon_x + \varepsilon_y + \varepsilon_z$. In terms of stresses, the strain equations become

$$\varepsilon_x = \frac{1}{E}(\sigma_x - \nu(\sigma_y + \sigma_z)) \qquad \gamma_{xy} = \frac{\tau_{xy}}{G}$$

$$\varepsilon_y = \frac{1}{E}(\sigma_y - \nu(\sigma_x + \sigma_z)) \qquad \gamma_{yz} = \frac{\tau_{yz}}{G} \qquad (4.27)$$

$$\varepsilon_z = \frac{1}{E}(\sigma_z - \nu(\sigma_x + \sigma_y)) \qquad \gamma_{zx} = \frac{\tau_{zx}}{G}$$

Composite Materials

Advanced composite materials combine two or more materials to utilize various desirable characteristics. They consist of high-strength, high-modulus-of-elasticity, and low-density filaments embedded in the matrix of essentially homogeneous materials. The filaments most commonly used are boron and graphite. The matrix materials most commonly used are epoxy in organic matrix resins and aluminum in metals. Since these materials are anisotropic, the stress–strain analysis becomes more complex.

There are three levels of composite material properties. The first level is the properties of the fibers and the matrix. The second level is the properties of the unidirectional composites, called plies. The third level is the multidirectional laminates, consisting of plies with arbitrary orientations. In the stress analysis, there are three levels of corresponding stresses. Micromechanical stresses are calculated at the level of the fiber and matrix. Ply stress is calculated on the basis of an assumed homogeneity within each ply where the fiber and matrix are smeared and no longer recognized as distinct phases. Laminate stress is an average of the ply stresses across the thickness of the laminate.

Unidirectional composites. Unidirectional composites have two orthogonal planes of symmetry as shown in Fig. 4.13. One plane is parallel to the fibers and the other is transverse to the fibers. When the axes x-y coincide with the longitudinal and transverse directions, the material is called orthotropic and on-axis. The on-axis stress–strain relations are

$$\varepsilon_x = \frac{1}{E_x}\sigma_x - \frac{\nu_y}{E_y}\sigma_y$$

$$\varepsilon_y = -\frac{\nu_x}{E_x}\sigma_x + \frac{1}{E_y}\sigma_y \tag{4.28}$$

$$\varepsilon_s = \frac{1}{E_s}\sigma_s$$

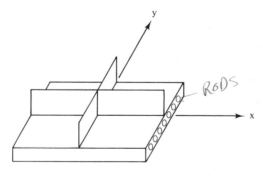

Figure 4.13 Unidirectional composites.

where E_x = longitudinal modules of elasticity
$\quad\quad v_x$ = longitudinal Poisson's ratio
$\quad\quad E_y$ = transverse modulus of elasticity
$\quad\quad v_y$ = transverse Poisson's ratio
$\quad\quad E_s$ = longitudinal shear modulus

The constants E_x, E_y, v_x, and v_y are related by

$$\frac{v_x}{v_y} = \frac{E_x}{E_y} \tag{4.29}$$

Hence an on-axis orthotropic unidirectional composite has four independent property constants. The transformation for off-axis stress is derived by using the balance of forces. The off-axis strain transformation is purely geometric. The transformation from off-axis stress to off-axis strain is as follows. Transform off-axis stress to on-axis stress, transform on-axis stress to on-axis strain by using the relationship above, and transform on-axis strain to off-axis strain. The off-axis modulus can be determined in a similar way.

Symmetric laminates. Multidirectional laminates consist of plies with arbitrary orientations. These laminates with midplane symmetry will behave as homogeneous anisotropic plates. The effective modulus of the laminates is simply the arithmetic average of the moduli of the constituent plies.

A multidirectional composite laminate, as shown in Fig. 4.14, is defined by the following code to designate the stacking sequence of the ply group.

$$[0_2/90_2/45_2/-45_2]_S$$

Starting from the bottom of the plate, at $z = -h/2$, the first ply group has two plies of 0° orientation, followed by two 90° plies, followed by two 45° plies, and finally two −45° plies. For a symmetric plate, the order of plies is symmetric about the midplane of the laminate. The subscript s

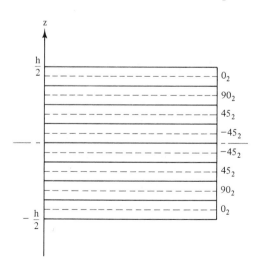

Figure 4.14 Multidirectional composite laminate.

denotes that the laminate is symmetric with respect to the midplane or the $z = 0$ plane.

To derive the stress–strain relation of a multidirectional laminate, two simplifying assumptions are made: the laminate is assumed to be symmetric (i.e., both the ply orientation and the ply material modulus are symmetric with respect to the midplane of the laminate) and the strain is assumed to be constant across the laminate thickness.

The stress distribution across the multidirectional laminate is not constant because the modulus varies from ply to ply. Therefore, it is more convenient to define an average stress across the laminate. The average stress can be used to define the stress–strain relationship. Once we know the strain in the laminate, the stress at any ply within the laminate can be determined.

As an example, average stress based on average strain is given by

$$\bar{\sigma}_x = \frac{1}{h} \int_{-h/2}^{h/2} (Q_{xx}\varepsilon_x^0 + Q_{xy}\varepsilon_y^0 + Q_s\varepsilon_s^0)\, dz \qquad (4.30)$$

where Q_{xx}, Q_{xy}, Q_s are the modulus components of the plies and ε_x^0, ε_y^0, and ε_s^0 are the average strain components. Equation (4.30) integrates over all the plies in the laminate.

Strain Energy

An elastic body when acted upon by external loads suffers deformation, so the points of application of the loads move. If the loads are applied gradually, the kinetic energy imparted to the body is negligible. All the work done by the loads is transformed into the potential energy of the strain, known as strain energy. The strain energy per unit volume is given by

$$U_i = \frac{1}{2} (\underbrace{\sigma_x\varepsilon_x + \sigma_y\varepsilon_y + \sigma_z\varepsilon_z}_{NORMAL} + \underbrace{\tau_{xy}\gamma_{xy} + \tau_{yz}\gamma_{yz} + \tau_{zx}\gamma_{zx}}_{SHEAR}) \qquad (4.31)$$

4.5 MATRIX METHODS OF STRUCTURAL ANALYSIS

For simple structural elements, such as a beam, the stress and strain due to external loads can be denoted by elementary stress–strain analysis methods, as discussed in the preceding section. However, a spacecraft structure is inherently complex. It consists of an assemblage of many structural elements of irregular shapes. Hence the true structure must be replaced by an idealized model, which is suitable for mathematical analysis.

For a structural analyst, establishing an idealized structural model is the major task. This requires experience, sound judgment, and the knowledge of structural analysis theory. Several general-purpose finite-element-method structural analysis computer programs are now available. However, to use them correctly, a basic knowledge of matrix methods of structural analysis is useful.

Flexibility and Stiffness Matrices

Let us consider an arbitrary elastic structure, shown in Fig. 4.15, supported against rigid-body motion and subjected to loads F_1, F_2, ..., F_n acting at nodes 1, 2, ..., n. The corresponding set of displacements is represented by δ_1, δ_2, ..., δ_n.

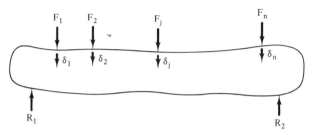

Figure 4.15 Elastic structure under loads.

The total displacement at any node, both linear and angular, can be expressed as the sum of the displacements at that node due to the individual loads. This is a principle of superposition, which is fundamental to the analysis of linear systems.

The total displacement, δ_i, at node i can be expressed as

$$\delta_i = C_{i1}F_1 + \cdots + C_{ij}F_j + \cdots + C_{in}F_n \tag{4.32}$$

By definition, C_{ij} is the deflection produced at node i due to a unit load at node $j (F_j = 1)$. These coefficients are known as deflection influence coefficients. Similarly, we can write the equations of deflection for other nodes. These equations can be written into a single matrix equation as

$$\begin{Bmatrix} \delta_1 \\ \vdots \\ \delta_i \\ \vdots \\ \delta_n \end{Bmatrix} = \begin{bmatrix} C_{11} & \cdots & C_{1j} & \cdots & C_{1n} \\ \vdots & & & & \\ C_{i1} & \cdots & C_{ij} & \cdots & C_{in} \\ \vdots & & & & \\ C_{n1} & \cdots & C_{nj} & \cdots & C_{nn} \end{bmatrix} \begin{Bmatrix} F_1 \\ \vdots \\ F_i \\ \vdots \\ F_n \end{Bmatrix} \tag{4.33}$$

This equation may be written in a compact matrix form:

$$\delta = CF \tag{4.34}$$

Matrix C is known as the matrix of flexibility influence coefficients. If C is known, the nodal displacements due to any set of prescribed nodal forces, F, can be calculated at once from Eq. (4.33). The forces in terms of deflections can be written as

$$F = C^{-1}\delta \tag{4.35}$$
$$= K\delta$$

where K is known as the stiffness matrix and is equal to C^{-1}, which is the inverse of matrix C.

$$
\begin{Bmatrix} F_1 \\ \vdots \\ F_i \\ \vdots \\ F_n \end{Bmatrix} = \begin{bmatrix} K_{11} & \cdots & K_{1j} & \cdots & K_{1n} \\ \vdots & & \vdots & & \vdots \\ K_{i1} & \cdots & K_{ij} & \cdots & K_{in} \\ \vdots & & \vdots & & \vdots \\ K_{n1} & \cdots & K_{nj} & \cdots & K_{nn} \end{bmatrix} \begin{Bmatrix} \delta_1 \\ \vdots \\ \delta_i \\ \vdots \\ \delta_n \end{Bmatrix}
\tag{4.36}
$$

If $\delta_1 = 1$, $\delta_2 = \delta_3$, ..., $\delta_n = 0$, then from Eq. (4.36),

$$
F_1 = K_{11}, \quad F_2 = K_{21}, \quad \ldots, \quad F_n = K_{n1}
\tag{4.37}
$$

which are the elements of the first column of the matrix $[K]$. Similarly, the jth column of $[K]$ represents the forces required to maintain the displacement state $\delta_j = 1$ and all other nodal displacements equal to zero.

By applying the reciprocal theorem, we get

$$
C_{ij} = C_{ji}
\tag{4.38}
$$

Hence the matrix C is symmetric. Since C is symmetric, Eq. (4.35) guarantees that the stiffness matrix K will also be symmetric. Consequently,

$$
K_{ij} = K_{ji}
\tag{4.39}
$$

Stiffness Matrix Assemblage

It is difficult to determine the stiffness matrix of a complex structure directly. A structure generally consists of basic structural elements: bars, thin plates, and so on. A structure is usually assumed to consist of basic structural elements and joined together at nodes. Consequently, it is important to be able to form the total structure stiffness matrix from the stiffness matrices of the basic elements. As an example, let us consider a simple example consisting of two springs, shown in Fig. 4.16. The springs, a and b, have individual stiffness constants K_a and K_b. They are connected at node 2. First we determine the stiffness matrices of the two individual springs shown in Fig. 4.17. The stiffness matrices of these springs are

$$
K_a = \begin{matrix} \;\;u_1 & \;\;u_2 \\ \begin{bmatrix} K_a & -K_a \\ -K_a & K_a \end{bmatrix} \end{matrix} \qquad K_b = \begin{matrix} \;\;u_2 & \;\;u_3 \\ \begin{bmatrix} K_b & -K_b \\ -K_b & K_b \end{bmatrix} \end{matrix}
\tag{4.40}
$$

The matrices above are not immediately subject to superposition since the columns in the two matrix are not compatible. To correct this situation, we expand each member stiffness matrix to the order of the total structural stiffness matrix by adding columns and rows of zeros for those nodal displacements which are irrelevant for the member in question. So we obtain

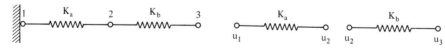

Figure 4.16 Spring system.

Figure 4.17 Spring elements.

$$K_a = \begin{array}{c} \begin{array}{ccc} u_1 & u_2 & u_3 \end{array} \\ \left[\begin{array}{ccc} K_a & -K_a & 0 \\ -K_a & K_a & 0 \\ 0 & 0 & 0 \end{array} \right] \end{array} \qquad K_b = \begin{array}{c} \begin{array}{ccc} u_1 & u_2 & u_3 \end{array} \\ \left[\begin{array}{ccc} 0 & 0 & 0 \\ 0 & K_b & -K_b \\ 0 & -K_b & K_b \end{array} \right] \end{array} \qquad (4.41)$$

By directly adding K_a and K_b, we obtain the total structure stiffness matrix,

$$K = \begin{array}{c} \begin{array}{ccc} u_1 & \quad u_2 & \quad u_3 \end{array} \\ \left[\begin{array}{ccc} K_a & -K_a & 0 \\ -K_a & K_a + K_b & -K_b \\ 0 & -K_b & K_b \end{array} \right] \end{array} \qquad (4.42)$$

It can be shown that the matrix K is singular and that its inverse does not exist. The system is free to move as a rigid body when external loading is applied. On specifying the sufficient boundary conditions to prevent rigid-body motion, the singularity in the matrix K will be removed. In our example, $u_1 = 0$. So

$$\begin{Bmatrix} F_1 \\ F_2 \\ F_3 \end{Bmatrix} = \begin{bmatrix} K_a & -K_a & 0 \\ -K_b & K_a + K_b & -K_b \\ 0 & -K_b & K_b \end{bmatrix} \begin{Bmatrix} u_1 = 0 \\ u_2 \\ u_3 \end{Bmatrix} \qquad (4.43)$$

In Eq. (4.43) F_2 and F_3 are applied loads and F_1 is the unknown reaction. Equation (4.43) can be written as

$$\begin{Bmatrix} F_2 \\ F_3 \end{Bmatrix} = \begin{bmatrix} K_a + K_b & -K_b \\ -K_b & K_b \end{bmatrix} \begin{Bmatrix} u_2 \\ u_3 \end{Bmatrix} \qquad (4.44)$$

and

$$\{F_1\} = [-K_a \quad 0] \begin{Bmatrix} u_2 \\ u_3 \end{Bmatrix} \qquad (4.45)$$

From Eq. (4.44), the displacements u_2 and u_3 can be determined in terms of the external loads F_2 and F_3. From Eq. (4.45), the reaction F_1 is determined from displacements u_2 and u_3.

4.6 FINITE ELEMENT ANALYSIS (Ref. 4)

A spacecraft structure consists of an assembly of different structural elements connected together either by discrete or by continuous attachments. If the structural elements are connected together by discrete joints, the assembled structure can be analyzed as discussed in the preceding section if the force–displacement relationship of the individual elements are known. If the structural elements are continuously attached, which is equivalent to infinite attached points, such as in plates and shells, we have a problem in its numerical solution. This problem is overcome by the finite element method where the continuum is divided into elements which are interconnected only at a finite number of nodal points at which some fictitious force, representative of the distributed stresses acting on the element boundaries, is introduced.

To calculate the stiffness matrix for a structural element by the finite element method, the following general method is used. The displacements at any point within the element are expressed in terms of nodal displacements as follows:

$$\{u\} = [N]\{\delta\} \tag{4.46}$$

where $\{u\}$ is a displacement vector at any point in the element and $\{\delta\}$ is a displacement vector of the nodes. $[N]$ is a transformation matrix with general functions. The strain vector $\{\varepsilon\}$ in terms of displacement vector at any point in the element, by using Eq. (4.25), can be expressed as

$$\{\varepsilon\} = [C]\{u\} \tag{4.47}$$

By substituting Eq. (4.46) into Eq. (4.47), we get

$$\{\varepsilon\} = [B]\{\delta\} \tag{4.48}$$

where

$$[B] = [C][N] \tag{4.49}$$

The stress vector $\{\sigma\}$ in terms of strain by using Eq. (4.26) can be expressed as

$$\{\sigma\} = [D]\{\varepsilon\} \tag{4.50}$$

Let us consider a virtual displacement $\{\delta^*\}$ at the nodes. This will result in displacements, strains, and stresses within the element given by Eqs. (4.48) and (4.50) as

$$\{\varepsilon^*\} = [B]\{\delta^*\}$$
$$\{\sigma^*\} = [D][B]\{\delta^*\} \tag{4.51}$$

The internal work done by internal forces per unit volume is

$$\{\varepsilon^*\}^T\{\sigma^*\} = \{\delta^*\}^T[B]^T[D][B]\{\delta^*\} \tag{4.52}$$

Equating the external work done by nodal forces with internal work obtained by integrating over the volume of the element,

$$\{\delta^*\}^T\{F\} = \{\delta^*\}^T\left(\int [B]^T[D][B]\,d(\text{vol})\right)\{\delta^*\} \tag{4.53}$$

The nodal forces $\{F\}$ are given by

$$\{F\} = \left(\int [B]^T[D][B]\,d(\text{vol})\right)\{\delta^*\} \tag{4.54}$$

From Eq. (4.54), the stiffness matrix becomes

$$[K] = \int [B]^T[D][B]\,d(\text{vol}) \tag{4.55}$$

Once the nodal displacements are determined by complete structure analysis, the stress at any point of the element can be found from the relation

$$\{\sigma\} = [D][B]\{\delta\}$$
$$= [S]\{\delta\} \tag{4.56}$$

where

$$[S] = [D][B] \qquad (4.57)$$

The matrix $[S]$ is known as the stress matrix. It defines the stresses in the element in terms of the nodal displacements.

Example 4.1

Let us consider a triangular element under plain stress as shown in Fig. 4.18. The displacements within the element can be represented by the equations

$$u = \alpha_1 + \alpha_2 x + \alpha_3 y \qquad (4.58)$$

$$v = \alpha_4 + \alpha_5 x + \alpha_6 y$$

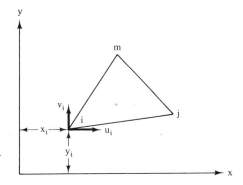

Figure 4.18 Triangular element under plain stress.

Since the nodal displacement will satisfy this equation, the six constants α can be determined in terms of nodal displacements. The displacement equations become

$$u = \frac{1}{2\Delta} [(a_i + b_i x + c_i y)u_i + (a_j + b_j x + c_j y)u_j + (a_m + b_m x + c_m y)\, u_m] \qquad (4.59)$$

$$v = \frac{1}{2\Delta} [(a_i + b_i x + c_i y)v_i + (a_j + b_j x + c_j y)v_j + (a_m + b_m x + c_m y)v_m]$$

where $a_i = x_j y_m - x_m y_j$, $b_i = y_j - y_m$, $c_i = x_m - x_j$ and the other coefficients are obtained by cyclic permutation of the subscripts in the order i, j, m, and Δ is the area of the triangle. The strain at any point, from Eq. (4.25), is

$$\{\varepsilon\} = \begin{Bmatrix} \varepsilon_x \\ \varepsilon_y \\ \gamma_{xy} \end{Bmatrix} = \begin{Bmatrix} \dfrac{\partial u}{\partial x} \\[2mm] \dfrac{\partial v}{\partial y} \\[2mm] \dfrac{\partial u}{\partial y} + \dfrac{\partial v}{\partial x} \end{Bmatrix} \qquad (4.60)$$

Using Eqs. (4.48) and (4.60), we have

$$\{\varepsilon\} = [B]\{\delta\}$$

where

$$[B] = \frac{1}{2\Delta} \begin{bmatrix} b_i & 0 & b_j & 0 & b_m & 0 \\ 0 & c_i & 0 & c_j & 0 & c_m \\ c_i & b_i & c_j & b_j & c_m & b_m \end{bmatrix} \qquad (4.61)$$

and

$$\{\delta\} = \begin{Bmatrix} u_i \\ v_i \\ u_j \\ v_j \\ u_m \\ v_m \end{Bmatrix} \tag{4.62}$$

From Eq. (4.26), for plane stress in an isotropic material, we have

$$\{\sigma\} = [D]\{\varepsilon\}$$

where

$$D = \frac{E}{1 - \nu^2} \begin{bmatrix} 1 & \nu & 0 \\ \nu & 1 & 0 \\ 0 & 0 & \dfrac{1 - \nu}{2} \end{bmatrix} \tag{4.63}$$

and

$$\{\sigma\} = \begin{Bmatrix} \sigma_x \\ \sigma_y \\ \tau_{xy} \end{Bmatrix} \tag{4.64}$$

From Eq. (4.55),

$$[K] = \int [B]^T[D][B]t \, dx \, dy \tag{4.65}$$

where t is the thickness of the element and the integration is taken over the area of the triangle. By substituting [B] from Eq. (4.61) and [D] from Eq. (4.63) into Eq. (4.65), the stiffness matrix [K] is determined. The chosen displacement function, Eq. (4.58), automatically guarantees continuity of displacements with adjacent elements because the displacements vary linearly along any side of the triangles and, with the same displacement at the nodes, the same displacements will be along the interface. The basic structural elements commonly used for spacecraft structural analysis are shown in Fig. 4.19.

A spacecraft structure is idealized into discretely connected basic structural elements as shown in Fig. 4.20 for Intelsat V. This idealized assembly is called the structural model. There are several general-purpose programs which can be used to analyze a spacecraft structure. The following are the commonly used computer programs:

NASTRAN
ASKA
STARDYNE
SAP
ANSYS
MARC

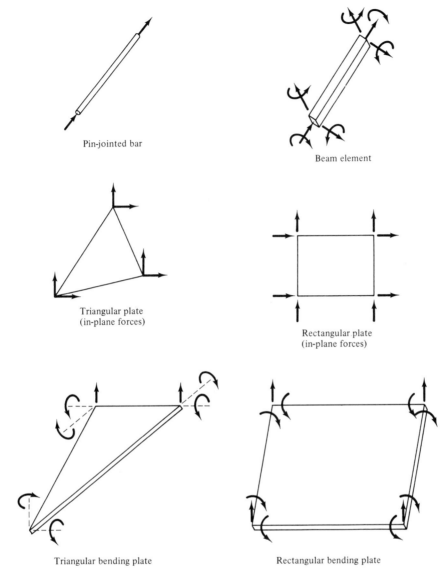

Pin-jointed bar

Beam element

Triangular plate
(in-plane forces)

Rectangular plate
(in-plane forces)

Triangular bending plate

Rectangular bending plate

Figure 4.19 Basic structural elements.

4.7 INSTABILITY OF STRUCTURES

An elastic body is called stable under given loads when an infinitesimal load added to the body would cause only infinitesimal changes in the displacements and the body recovers if the added loads are removed. When the displacements continuously increase with little or no further increment of the loads, the body is considered to be unstable. If the body will remain

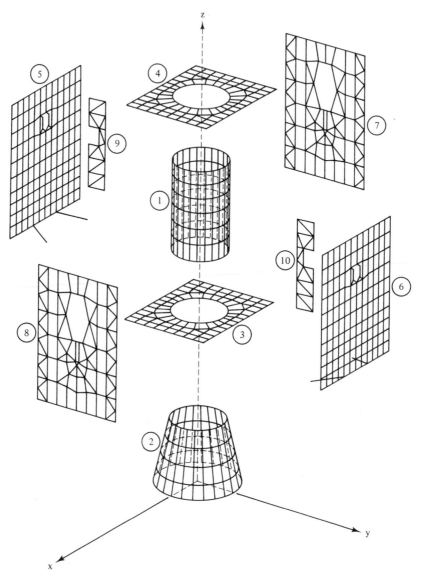

Figure 4.20 Finite element model of Intelsat V main body. (Courtesy of INTELSAT and FACC)

in the displaced position after removal of the disturbance, the body is said to be in neutral equilibrium. In structural elements, the columns, shells, and panels are subjected to the phenomenon of instability.

Columns

A column under compressive load will have elastic instability. Let us consider a column of length, L, with hinged ends, which are free to rotate, is subjected to an axial load P. The column is assumed to be straight before the load is applied. As the load is increased, the column remains straight until the load exceeds a certain value called critical load, P_{cr}. This is unstable equilibrium and if there is a slight disturbance or initial eccentricity, the column buckles and assumes the shape shown in Fig. 4.21. This critical load is called the buckling load.

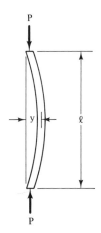

Figure 4.21 Column under axial compressive load.

Euler derived this critical load analytically and it is given by the formula

$$P_{cr} = \frac{\pi^2 EI}{L^2} \qquad load \qquad (4.66)$$

or

$$f_c = \frac{\pi^2 E}{(L/\rho)^2} \qquad stress \qquad (4.67)$$

where P_{cr} = critical buckling load
$f_c = P_{cr}/A$, critical buckling stress
A = area cross section
E = modulus of elasticity
L = column length

$$I = \frac{\pi r^4}{4}$$

I = moment of the inertia of the cross section about the axis about which the column tends to buckle, usually the minimum moment of inertia
$\rho = \sqrt{I/A}$, radius of gyration of the cross section

It should be noted from Eqs. (4.66) and (4.67) that the critical buckling load or stress is not a function of the material strength but of the modulus of elasticity of the material, E.

The critical buckling load will depend on the end condition of the column. The general formula for buckling stress can be written as

$$f_c = \frac{C\pi^2 E}{(L/\rho)^2} \qquad (4.68)$$

$$= \frac{\pi^2 E}{(L'/\rho)^2} \qquad (4.69)$$

where C = constant, dependent on the end conditions
L' = $L/\sqrt{C}$, effective length of the column
L'/ρ = slenderness ratio

Figure 4.22 shows the end-condition constant, C, of a column under different end conditions. In the hinged-end condition, the ends are allowed to rotate, while in the fixed-end condition, the ends are not allowed to rotate.

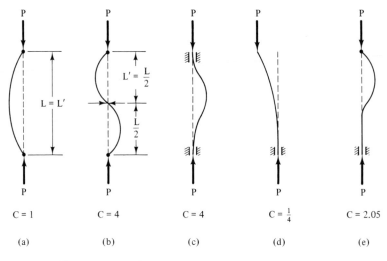

Figure 4.22 Buckling stress as a function of slenderness.

Figure 4.23 shows the plot of the buckling stress, f_c, against the slenderness ratio, L'/ρ. For a stable cross section, such as a round tube of relatively heavy wall thickness, the test results will closely follow the curve *ABFC*.

The Euler formula [Eq. (4.69)] is valid only for the elastic deflection. Between points *B* and *C*, the failure will be an elastic overall bending instability, and the Euler formula, Eq. (4.69), can be used. There will be an inelastic bending instability for most of the stress range *B* to *A*, and the Euler formula can be used to calculate the critical stress by replacing the modulus of elasticity E in Eq. (4.69) by the tangent modulus E_t.

Let us consider a case were the column has an open cross section, such as a channel or a hat section, with relatively small material thickness. The test results for such a column will follow the curve *DEFC*. The critical stresses for the region *DEF* are much lower than those given by the Euler

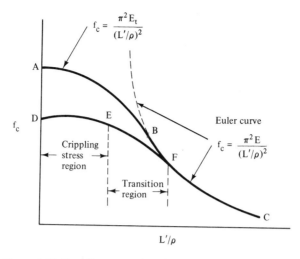

Figure 4.23 Buckling stress as a function of slenderness ratio.

formula. For region DE, the failure mode is the local crippling of the section. In region FC, there is overall elastic buckling and therefore the Euler formula is applicable. Region EF consists of a combination of local and overall buckling.

Crippling strength under compression. The methods to calculate crippling stress are semiempirical and have been sufficiently proven by testing. The angle method or the Needham method is as follows:

$$\sigma_c = C_e \frac{(\sigma_y E)^{1/2}}{(b'/t)^{0.75}} \qquad (4.70)$$

where σ_c = crippling stress
  C_e = coefficient that depends on the edge conditions
  = 0.316 (two edges free)
  = 0.342 (one edge free)
  = 0.366 (no edge free)
  σ_y = compression yield stress
  t = thickness
  $b'/t = (a + b)/2t$

The crippling load on an angle unit is

$$P_c = \sigma_c A \qquad (4.71)$$

where A is the area of the angle.

The crippling stress for channels and Z-shaped and rectangular tubes can be divided into equal angles as shown in Fig. 4.24. The basic angle unit for a channel section has one edge free. The crippling stress for other structural shapes can be determined by dividing the shape into a series of angle units and computing the crippling loads for each section by using Eq. (4.71). The average crippling stress is determined by summing the crippling

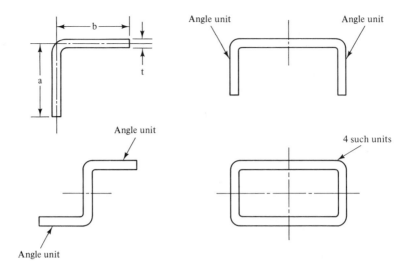

Figure 4.24 Basic angle units for different cross sections.

load for each section and dividing the total load by the total area of the cross section as follows:

$$\sigma_c = \frac{\Sigma \text{ (crippling loads of angles)}}{\Sigma \text{ (area of angles)}} \qquad (4.72)$$

The crippling stress formulas for channel, H, rectangular tube, and hat sections are given in Ref. 3.

Shells

For shells, there are discrepancies between the theoretical analyses and the experimental results. The causes of such discrepancies are generally the dependence of the buckling load on the initial imperfections of the shell, small deviations from nominal circular cylindrical or conical shape, and the edge conditions. The current methods of establishing design data tend to treat both initial imperfections and edge conditions as random effects. Results from all available tests are lumped together without regard to specimen construction or the method of testing, and are analyzed to yield lower-bound statistical correction factors to be applied to simplified versions of the theoretical results.

The shells which are commonly used in a spacecraft structure can be divided into two types: cylindrical and conical.

Cylindrical shells. Cylindrical shells can be further divided into three categories: monocoque, sandwich, and stiffened. This section will give buckling loads for these cylindrical shells under different load conditons.

Monocoque Shells

1. Axial compression. A theoretical analysis indicates that the critical buckling stress under axial compression is given by

$$\sigma_c = \gamma \frac{E}{\sqrt{3(1 - \nu^2)}} \frac{t}{r} \tag{4.73}$$

$$= 0.6 \, \gamma \, E \frac{t}{r} \quad \text{(for } \nu = 0.3\text{)} \tag{4.74}$$

where σ_c = critical buckling stress
E = modulus of elasticity
t = thickness of the shell
r = radius of the shell

The theoretical value of γ in Eq. (4.74) is unity. On the basis of experimental data, however, it is recommended (Ref. 9) that the following values of γ be used:

$$\gamma = 1 - 0.9(1 - e^{-\phi}) \quad \text{Theoretically } \gamma = 1 \tag{4.75}$$

where

$$\phi = \frac{1}{16} \sqrt{\frac{r}{t}} \quad \text{for } \left(\frac{r}{t} < 1500\right)$$

The critical buckling load, P_c, is given by critical buckling stress times the area of the cross section as

$$P_c = \sigma_c 2\pi rt \tag{4.76}$$

$$= 1.2\gamma\pi Et^2$$

2. Bending. The theoretical critical stress formula for bending is the same as that for axial compression and is given as

$$\sigma_b = 0.6\gamma E \frac{t}{r} \; . \tag{4.77}$$

The theoretical value of γ is unity. However, the correlation factor for bending is greater than that for compression and is given as (Ref. 9)

$$\gamma = 1 - 0.731(1 - e^{-\phi}) \tag{4.78}$$

The primary reason that the correlation factor for bending is higher than that for compression is that in compression the buckling can be triggered by any imperfection on the shell surface, whereas in bending, the buckling is generally initiated in the region of the greatest compressive stress.

The critical bending moment, M_c, is given by

$$M_c = \sigma_b \pi r^2 t \tag{4.79}$$

Substituting Eq. (4.77) into Eq. (4.79), we get

$$M_c = 0.6\gamma\pi Ert^2 \tag{4.80}$$

3. Torsion. The critical shear stress due to torsion is given in Ref. 15 by

$$\tau_c = \frac{K_t \pi^2 E}{12(1 - v^2)} \left(\frac{t}{L}\right)^2 \tag{4.81}$$

where L is the length of the cylinder and K_t is given in Fig. 4.25. The critical torque, T_c, is given by

$$T_c = \tau_c \cdot 2\pi r^2 t = \frac{K_t \pi^3 E r^2 t}{6(1 - v^2)} \left(\frac{t}{L}\right)^2 \tag{4.82}$$

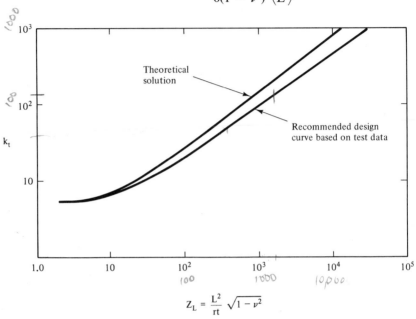

$$Z_L = \frac{L^2}{rt} \sqrt{1 - v^2}$$

Figure 4.25 Buckling of simply supported circular cylinders in torsion or transverse shear.

4. Transverse shear. Shear stresses are also produced under transverse loads. These shear stresses are a maximum at the neutral axis and zero at the extreme distance from the neutral axis. The torsional shear stresses are uniform over the entire cylindrical wall. A recommended approach in Ref. 3 is to determine buckling stress under transverse shear by replacing K_t by $1.25 K_t$ in Eq. (4.81), that is,

$$\tau_c' = \frac{1.25 K_t \pi^2 E}{12(1 - v^2)} \left(\frac{t}{L}\right)^2 \tag{4.83}$$

where τ_c' is transverse shear stress and K_t is determined from Fig. 4.25. The critical transverse shear force, V_c, is given by

$$V_c = \frac{1.25 K_t \pi^3 E r t}{12(1 - v^2)} \left(\frac{t}{L}\right)^2 \tag{4.84}$$

Combined Load. Let us consider a cylindrical shell subjected to compressive load, P, bending moment, M, torque, T, and transverse shear force, V. The critical compressive load, P_c, the critical bending moment, M_c, the critical torque, T_c, and the critical transverse shear force, V_c, are calculated from Eqs. (4.76), (4.80), (4.82), and (4.84), respectively. The critical load ratios are calculated as follows:

$$R_c = \frac{P}{P_c} \qquad R_b = \frac{M}{M_c} \qquad R_s = \frac{V}{V_c} \qquad R_{st} = \frac{T}{T_c}$$

The margin of safety, M.S., is given by

$$\text{M.S.} = \frac{1}{R_c + R_{st}^2 + (R_s^3 + R_b^3)^{1/3}} - 1 \tag{4.85}$$

Sandwich Shells. A sandwich shell consists of two thin-face skins bonded to a thick core. The face skins resist nearly all of the applied edgewise loads and the bending moment. The core spaces the face skins and transmits shear forces between them so that they are effective about a common neutral axis. The primary difference between sandwich shells and monocoque shells is the relatively low transverse shear stiffness of the sandwich construction. The instability of a sandwich shell is analyzed, generally, for two modes of failures: (1) local instability failure, in which the facing skin fails because of insufficient stabilization by the core, and (2) general instability failure, in which the entire shell fails with the core and face skins acting together.

1. Local instability. For a honeycomb core, it is possible for the face skin to buckle or dimple into the spaces between core walls. This is called *intercell buckling*. The critical stress (Ref. 10) is

$$\sigma_s = 2.5 E_R \frac{t_f}{S} \tag{4.86}$$

where $E_R = \dfrac{4 E_f E_{\text{tan}}}{(\sqrt{E_f} + \sqrt{E_{\text{tan}}})^2}$

    E_f = modulus of elasticity of the face skin
    E_{tan} = tangent modulus of elasticity of the face skin
    t_f = thickness of the face skin
    S = core shell size expressed as the diameter of the largest inscribed circle

When a face skin of a sandwich element is subjected to axial compression, *face-skin wrinkling* may occur. Typical wrinkling failures are shown in Fig. 4.26. The critical uniaxial wrinkling stress, σ_c, can be obtained from

$$\sigma_c = 0.43 (E_f E_c G_c)^{1/3} \tag{4.87}$$

where E_f = modulus of elasticity of the face skin
    E_c = modulus of elasticity of the core in the direction normal to the surface of the core
    G_c = transverse shear modulus of the core in the direction of the maximum compressive stress

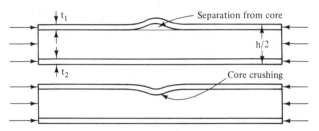

Figure 4.26 Face-skin wrinkling.

2. General instability. Reference 3 gives the equations for the buckling stresses in a sandwich cylinder under general instability by taking into account the low transverse shear stiffness of the core. However, if it is assumed that the core has infinite transverse shear stiffness and no load-carrying capacity in the meridional and circumferential directions, the analysis for the monocoque cylinder can be used for a sandwich cylinder by determining its equivalent thickness and modulus of elasticity. The face skins may be of different materials, subject to the restriction that the Poisson's ratios of the two materials are identical. Longitudinal and bending stiffnesses of a sandwich cylinder are equated to the longitudinal and bending stiffnesses of an equivalent monocoque cylinder. So we have

$$\bar{E}\bar{t} = E_1 t_1 + E_2 t_2 \tag{4.88}$$

$$\frac{\bar{E}(\bar{t})^3}{12} = \frac{h^2}{1/E_1 t_1 + 1/E_2 t_2}$$

By solving Eq. (4.88) for $\bar{E}$ and $\bar{t}$, we get

$$\bar{t} = \frac{\sqrt{12}h}{\sqrt{E_1 t_1/E_2 t_2} + \sqrt{E_2 t_2/E_1 t_1}} \tag{4.89}$$

$$\bar{E} = \frac{E_1 t_1 + E_2 t_2}{\bar{t}}$$

Using these equivalent thickness $\bar{t}$ and modulus of elasticity parameters $\bar{E}$, the critical loads for general instability can be determined by using Eqs. (4.76), (4.80), (4.82), and (4.84).

Stiffened Shells. The stiffened shell for a spacecraft structure normally consists of a monocoque shell with longitudinal stiffeners as shown in Fig. 4.27. The commonly used stiffeners are of hat, channel, Z-, and H-shaped sections. The instability of a stiffened shell is examined for the following types of failures:

1. Buckling of the sheet between the stiffeners
2. Crippling of the stiffeners
3. Lateral buckling of the stiffeners along with the sheet

The design criterion for a spacecraft structure does not normally permit

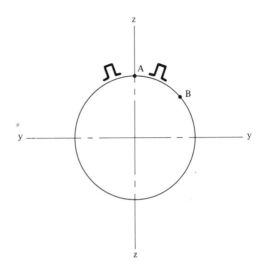

Figure 4.27 Stiffened shell.

either buckling or inelastic deformation of the sheet and the stiffeners. In such cases, stresses in the shell can be calculated by using the elementary beam theory.

The axial stress σ is given by

$$\sigma = \frac{P}{A} + \frac{M}{I} z \tag{4.90}$$

where P = axial load
  A = total cross-sectional area, both sheet and stiffeners
  M = bending moment
  I = moment of inertia of the cross section
  z = distance of the section from the neutral axis

The shear flow q at the cross-section B, as shown in Fig. 4.27, is

$$q = \frac{V_z}{I_y} \int z \, da \tag{4.91}$$

where V_z = shear force
  I_y = moment of inertia about y-y axis
  $\int z \, da$ = moment of the area between points A and B about the neutral axis yy

The shear stress at B is given by

$$\tau = \frac{V_z}{I_y t} \int z \, da \tag{4.92}$$

where t is the thickness of the section at B. The stiffeners and the panels between the stiffeners are analyzed for instability by using the design formulas given in Reference 3.

Conical shells.   The equivalent cylinder approach can be used for conical shells. The equivalent radius for cylinder r_e, as shown in Fig. 4.28

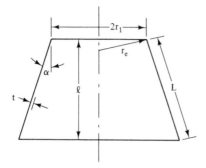

Figure 4.28 Conical shell.

is given by

$$r_e = \frac{r_1}{\cos \alpha} \tag{4.93}$$

$$L = \frac{l}{\cos \alpha} \tag{4.94}$$

Monocoque Shells

1. Axial compression (Ref. 10). The critical buckling stress for a conical shell under compression is given by

$$\sigma_c = \gamma \frac{E}{\sqrt{3(1 - \nu^2)}} \frac{t}{r_e} \tag{4.95}$$

The theoretical value of γ is unity. However, based on experiments, the recommended value of γ (Ref. 10) is

$$\gamma = 0.33 \ (10° < \alpha < 75°)$$

This gives a lower bound to the experimental data. For $\alpha < 10°$, the formulas for the cylinder shells can be used with the equivalent radius and length of the cylinder.

By substituting Eq. (4.93) into Eq. (4.95) and using $\nu = 0.3$, we get the equation for the critical buckling stress at the smaller end of the cone as

$$\sigma_c = 0.19972 \frac{Et \cos \alpha}{r_1} \tag{4.96}$$

The critical buckling axial load P_{cr} is given by

$$P_{cr} = 2\pi r_1 t \sigma_c \cos \alpha \tag{4.97}$$
$$= 0.3994\pi Et^2 \cos^2\alpha$$

2. Bending. The critical stress for buckling is

$$\sigma_b = \gamma \frac{E}{\sqrt{3(1 - \mu^2)}} \frac{t}{r_e} \tag{4.98}$$

The theoretical value of γ is unity. However, the recommended value (Ref. 10) is

$$\gamma = 0.41 \ (10° < \alpha < 60°) \tag{4.99}$$

The critical bending moment is

$$M_{cr} = \pi r_1^2 \sigma_b t \cos \alpha \tag{4.100}$$

Substituting Eqs. (4.93), (4.98), and (4.99) into Eq. (4.100), we get the critical bending moment at the short end of the cone as

$$M_{cr} = 0.24814 E \pi r_1 t^2 \cos^2 \alpha \tag{4.101}$$

3. *Torsion.* The critical torque for a conical shell is

$$T_c = 52.8 \gamma D \left(\frac{t}{l} \right)^{1/2} \left(\frac{r}{t} \right)^{5/4} \tag{4.102}$$

where $r = r_2 \cos \alpha \left\{ 1 + \left[\frac{1}{2} \left(1 + \frac{r_2}{r_1} \right) \right]^{1/2} - \left[\frac{1}{2} \left(1 + \frac{r_2}{r_1} \right) \right]^{-1/2} \right\} \frac{r_1}{r_2}$

$$D = \frac{E t^3}{12(1 - v^2)}$$

The theoretical value of γ is unity. For design purposes it is recommended that γ be assumed as

$$\gamma = 0.67 \tag{4.103}$$

4. *Transverse shear.* The critical transverse shear force, similar to the cylindrical shell, Eq. (4.84), is given by

$$V_{cr} = C_S \frac{\pi^3 E r t}{12(1 - v^2)} \left(\frac{t}{L} \right)^2 \tag{4.104}$$

For combined load conditions, the margin of safety can be determined by Eq. (4.85). Using the approach similar to the cylindrical shell, the sandwich shell is analyzed for local buckling and general instability. The equivalent monocoque shell approach can also be used.

The stiffened conical shells can be analyzed in a manner similar to that for the stiffened cylindrical shells. The compressive stress is given by

$$\sigma_c = \left(\frac{P}{A} + \frac{M}{I} z \right) \frac{1}{\cos \alpha} \tag{4.105}$$

The terms in Eq. (4.105) are defined in Eq. (4.90).

4.8 DYNAMIC ANALYSIS

For a spacecraft, dynamic launch loads are introduced through acceleration at the interface with the launch vehicle and direct acoustic noise through the shroud. Due to dynamic interaction between the spacecraft and the launch vehicle, the launch loads are functions of spacecraft dynamic characteristics, such as natural frequencies and mode shapes. Hence determination of spacecraft dynamic characteristics and performing dynamic analysis to calculate launch loads constitute an important part of structural analysis.

In this section the basics of dynamic analysis of a spacecraft structure are given. The equations for the response of a structure due to sinusoidal, random, and acoustic excitation are given.

Single Degree of Freedom

A mechanical system is said to have a single degree of freedom if its geometrical position at any instant can be expressed by one coordinate. The mechanical system shown in Fig. 4.29 consists of a mass, m, attached by means of spring, K, and a dashpot, C, to an immovable support. The mass is constrained to translational motion in the x direction. The position of the mass at any instant is fully described by a single parameter, x, which is the displacement of the mass from the position of static equilibrium. Figure 4.29(b) represents the free-body diagram of the mass, m. The equation of motion of the mass, m, is $\quad$ *for free vibration*

$$F(t) - m\ddot{x} - c\dot{x} - kx = 0 \tag{4.106}$$

which can be rearranged in the form

$$m\ddot{x} + c\dot{x} + kx = F(t) \tag{4.107}$$

where $\quad m$ = mass
$\quad\quad c$ = viscous damping constant
$\quad\quad k$ = spring stiffness
$\quad\quad F(t)$ = external force

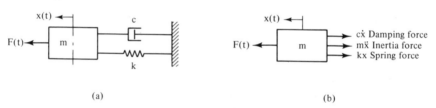

(a) (b)

Figure 4.29 Single degree of freedom.

Free Vibration

The vibration of a mechanical system in the absence of an external force is called free vibration. Assuming the external force, $F(t)$, and damping constant, c, to be zero, the equation of the motion of the undamped system becomes

$$m\ddot{x} + kx = 0 \tag{4.108}$$

The solution of Eq. (4.108) is

$$x = A \sin(\omega_n t + \theta) \tag{4.109}$$

where A, θ = constants and dependent on the initial conditions
$\quad\quad \omega_n = \sqrt{k/m}$, natural frequency, rad/s

Hence if the mass, m, is displaced from its equilibrium position and released, in the absence of damping it will vibrate with constant amplitude A and ω_n frequency. The frequency, ω_n, is called the natural frequency of the system.

In the presence of damping, $c \neq 0$, the equation of motion is

$$m\ddot{x} + c\dot{x} + kx = 0 \tag{4.110}$$

The general solution of Eq. (4.110) is

$$x = A_1 e^{s_1 t} + A_2 e^{s_2 t} \qquad (4.111)$$

where

$$s_{1,2} = -\frac{c}{2m} \pm \sqrt{\left(\frac{c}{2m}\right)^2 - \frac{k}{m}} \qquad (4.112)$$

In discussing the physical significance of Eq. (4.112), three cases have to be distinguished, depending on whether the expression within the square root is zero, positive, or negative. The damping at which the expression within the square root of zero is called critical damping c_c:

$$c_c = 2\sqrt{mk} = 2m\omega_n \qquad (4.113)$$

The ratio $\xi = c/c_c$ is defined as the fraction of critical damping and is

$$\xi = \frac{c}{2\sqrt{mk}} = \frac{c}{2m\omega_n} \qquad (4.114)$$

Case 1 $\xi > 1$. For this case,

$$s_{1,2} = -\xi\omega_n \pm \omega_n\sqrt{\xi^2 - 1} \qquad (4.115)$$

The solution of Eq. (4.110) becomes

$$x = e^{-\xi\omega_n t}(A_1 e^{\omega_n\sqrt{\xi^2 - 1}t} + A_2 e^{-\omega_n\sqrt{\xi^2 - 1}t}) \qquad (4.116)$$

Thus free vibration of a system with damping greater than critical is a combination of two motions which decrease exponentially.

Case 2 $\xi = 1$, *critically damped.* For this case

$$s_{1,2} = -\omega_n \qquad (4.117)$$

Since this is a case of repeated roots, the solution for this case will be

$$x = (A_1 + A_2 t)e^{-\omega_n t} \qquad (4.118)$$

Case 3 $\xi < 1$

$$s_{1,2} = -\xi\omega_n \pm j\omega_n\sqrt{1 - \xi^2} \qquad (4.119)$$

The solution of Eq. (4.110) becomes

$$x = Ae^{-\xi\omega_n t}\sin(\omega_d t + \theta) \qquad (4.120)$$

where $\omega_d = \omega_n\sqrt{1 - \xi^2}$, damped frequency.

The solution is a sinusoidal motion with exponentially decreasing amplitude. The time constant, ΔT, is defined as the time required to decrease the amplitude by a factor e. So

$$e = \frac{e^{-\xi\omega_n t}}{e^{-\xi\omega_n(t + \Delta T)}} = e^{\xi\omega_n \Delta T}$$

or $\qquad (4.121)$

$$\Delta T = \frac{1}{\xi\omega_n} \qquad reduce \ Amplitude \ by \ e$$

$$P_2 = \frac{P_1}{e}$$

Forced Vibration

The equation of motion of the system with sinusoidal external force is

$$m\ddot{x} + c\dot{x} + kx = F_0 \sin \omega t \tag{4.122}$$

Using Eq. (4.114), Eq. (4.122) can be written as

$$\ddot{x} + 2\xi\omega_n\dot{x} + \omega_n^2 x = \frac{F_0}{m} \sin \omega t \tag{4.123}$$

The solution of Eq. (4.123) for $\xi < 1$ will be a combination of free vibration given by Eq. (4.120) and a forced vibration at the forcing frequency, ω. The free vibration will damp out. Therefore, the steady-state solution is

$$x = a \sin (\omega t - \theta) \tag{4.124}$$

where

$$a = \frac{a_s}{\sqrt{(1 - \omega^2/\omega_n^2)^2 + [2\xi(\omega/\omega_n)]^2}}$$

$$\tan \theta = \frac{2\xi(\omega/\omega_n)}{1 - (\omega/\omega_n)^2} \tag{4.125}$$

$$a_s = \frac{F_0}{k}$$

The amplitude ratio, a_0/a_s, and phase angle, θ, are given as functions of damping ratio and frequency ratio in Fig. 4.30. From Fig. 4.30 it can be seen that the response of the system is maximum when the forcing frequency, ω, is close to ω_n (i.e., $\omega/\omega_n \cong 1$). The system under this condition is said to be at resonance.

Base Excitation

Let us consider the second type of forced vibration, where there is no external force on the mass but the support is moving as shown in Fig. 4.31. The equation of motion of the mass is

$$m\ddot{x} + c(\dot{x} - \dot{u}) + k(x - u) = 0 \tag{4.126}$$

Let

$$y = x - u \tag{4.127}$$

Substituting Eq. (4.127) into Eq. (4.126), we get

$$m\ddot{y} + c\dot{y} + ky = -m\ddot{u} \tag{4.128}$$

Equation (4.128) is similar to Eq. (4.122) with the external force equivalent to $-m\ddot{u}$; that is,

$$F(t) = -m\ddot{u} \tag{4.129}$$

Assuming that the base excitation is sinusoidal, that is,

$$u = u_0 \sin \omega t \qquad \ddot{u} = -u_0\omega^2 \sin \omega t \tag{4.130}$$

Spacecraft Structures   Chap. 4

u is the displacement of the base

Substituting Eq. (4.130) into Eq. (4.128) and rearranging it, we get

$$\ddot{y} + 2\xi\omega_n\dot{y} + \omega_n^2 y = \omega^2 u_0 \sin \omega t \qquad (4.131)$$

The solution of Eq. (4.131) will be similar to Eq. (4.124) as

$$y = a \sin (\omega t - \theta) \qquad (4.132)$$

where

$$a = \frac{u_0\omega^2/\omega_n^2}{\sqrt{(1 - \omega^2/\omega_n^2)^2 + [2\xi(\omega/\omega_n)]^2}} \qquad (4.133)$$

$$\tan \theta = \frac{2\xi(\omega/\omega_n)}{1 - (\omega/\omega_n)^2} \qquad (4.134)$$

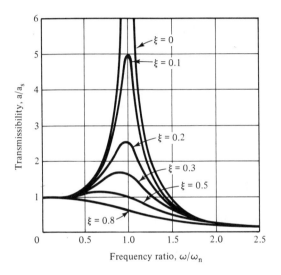

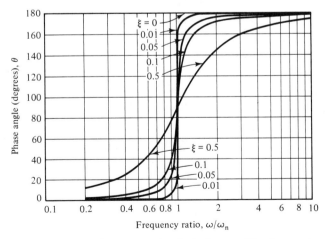

Figure 4.30 Forced response of a single-degree-of-freedom system.

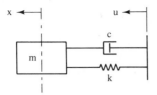

Figure 4.31 Single-degree-of-freedom system under base excitation.

4.9 MULTI-DEGREE-OF-FREEDOM SYSTEM

In the last example, the system consisted of a concentrated mass and a massless spring. A real spacecraft structure is a much more complex system. It will require infinite coordinates and degrees of freedom to describe its motion. For the structural analysis, a spacecraft structure is represented by an idealized structural model with finite degrees of freedom. The structural model consists of an assemblage of basic structural elements, such as bars, beams, and plates, interconnected only at a finite number of nodal points. Hence we determine deflections due to static loads and responses due to dynamic loads only at the nodal points. A method to determine the stiffness matrix of the structural model is described in Section 4.6. To determine the mass matrix, $[M]$, first the mass matrix for each structural element is determined. These mass matrices of the elements are assembled in a similar manner as the stiffness matrices. There are two commonly used methods to determine mass matrices.

Lumped-Mass Representation

In this method, concentrated masses are placed at the nodal points in the direction of the assumed element degree of freedom. These masses refer to the translational and rotational inertia of the element. The resulting element mass matrix is purely diagonal.

Consistent Mass Matrix

The Archer consistent mass matrix technique is commonly used to formulate a coupled mass matrix. The consistent mass matrix for a structural element is obtained from kinetic energy under the assumption that the inertia loading does not change the displacements at the interior points. For a plate element, the kinetic energy for sinusoidal transverse motion at ω frequency is

$$T = \frac{1}{2} \omega^2 \int mu^2 \, dA \qquad (4.135)$$

where m is the mass per unit area. The translational displacement function, u, is related to nodal displacement, u_k, by a finite element analysis as given by Eq. (4.59), as

$$u = \sum_k c_k u_k \qquad (4.136)$$

c - damping

So the kinetic energy expression becomes

$$T = \frac{1}{2} \omega^2 \int m \left[\sum_k \sum_l c_k c_l u_k u_l \right] dA$$

$$= \frac{1}{2} \omega^2 \{u\}^T [M]\{u\} \qquad (4.137)$$

where $[M]$ is mass matrix and the elements are given by

$$M_{kl} = \int m c_k c_l \, dA \qquad (4.138)$$

Basic Equations

The equations of the motion of an n-degree-of-freedom linear system under an external force can be written as follows:

$$[M]\{\ddot{x}\} + [C]\{\dot{x}\} + [K]\{x\} = \{F(t)\} \qquad (4.139)$$

equation

where $[M] = n \times n$ symmetric mass matrix
$[C] = n \times n$ symmetric damping matrix
$[K] = n \times n$ symmetric stiffness matrix
$\{F(t)\} = $ forcing function
$\{x\} = $ coordinate vector

Let

eigenvalue

$$\{x\} = [M]^{-1/2}\{X\} \qquad (4.140)$$

This transformation will result in the eigenvalue solution of a symmetric matrix instead of an asymmetric matrix and hence a significant reduction in the computation time. Substituting Eq. (4.140) into Eq. (4.139), and premultiplying the result by $[M]^{-1/2}$, we obtain

$$[I]\{\ddot{X}\} + [\bar{C}]\{\dot{X}\} + [\bar{K}]\{X\} = [M]^{-1/2}\{F(t)\} \qquad (4.141)$$

where $[I]$ is the identity matrix and

$$[\bar{C}] = [M]^{-1/2}[C][M]^{-1/2} \text{ and } [\bar{K}] = [M]^{-1/2}[K][M]^{-1/2} \qquad (4.142)$$

are symmetric and positive definite matrices, respectively.

Let us consider the free vibration of an undamped multi-degree-of-freedom system, that is, assuming that $[\bar{C}] = \{F(t)\} = 0$, in Eq. (1.141), i.e.

$$[I]\{\ddot{X}\} + [\bar{K}]\{X\} = 0 \qquad (4.143)$$

Let the solution of Eq. (4.143) be

$$\{X\} = \{\phi\}e^{i\omega t} \qquad (4.144)$$

Substituting Eq. (4.144) into Eq. (4.143), we get

$$[\overline{K}]\{\phi\} = \omega^2\{\phi\} \qquad (4.145)$$

Equation (4.145) is an eigenvalue equation with eigenvalue equal to ω^2 and eigenvector equal to $\{\phi\}$. For the symmetric matrix, $[K]$, the eigenvalues, ω_r^2, and corresponding eigenvectors, $\{\phi_r\}$, can be determined by using standard computer programs. Here ω_r corresponds to the rth natural frequency and $\{\phi_r\}$ corresponds to the rth modal vector.

The normal modes have a useful property called orthogonality, such that

$$\{\phi_s\}^T\{\phi_r\} = 0 \qquad r \ne s \qquad (4.146)$$

and

$$\{\phi_s\}^T[\overline{K}]\{\phi_r\} = 0 \qquad r \ne s \qquad 4.147)$$

The modal vectors can be arranged in a square modal matrix of order n as

$$[\phi] = [\{\phi_1\} \cdots \{\phi_n\}] \qquad (4.148)$$

Using orthogonality conditions, from Eqs. (4.146) and (4.147) we obtain

$$[\phi]^T[\phi] = [I] \qquad (4.149)$$

$$[\phi]^T[K][\phi] = \begin{bmatrix} \omega_1^2 & & \\ & \ddots & \\ & & \omega_n^2 \end{bmatrix} \qquad (4.150)$$

Let

$$\{X\} = [\phi]\{q\} \qquad (4.151)$$

Substituting Eq. (4.151) into Eq. (4.143), premultiplying by $[\phi]^T$, and using Eqs. (4.149) and (4.150), we get a set of uncoupled equations of the type

$$\ddot{q}_r(t) + \omega_r^2 q_r(t) = 0 \qquad r = 1, 2, \dots, n \qquad (4.152)$$

The solution of Eq. (4.152) can be written as

$$q_r = A_r \sin(\omega_r t - \theta_r) \qquad r = 1, 2, \dots, n \qquad (4.153)$$

Substituting Eq. (4.153) into Eq. (4.151) and using Eq. (4.140), we get

$$\{x\} = [M]^{-1/2}[\phi]\{q\} = [\psi]\{q\} = \sum_{r=1}^{n} A_r\{\psi_r\} \sin(\omega_r t - \theta_r) \qquad (4.154)$$

where $\{\psi_r\}$ is the mode shape corresponding to the rth mode of the natural frequency, ω_r. Hence the free vibration of a system can be considered as the superimposition of n harmonic motions with frequencies equal to the natural frequencies of the system, and amplitudes and phase angles determined by initial conditions.

Substituting Eq. (4.151) into Eq. (4.141) and premultiplying the result by $[\phi]^T$, we obtain

$$[I]\{\ddot{q}\} + [\overline{\overline{C}}]\{\dot{q}\} + \begin{bmatrix} \omega_1^2 & & \\ & \ddots & \\ & & \omega_n^2 \end{bmatrix}\{q\} = [\phi]^T[M]^{-1/2}\{F(t)\} \qquad (4.155)$$

where

$$[\overline{\overline{C}}] = [\phi]^T[\overline{C}][\phi] \qquad (4.156)$$

Normally, $[\overline{\overline{C}}]$ will not be a diagonal matrix except for classical damping. Neglecting the nondiagonal elements in the damping matrix, Eq. (4.155) becomes

$$\ddot{q}_i + 2\xi_i\omega_i\dot{q}_i + \omega_i^2 q_i = Q_i(t) \qquad (4.157)$$

where ξ_i = fraction of critical damping for the ith mode
     $\{Q\} = [\phi]^T[M]^{-1/2}\{F(t)\}$
     q_i = modal coordinate of the ith mode

The modal coordinate for each mode is obtained by solving Eq. (4.157) and the response coordinate vector (x) is obtained from Eqs. (4.140) and (4.141) as

$$\{x\} = [M]^{-1/2}[\phi]\{q\}$$

and $\qquad\qquad\qquad\qquad\qquad\qquad\qquad\qquad$ (4.158)

$$\{\ddot{x}\} = [M]^{-1/2}[\phi]\{\ddot{q}\}$$

The forcing function $\{F(t)\}$ can be a general time-dependent function. It should be noted that by using the transformation above, an equation of motion of a multi-degree-of-freedom system is reduced to n uncoupled equations similar to a single-degree-of-freedom system. This simplifies the computation of the response significantly. Similar to a single-degree-of-freedom system, if a multiple-degree-of-freedom system is subjected to a base excitation, the equations of motion become

$$[M]\{\ddot{y}\} + [C]\{\dot{y}\} + [K]\{y\} = -[M]\begin{Bmatrix} 1 \\ 1 \\ \vdots \\ 1 \end{Bmatrix}\ddot{u} \qquad (4.159)$$

where $\{u\}$ = base motion
     $\{y\}$ = relative response vector; response relative to the base

Sinusoidal Excitation

Let the excitation force be

$$\{F\} = \{P\} \sin \omega t \qquad (4.160)$$

The modal force vector is

$$\{Q\} = [\phi]^T[M]^{-1/2}\{P\} \sin \omega t = \{\overline{Q}\} \sin \omega t \qquad (4.161)$$

By using Eqs. (4.124) and (4.125), the modal response becomes

$$q_i = a_i \sin (\omega t - \theta_i) \qquad (4.162)$$

where

$$a_i = \frac{\overline{Q}_i}{\sqrt{(\omega_i^2 - \omega^2)^2 + (2\xi_i \omega \omega_i)^2}} \qquad (4.163)$$

$$\theta_i = \tan^{-1} \frac{2\xi(\omega/\omega_i)}{1 - (\omega/\omega_i)^2}$$

The structure response is

$$\{x\} = [M]^{-1/2}[\phi] \begin{Bmatrix} a_1 \sin (\omega t - \theta_1) \\ \vdots \\ a_n \sin (\omega t - \theta_n) \end{Bmatrix}$$

$$= [\psi] \begin{Bmatrix} a_1 \sin (\omega t - \phi_1) \\ \vdots \\ a_n \sin (\omega t - \phi_n) \end{Bmatrix} \qquad (4.164)$$

where $[\psi] = [M]^{-1/2}[\phi]$ is the modal vector matrix of the structure. Equation (4.164) can be written as

$$\{x\} = a_1 \sin (\omega t - \theta_1)\{\psi_1\} + \cdots + a_i \sin (\omega t - \theta_i)\{\psi_i\}$$
$$+ \cdots + a_n \sin (\omega t - \theta_n)\{\psi_n\} \qquad (4.165)$$

In Eq. (4.165), $\{\psi_i\}$ is the mode shape corresponding to the ith mode, ω_i is its natural frequency, a_i is the amplitude of the ith mode, and its magnitude is dependent on the frequency ratio, ω/ω_i, and the damping ratio, ξ_i. For example, for $\omega/\omega_i \approx 1$, a_i, the amplitude of the ith modal response, will be significantly higher than the other modal response amplitudes and the response of the structure will have essentially the ith mode shape.

The determination of natural frequencies and natural mode shapes is an important part of dynamic analysis. The natural frequencies of a complex spacecraft structure are determined by the digital computer programs. However, for preliminary design, it is desired to approximate the structure by a simple mechanical system to determine its natural frequencies. The expressions generally used for the natural frequencies of simple mechanical systems are given in Tables 4.7, 4.8, and 4.9.

4.10 RANDOM EXCITATION

There are many physical phenomena which produce excitation that are not deterministic. It is not possible to predict an exact value at a future instant of time. As an example, Fig. 4.32 shows the response of a launch vehicle due to engine noise at lift-off for different launches. These excitations are random in character and must be described in terms of statistical averages rather than by explicit equations.

The mean value of the random response at some time t_1 can be computed

TABLE 4.7 FUNDAMENTAL NATURAL FREQUENCIES OF SPRINGS AND BEAMS WITH CONCENTRATED MASS.

System:	Natural frequency:
Mass-helical spring:	$\omega_n = \sqrt{\dfrac{K}{(m + m_s/3)}}$
Fixed-fixed center load: $\|{\leftarrow}\ell/2{\rightarrow}\|{\leftarrow}\ell/2{\rightarrow}\|$	$\omega_n = 14\sqrt{\dfrac{3EI}{\ell^3(m + 0.375m_b)}}$
Simply supported center load: $\|{\leftarrow}\ell/2{\rightarrow}\|{\leftarrow}\ell/2{\rightarrow}\|$	$\omega_n = \sqrt{\dfrac{48EI}{\ell^3(m + 0.5m_b)}}$
Cantilever with end load: $\|{\leftarrow}\ \ell\ {\rightarrow}\|$	$\omega_n = \sqrt{\dfrac{3EI}{\ell^3(m + 0.23m_b)}}$

where:

ω_n = natural frequency, rad/s
K = stiffness of the spring, N/m
m_s = mass of the spring, kg
m_b = mass of the beam, kg
m = mass of the load, kg
E = modulus of elasticity, N/m^2
I = area moment of inertia of beam cross-section, m^4

(margin notes)
$S = db/oct$

$\beta^2 = \dfrac{S\,(\#oct)}{\sqrt{2}}$

$\#oct = \dfrac{\log\left(\dfrac{f\,upper}{f\,lower}\right)}{\log 2}$

$P^2 = 10^{\left(\frac{P_{db}^2}{10}\right)}$

$P^2 = (B-A)\,amp$

$g_{rms} = \sqrt{P_1^2 + P_2^2 + \cdots}$

by taking the instantaneous value of response at time t_1 for all the launches, summing the values, and dividing by the number of launches.

$$\mu_x(t_1) = \lim_{n \to \infty} \frac{1}{n} \sum_{k=1}^{n} x_k(t_1) \qquad (4.166)$$

The random vibration is called stationary if the mean is independent of time t_1 or $\mu_x(t_1) = \mu_x(t_1 + t)$ for all values of t. Hence we see that the statistical properties of a stationary random process can be determined by computing them at a specific instant of time.

Next consider whether it is possible to determine the statistical prop-

TABLE 4.8 NATURAL FREQUENCIES OF BEAMS OF UNIFORM SECTION AND UNIFORM DISTRIBUTED MASS. (REF. 25)

The natural frequency ω_n is given by

$$\omega_n = K \sqrt{\frac{EI}{\mu \ell^4}}$$

where ω_n = natural frequency, rad/s
  E = modulus of elasticity, N/m^2
  I = area moment of inertia of beam cross section, m^4
  ℓ = length of beam, m
  μ = mass per unit length of beam, kg/m
  K = coefficient from the table below

Nodes are shown as a fraction of length ℓ from left end.

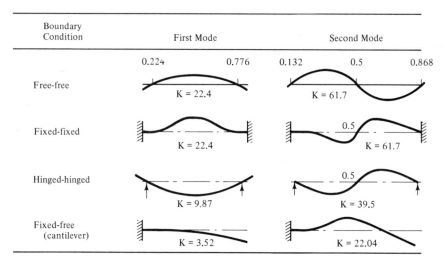

Boundary Condition	First Mode	Second Mode
Free-free	0.224 0.776 K = 22.4	0.132 0.5 0.868 K = 61.7
Fixed-fixed	K = 22.4	0.5 K = 61.7
Hinged-hinged	K = 9.87	0.5 K = 39.5
Fixed-free (cantilever)	K = 3.52	K = 22.04

erties—mean, mean square, and so on—by computing these properties for a single sample or a single launch in the example. The mean value for k launch will be

$$\mu_x(k) = \lim_{T \to \infty} \frac{1}{T} \int_0^T x_k(t)\, dt \tag{4.167}$$

If the random process is stationary and $\mu_x(k)$ does not differ when computed over different sample functions, the random process is called ergodic. An ergodic random process is an important class of random processes since all properties can be determined by performing time averages over a single sample function. Random excitations are normally assumed to be ergodic. *if you know one, can define rest of ergodic*

Basic Properties of Random Data

For random data, the basic properties are: mean-square values, probability density functions, autocorrelation functions, and power spectral density functions.

TABLE 4.9 NATURAL FREQUENCIES OF RECTANGULAR PLATES. (REF. 25)

b/a	1.0	1.5	2.0	2.5	3.0	∞
$\omega_n/\sqrt{D/\gamma a^4}$	19.74	14.26	12.34	11.45	10.97	9.87

b/a	1.0	1.5	2.0	2.5	3.0	∞
$\omega_n/\sqrt{D/\gamma a^4}$	23.65	18.90	17.33	16.63	16.26	15.43
a/b	1.0	1.5	2.0	2.5	3.0	∞
$\omega_n/\sqrt{D/\gamma b^4}$	23.65	15.57	12.92	11.75	11.14	9.87

b/a	1.0	1.5	2.0	2.5	3.0	∞
$\omega_n/\sqrt{D/\gamma a^4}$	28.95	25.05	23.82	23.27	22.99	22.37
a/b	1.0	1.5	2.0	2.5	3.0	∞
$\omega_n/\sqrt{D/\gamma b^4}$	28.95	17.37	13.69	12.13	11.36	9.87

b/a	1.0	1.5	2.0	2.5	3.0	∞
$\omega_n/\sqrt{D/\gamma a^4}$	35.98	27.00	24.57	23.77	23.19	22.37

where ω_n = natural frequency, rad/s
D = plate stiffness = $Eh^3/12(1 - \nu^2)$, N · m
h = plate thickness, m
a = plate length, m
b = plate width, m
γ = mass density per unit area, kg/m²
c = denotes clamped or built-in edge
s = denotes simply supported edge

Mean-square values. The mean-square value of $x(t)$ is defined as

$$\psi_x = \lim_{T\to\infty} \frac{1}{T} \int_0^T x^2(t)\, dt \tag{4.168}$$

RMS $= \sqrt{\psi_x}$ positive

The positive square root of the mean square is called the root mean square (rms). The variance σ_x^2 is simply the mean square value about the mean μ_x as follows:

$$\sigma_x^2 = \lim_{T\to\infty} \frac{1}{T} \int_0^T [x(t) - \mu_x]^2\, dt \tag{4.169}$$

where σ_x is called the standard deviation. *more accurate with high # of samples.*

Sec. 4.10 Random Excitation

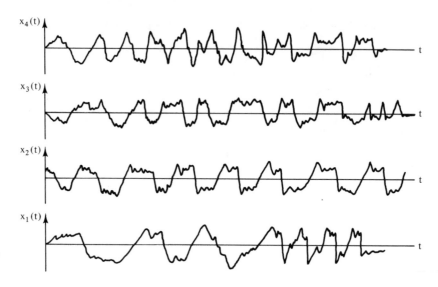

Figure 4.32 Random vibration.

Probability density function. The probability that $x(t)$ assumes a value within the range between x and $(x + \Delta x)$ from Fig. 4.33 may be obtained as

$$\text{Prob}\,[x < x(t) \leqslant x + \Delta x] = \lim_{T \to \infty} \frac{T_x}{T} \qquad (4.170)$$

where

$$T_x = \sum_{i=1}^{k} \Delta t_i$$

The probability density function $p(x)$ is defined as

$$p(x) = \lim_{\Delta x \to 0} \frac{\text{Prob}\,[x < x(t) < x + \Delta x]}{\Delta x} \qquad (4.171)$$

$$= \lim_{\Delta x \to 0} \lim_{T \to \infty} \frac{1}{T} \left(\frac{T_x}{\Delta x} \right)$$

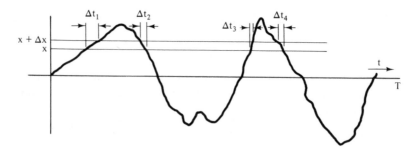

Figure 4.33 Random function.

The probability that $x(t)$ lies between x_1 and x_2 is given as

$$P = \int_{x_1}^{x_2} p(x)\, dx \qquad (4.172)$$

In terms of probability density function $p(x)$, the mean value μ_x and mean-square value ψ_x^2 of $x(t)$ are given by

$$\mu_x = \int_{-\infty}^{\infty} x p(x)\, dx \qquad (4.173)$$

and

$$\psi_x = \int_{-\infty}^{\infty} x^2 p(x)\, dx \qquad (4.174)$$

For a classical Gaussian form, also known as a normal distribution, the probability distribution is given by

$$p(x) = \frac{1}{\sigma_x \sqrt{2\pi}}\, e^{-1/2 \frac{(x - \mu_x)^2}{\sigma_x^2}} \qquad (4.175)$$

Let us consider the sum

$$x = \sum_{i=1}^{n} x_i \qquad (4.176)$$

where x_i are statistically random variables with corresponding probability densities $p_i(x)$. The central limit theorem states that under very minor restrictions which can always be assumed to be satisfied in practice, the distribution of the sum of random variable x_i approaches the normal distribution function as $n \to \infty$ regardless of the shape of densities, $p_i(x)$.

Autocorrelation function. The autocorrelation function for random data describes the general dependence on the values of the data at one time on the values at another time. The autocorrelation function $R_x(\tau)$ is defined as

$$R_x(\tau) = \lim_{T \to \infty} \frac{1}{T} \int_0^T x(t) x(t + \tau)\, dt \qquad (4.177)$$

The mean-square value is equal to the autocorrelation at zero time displacement as

$$\psi_x = R(0) \qquad (4.178)$$

Power spectral density function. The power spectral density function for a random data describes the spectral density of its mean-square value. The mean-square value of a random data in frequency range between f and $f + \Delta f$ may be obtained by filtering the sample record with a bandpass filter of sharp cutoff characteristics, and computing the average of the squared output from the filter.

$$\psi_x[f, f + \Delta f] = \lim_{T \to \infty} \frac{1}{T} \int_0^T x^2(t, f, f + \Delta f)\, dt \qquad (4.179)$$

The power spectral density $S_x(f)$ is defined as

$$S_x(f) = \lim_{\Delta f \to 0} \frac{\psi_x[f, f + \Delta f]}{\Delta f} \quad\quad = \frac{\text{Mean Square Value}}{\text{freq Range}} = \frac{g^2}{Hz}$$

$$\tag{4.180}$$

$$= \lim_{\Delta f \to 0} \lim_{T \to \infty} \frac{1}{(\Delta f)T} \int_0^T x^2(t, f, f + \Delta f)\, dt$$

The mean-square value of $x(t)$ is given by

$$\psi_x = \int_0^\infty S_x(f)\, df \tag{4.181}$$

$$\psi_{xRMS} = \sqrt{\psi_x}$$

Relationship Between Autocorrelation and Power Spectral Density

For stationary data, it can be shown by using the theory of the Fourier integral that the power spectral density functions are Fourier transforms of the autocorrelation functions. Therefore,

$$S_x(f) = \int_{-\infty}^{\infty} R_x(\tau)e^{-i2\pi f\tau}\, d\tau = 2\int_0^\infty R_x(\tau)e^{-i2\pi f\tau}\, d\tau \tag{4.182}$$

Transfer Function

The transfer function theorem states that if for a linear system, the response, x, due to excitation, $F(t)$, is related in such a way that

$$X(f) = H(f)F(f) \tag{4.183}$$

where $H(f)$ is the Fourier transform of the transfer function of the linear system and $X(f)$ and $F(f)$ are Fourier transforms of $x(t)$ and $F(t)$, respectively, then the power spectral density of the response, $S_x(f)$, is related to the power spectral density of the excitation, $S_F(f)$, by the relation

$$S_x(f) = |H(f)|^2 S_F(f) \tag{4.184}$$

Another important result is that if the excitation is a combination of excitations $F_1, F_2, ..., F_n$ which are statistically independent, that is, if the cross-correlation function between any pair of excitations

$$R_{kr}(\tau) = \lim_{T \to \infty} \frac{1}{T} \int_0^T F_k(t)F_r(t - \tau)\, dt \tag{4.185}$$

is zero, the spectral density of the total response is equal to the sum of power spectral densities of the responses due to individual sources. Thus

$$S_x(f) = \sum_{r=1}^n |H_r(f)|^2 S_r(f) \tag{4.186}$$

where S_r is the power spectral density of F_r.

If the sources are statistically correlated, the degree of correlation can be expressed by a cross-spectral density, S_{kr} [Fourier transform of cross-correlation $R_{kr}(\tau)$] and the spectral density of the response may be evaluated from

$$S_x(f) = \sum_{k=1}^{n} \sum_{r=1}^{n} H_k H_r^* S_{kr}(f) \qquad (4.187)$$

where H_r^* is the complex conjugate of H_r.

Response Due to Random Excitation

The modal equations for a multi-degree-of-freedom system, Eq. (4.157), are rewritten as

$$\ddot{q}_i + 2\xi_i \omega_i \dot{q}_i + \omega_i^2 q_i = F_i(t) \qquad (4.188)$$

The response coordinate vector $\{x\}$ is obtained from the modal coordinate vector $\{q\}$ by using Eq. (4.158) as follows:

$$\{x\} = [M]^{-1/2}[\phi]\{q\}$$
$$= [u]\{q\} \qquad (4.189)$$

where

$$[u] = [M]^{-1/2}[\phi]$$

Let

$$Q_i(f) = \int_{-\infty}^{\infty} q_i(t)e^{-i2\pi ft}\, dt \qquad (4.190)$$

$$F_i(f) = \int_{-\infty}^{\infty} F_i(t)e^{-i2\pi ft}\, dt \qquad (4.191)$$

be the Fourier transforms of $q_i(t)$ and $F(t)$, respectively. By taking the Fourier transform of the both sides of Eq. (4.188), we obtain

$$Q_i(f) = H_i(f)F_i(f) \qquad (4.192)$$

where

$$H_i(f) = \frac{1}{f_i^2 - f^2 + 2i\xi_i f_i f} \cdot \frac{1}{4\pi^2} \qquad (4.193)$$

is the complex frequency response of the ith mode.

From Eq. (4.189), the response of the structure at the rth point, x_r, can be written as

$$x_r = \sum_{i=1}^{n} u_r^i q_i(t) \qquad (4.194)$$

where u_r^i is the element in the matrix $[u]$ in the rth row and ith column. Taking the Fourier transformation of Eq. (4.194), we get

$$X_r(f) = \sum_{i=1}^{n} u_r^i Q_i(f)$$
$$= \sum_{i=1}^{n} u_r^i H_i(f)F_i(f) \qquad (4.195)$$

Using Eq. (4.195), the power spectral density of the response x_r, S_{xr}, will be given as

$$S_{xr}(f) = \sum_{i=1}^{n} \sum_{j=1}^{n} u_r^i u_r^j H_i(f) H_j^*(f) S_{Fij}(f) \qquad (4.196)$$

where S_{Fij} is the cross-spectral density of the excitation $F_i(t)$ and $F_j(t)$. The mean-square value of the response x_r, ψ_{xr}, is given as

$$\psi_{xr} = \int_0^\infty S_{xr}(f) \, df \qquad (4.197)$$

$$= \sum_{i=1}^{n} \sum_{j=1}^{n} u_r^i u_r^j \int_0^\infty H_i(f) H_j^*(f) S_{Fij}(f) \, df \qquad (4.198)$$

If it is assumed that

$$F_i = P_i F(t) \qquad (4.199)$$

which will be the case when a spacecraft is subjected to base excitation, then Eq. (4.198) can be written as

$$\bar{x}_r^2 = \psi_{xr} = \sum_{i=1}^{n} \sum_{j=1}^{n} P_i P_j u_r^i u_r^j \int_0^\infty H_i(f) H_j^*(f) S_F(f) \, df \qquad (4.200)$$

where S_F is the power spectral density of $F(t)$. For a Gaussian random excitation, it is sufficient to know $\bar{x}_r^2$ ($r = 1, ..., n$) to determine the probability that the response exceeds a certain value if the excitations' mean values are zero.

Acoustic Excitation

The acoustic excitation is normally given in terms of the sound pressure levels in the octave bands, as given in Table 4.3. The sound pressure level for an octave band is given in terms of decibels as follows:

$$K = 20 \log \frac{P}{P_0} \qquad (4.201)$$

or

$$P = P_0 \, 10^{K/20} \qquad (4.202)$$

where P = sound pressure level
P_0 = reference pressure level ($2 \times 10^{-5} \text{N/m}^2$)
K = sound pressure level, dB

Let P be the sound pressure level for a band with center frequency, f. So P is the rms value of the sound pressure in the frequency range $f/\sqrt{2}$ and $\sqrt{2}f$. If S is the mean power spectral density for this range, we obtain

$$P_{RMS} = \sqrt{S\left(\sqrt{2}f - \frac{f}{\sqrt{2}}\right)} = \sqrt{\frac{Sf}{\sqrt{2}}} \qquad (4.203)$$

In terms of sound pressure level, the power spectral density can be written as

$$S = \frac{\sqrt{2}\,P^2}{f} \tag{4.204}$$

Substituting Eq. (4.202) into Eq. (4.204), we obtain

$$S = \frac{\sqrt{2}\,P_0^2\ 10^{K/10}}{f} \tag{4.205}$$

From Eq. (4.205), the sound pressure levels are converted into sound pressure power spectral density. Knowing the power spectral density, the analysis developed in the preceding section for random excitation can be used for acoustic excitation.

4.11 MODE SYNTHESIS

A complex structure is often designed and fabricated in parts by different organizations. Therefore, it is often difficult to assemble the entire finite element model for dynamic analysis. In addition, many finite element models may contain so many degrees of freedom that they cannot be assembled and analyzed directly on the available computer. For these reasons, it is desirable to analyze the substructures independently and then synthesize their dynamic characteristics for the analysis of the complete structure. A number of substructure synthesis methods known as mode synthesis have been proposed.

The mode synthesis methods can be classified on the basis of the boundary conditions imposed at the interfaces of the substructures when their mode shapes are determined. One class is called the fixed-interface method and the second is called the free-interface method.

Mode synthesis methods are used to synthesize the dynamic characteristics of a booster and the spacecraft for coupled booster/spacecraft dynamic analysis. There are two commonly accepted mode synthesis methods for coupled booster/spacecraft dynamic analysis. In the first method, known as the Craig and Bampton method (Ref. 16), the fixed-interface normal modes and constraint modes of the booster and the spacecraft are used. The second method uses the free-interface normal modes plus the residual attachment modes of the booster and the fixed-interface normal modes and constraint modes of the spacecraft. The constraint modes for a substructure are defined as the mode shapes of its interior points due to successive unit displacements of interface points, all other interface points being totally constrained. The residual attachment modes, also known as residual flexibility modes, as proposed by MacNeal (Ref. 17) and Rubin (Ref. 18), are used to retain the static contribution of the higher-frequency truncated modes. In this section the first method will be discussed.

Let the equation of motion of a substructure be partitioned as

$$\begin{bmatrix} M_{II} & M_{IB} \\ M_{BI} & M_{BB} \end{bmatrix} \begin{Bmatrix} \ddot{X}_I \\ \ddot{X}_B \end{Bmatrix} + \begin{bmatrix} K_{II} & K_{IB} \\ K_{BI} & K_{BB} \end{bmatrix} \begin{Bmatrix} X_I \\ X_B \end{Bmatrix} = 0 \qquad (4.206)$$

where the subscript B refers to the coordinates which are on a boundary to another substructure and I refers to the coordinates which are interior. The constraint modes are given by

$$[\phi_c] = -[K_{II}]^{-1}[K_{IB}] \qquad (4.207)$$

The fixed-interface normal modes of the substructure are defined as

$$[K_{II}]^{-1}[M_{II}][\phi_N] = -\begin{bmatrix} \ddots & & \\ & 1/\omega_N^2 & \\ & & \ddots \end{bmatrix}[\phi_N] = -[\omega_N^2]^{-1}[\phi_N] \qquad (4.208)$$

where $\quad [\phi_N] =$ normal-mode matrix

$\begin{bmatrix} \ddots & \\ & 1/\omega^2 \\ & & \ddots \end{bmatrix} = $ diagonal matrix of eigenvalues or reciprocal of natural frequencies squared

The displacement of the substructure can now be written as

$$\begin{Bmatrix} X_I \\ X_B \end{Bmatrix} = \begin{bmatrix} \phi_N & \phi_c \\ 0 & I \end{bmatrix} \begin{Bmatrix} q_I \\ X_B \end{Bmatrix} = [T] \begin{Bmatrix} q_I \\ X_B \end{Bmatrix} \qquad (4.209)$$

where $\{q\}$ is a normal-mode coordinate vector. Substituting Eq. (4.209) into Eq. (4.206) and premultiplying the resulting equation by $[T]^T$, we get

$$\begin{bmatrix} \overline{M}_{NN} & \overline{M}_{NB} \\ \overline{M}_{BN} & \overline{M}_{BB} \end{bmatrix} \begin{Bmatrix} \ddot{q}_I \\ \ddot{X}_B \end{Bmatrix} + \begin{bmatrix} \omega_N^2 \overline{M}_{NN} & 0 \\ 0 & \overline{K}_{BB} \end{bmatrix} \begin{Bmatrix} q_I \\ X_B \end{Bmatrix} = \begin{Bmatrix} 0 \\ 0 \end{Bmatrix} \qquad (4.210)$$

where $\overline{M}_{NN} = \phi_N^T M_{II} \phi_N$

$\overline{M}_{NB} = \phi_N^T M_{II} \phi_c + \phi_N^T M_{IB}$

$\overline{M}_{BN} = \overline{M}_{NB}^T$

$\overline{M}_{BB} = \phi_c^T M_{II} \phi_c + M_{BI} \phi_c + \phi_c^T M_{IB} + M_{BB}$

$\overline{K}_{BB} = K_{BB} - K_{BI} K_{II}^{-1} K_{IB}$

For each substructure, Eq. (4.210) is generated. The sets of equations for the substructures are assembled in an uncoupled form as follows:

$$\begin{bmatrix} \overline{M}_1 & & & \\ & \overline{M}_2 & & \\ & & \ddots & \\ & & & \overline{M}_N \end{bmatrix} \begin{Bmatrix} \ddot{X}_1 \\ \ddot{X}_2 \\ \vdots \\ \ddot{X}_N \end{Bmatrix} + \begin{bmatrix} \overline{K}_1 & & & \\ & \overline{K}_2 & & \\ & & \ddots & \\ & & & \overline{K}_N \end{bmatrix} \begin{Bmatrix} X_1 \\ X_2 \\ \vdots \\ X_N \end{Bmatrix} = \begin{Bmatrix} 0 \\ 0 \\ \vdots \\ 0 \end{Bmatrix} \qquad (4.211)$$

The vectors $X_1, X_2 \ldots, X_N$ contain the normal modal coordinates of each substructure plus the discrete boundary coordinates. To reduce the order of the coupled equation, only a fraction of the total normal modes of the substructures are included in the coupled equation. Generally, the modes above a certain frequency, known as the cutoff frequency, are neglected. This is called modal truncation.

To couple these substructures, compatibility conditions are needed to ensure that the displacements on the interfaces of the substructures match those of the adjoining substructures. Using the compatibility conditions, the following coordinate transformation is made:

$$\begin{Bmatrix} X_1 \\ X_2 \\ \vdots \\ X_N \end{Bmatrix} = [TC]\{X_s\} \tag{4.212}$$

Substituting Eq. (4.212) into Eq. (4.211) and premultiplying the resulting equation by $[TC]^T$, we get

$$[M_c]\{\ddot{X}_s\} + [K_c]\{X_s\} = 0 \tag{4.213}$$

Equation (4.213) represents the free-vibration equation of the motion of coupled structures. If other substructures remain to be coupled, a similar procedure is used to couple those substructures. Once the mode shapes and responses are calculated in the transformed coordinates, the original coordinates of the substructures are calculated by the transformation

$$\begin{Bmatrix} X_I \\ X_B \end{Bmatrix} = [T][TC]\{X_s\} \tag{4.214}$$

Coupled Booster/Spacecraft Dynamic Analysis

A coupled booster/spacecraft dynamic analysis is performed to determine the dynamic response and loads in the spacecraft. A STS/PAM dynamic load analysis cycle is shown in Fig. 4.34. Two dynamic load

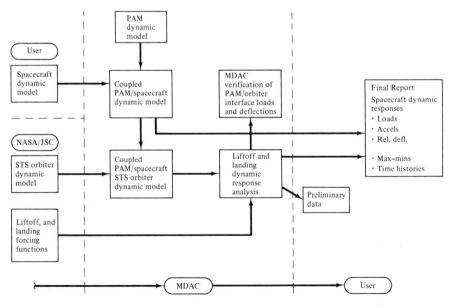

Figure 4.34 STS PAM dynamic loads analysis cycle (Ref. 19).

analysis cycles are performed to calculate the maximum expected dynamic responses of spacecraft structures for the STS orbiter lift-off and abort-landing conditions. The first load cycle is performed to determine the spacecraft launch loads for design evaluation. The second load cycle is performed prior to launch using the updated spacecraft structural model to satisfy all spacecraft, PAM, and STS orbiter design requirements.

Each load cycle involves coupling the structural models of the spacecraft and the PAM with the structural model STS which is supplied by NASA/JSC. Coupled spacecraft/PAM/Orbiter transient response analyses are performed using 16 lift-off and five landing forcing functions, also supplied by NASA/JSC. The time histories of spacecraft dynamic response in the form of loads, accelerations, and relative deflections are determined.

Example 4.2

The structural configuration of a spacecraft is shown in Fig. 4.35. The structural configuration is similar to the Intelsat V structure, as discussed in Section 4.2. The

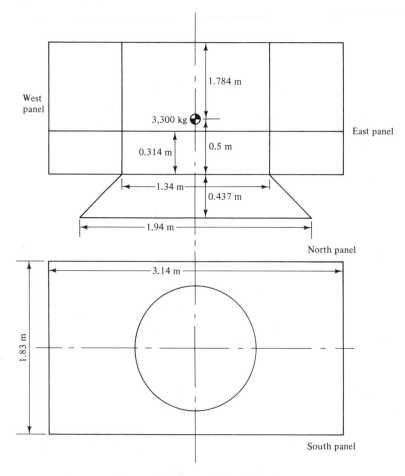

Figure 4.35 Structural configuration.

spacecraft mass is 3300 kg and the launch vehicle is Ariane. Perform a preliminary structural design of the adapter and the north-south panels.

Solution

Adapter. The center tube structure, including the adapter, is designed primarily for ultimate quasi-static loads during the launch. For Ariane, as given in Table 4.5, the ultimate loads are 1.5 times limit loads; for a second stage burnout during flight are 11.85g axial compression and 1.5g lateral.

Assuming the spacecraft mass to be concentrated at the center of mass of the spacecraft, the axial load, P, and bending moment, M, at the top radius of adapter, as shown in Fig. 4.36, are

$$P = 3300 \times 11.85 \times 9.81 = 3.84 \times 10^5 \, N$$

$$M = 3300 \times 1.5 \times 9.81 \times 0.5 = 2.43 \times 10^4 \, N \cdot m$$

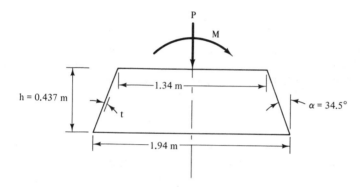

Figure 4.36 Thrust cone loads.

Let us assume the adapter to be a monocoque aluminum conical shell structure. The thickness of the shell is assumed to be $t = 3$ mm. The critical buckling axial load, P_{cr}, from Eq. (4.97) is given by

$$P_{cr} = 0.399\pi E t^2 \cos^2\alpha$$
$$= 0.399\pi \times 7 \times 10^{10} \times (3 \times 10^{-3})^2 \cos^2 34.5°$$
$$= 5.36 \times 10^5 \, N$$

The critical bending moment at the short end of the cone, using Eq. (4.101), is given by

$$M_{cr} = 0.248\pi E r_1 t^2 \cos^2\alpha$$
$$= 0.248\pi \times 7 \times 10^{10} \times 0.67 \times (3 \times 10^{-3})^2 \cos^2 34.5°$$
$$= 2.23 \times 10^5 \, N \cdot m$$

The load ratios, R_c and R_b, are given by

$$R_c = \frac{P}{P_{cr}} = \frac{3.84 \times 10^5}{5.36 \times 10^5} = 0.716$$

and

$$R_b = \frac{M}{M_{cr}} = \frac{2.42 \times 10^4}{2.23 \times 10^5} = 0.1$$

The margin of safety, M.S., is given by

$$\text{M.S.} = \frac{1}{R_c + R_b} - 1 = \frac{1}{0.716 + 0.1} - 1 = 0.22$$

Normally, a margin of safety of 10% is required. Hence the thickness of the shell can be further reduced and the calculations above repeated.

Upper north and south panels. The north and south panels are supported at the middle by the vertical webs. The boundary conditions for the half of the north-south panels are assumed to be clamped at the vertical web joint and simply supported at the other ends, as shown in Fig. 4.37. The panels are assumed to be sandwich panels with an aluminum honeycomb core and aluminum face skins.

The total mass of the panel, shown in Fig. 4.37, including the equipment mass and the assumed panel structural mass, is 77 kg. The mass per unit area,

$$\gamma = \frac{77}{1.97 \times 1.57}$$
$$= 24.9 \text{ kg/m}^2$$

The panels are designed for stiffness to meet design criteria for minimum natural frequency and for stress due to dynamic loads. To avoid coupling with the primary structure, the fundamental natural frequency is assumed to be 30 Hz. The fundamental natural frequency of the panel from Table 4.9 is given by

$$f = \frac{1}{2\pi} \beta \sqrt{\frac{D}{\gamma a^4}}$$

For the panel, $a = 1.57$ m, $b = 1.97$ m, and $b/a = 1.25$. For the given boundary condition and aspect ratio, the value of β from Table 4.9 is

$$\beta = 21.27$$

The panel stiffness, D, is given by

$$D = \frac{Eth^2}{2(1 - \nu^2)} = \frac{7 \times 10^{10}th^2}{2 \times 0.91} = 3.846 \times 10^{10}th^2$$

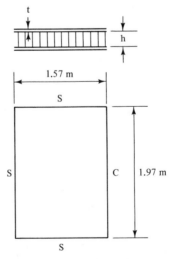

Figure 4.37 Upper north panel.

Substituting these panel parameters in the natural frequency equation

$$30 = \frac{1}{2\pi} \times 21.27 \sqrt{\frac{3.846 \times 10^{10} th^2}{24.9(1.57)^4}}$$

or

$$th^2 = 3.08 \times 10^{-7}$$

Assuming that $h = 0.0254$ m (1 in.),

$$t = \frac{3.08 \times 10^{-7}}{(2.54 \times 10^{-2})^2} = 4.78 \times 10^{-4}$$

$$\cong 5 \times 10^{-4} \text{m} = 0.5 \text{ mm}$$

Stress. Let us assume a uniform dynamic acceleration of 36g across the panel. The maximum stress in the face skin at the center of the panel from Table 4.10 is given by

$$\sigma_{max} = \beta \frac{wa^2}{6th}$$

For the panel, the limit load per unit area, w, is

$$w = \frac{77 \times 36 \times 9.81}{1.57 \times 1.97} = 8.792 \times 10^3 \text{ N/m}^2$$

For $b/a = 1.25$, the parameter β from Table 4.10 is 0.39. Substituting these parameters in the equation of stress gives us

$$\sigma_{max} = \frac{0.39 \times 8792 \times (1.57)^2}{6 \times 5 \times 10^{-4} \times 2.54 \times 10^{-2}}$$

$$= 111 \times 10^6 \text{ N/m}^2 = 111 \text{ N/mm}^2$$

TABLE 4.10

Maximum stress in a uniformly loaded panel is given by

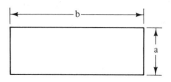

$$\sigma_{max} = \beta \frac{wa^2 c}{6I}$$

where σ_{max} = maximum stress, N/m²
     w = uniform load/unit area, N/m²
     c = distance from the center of the cross-section to the top skin, m
     a = smaller side dimension, m
     I = cross-sectional moment per unit length, m³
     β = coefficient as the function of the panel edge conditions and aspect ratio

b/a	1	1.4	1.6	2	4	5	∞
β	0.2874	0.4530	0.5172	0.6102	0.7410	0.7476	0.75

The allowable yield stress for the aluminum is 240 N/mm². Hence the panel design is acceptable.

4.12 MATERIALS

The materials for spacecraft structures are selected on the basis of several considerations, such as the structural mass, the loads, the dimensional stability, the cost, and the availability of the materials. Aluminum has been most commonly used as a spacecraft material. To meet the stringent requirements of lower thermal distortion, light weight, and high stiffness, advanced composite materials, such as graphite/epoxy, are finding increased applications. This section discusses the mechanical properties of spacecraft materials and the design considerations in the selection of these materials.

The mechanical and thermal properties of spacecraft materials are given in Table 4.11. Besides these properties, there are other properties, such as ductility, brittleness, creep, and fatigue strength.

Ductility/Brittleness

Ductility measures the capacity of a material for inelastic deformation without rupture. Brittleness indicates little capacity for plastic deformation without failure. Ductility is usually measured by the percentage elongation of tensile test specimen after failure for a specified gauge length. Usually, a material having less than 5% elongation at fracture is said to be brittle and one having more is said to be ductile.

Creep

Creep is defined as the time-dependent deformation of a material under an applied load. It is usually regarded as an elevated temperature phenomenon, although some materials also creep at room temperature. The results of tests of materials under a constant load and temperature are usually plotted as strain versus time up to rupture, as shown in Fig. 4.38. The curve exhibits three distinct regions. The first stage includes both elastic and plastic de-

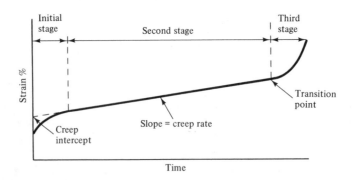

Figure 4.38 Creep diagram.

TABLE 4.11 PROPERTIES OF SPACECRAFT MATERIALS

Material	Material Type	Density, f (kg/m³ × 10³)	Longitudinal Ultimate Tensile Strength (N/m² × 10⁶)	Transverse Ultimate Tensile Strength (N/m² × 10⁶)	Longitudinal Tensile Yield Strength (N/m² × 10⁶)	Young's Modulus (N/m² × 10⁹)	Shear Modulus (N/m² × 10⁹)	Specific Longitudinal Ultimate Strength (N·m/kg × 10³)	Specific Stiffness, E/f (N·m/kg × 10³)	Specific Heat, C (J/kg·K)	Thermal Expansion α (10⁻⁶/K)	Thermal Conductivity K (W/m·K)
Aluminum, sheet	2014-T6	2.80	441	—	386	72	27.6	157.6	25.9	962	22.5	155
	2024-T36	2.77	482	—	413	72	27.6	174.2	26.1	879	22.5	121
	6061-T6	2.71	289	—	241	67	26.2	106.8	24.9	962	23.4	166
	7075-T6	2.80	523	—	448	71	26.9	187.1	25.4	837	28.9	134
Beryllium	Extrusion	1.85	620	—	413	293	138	335.4	158.4	1862	11.5	179
Be-38% Al	Lockalloy	2.10	426	—	431	186		203.2	88.6		17.0	212
Cross rolled	Sheet	1.85	448	—	289	293	138	242.2	158.4	1862	11.5	179
Hot pressed	Wrought	1.83	275	—	179	293	138	150.6	160.1	1862	11.5	179
Boron epoxy |0|		2.01	1337	71		206	4.8	665.4	102.9	920	4.2	1.9
|O2/±45|		2.01	717	107		115		356.7	57.6		4.6	0.4
Graphite/epoxy V_f 55% |0|	HTS	1.49	1337	66		151	5.9	897.6	101.7		−0.36	
|O2/±45|	HTS	1.49	641	289		82		430.3	55.5			
|0|	HM	1.61	675	29		186	5.9	419.6	115.6			
|0|	UHM	1.69	620	20		289	4.1	367.1	171.3		−1.0	
|O2/±45|	UHM											
Invar 36	Annealed	8.08	489		257	144	55	60.6	17.9	514	1.26	13.5
Magnesium Extrusion tubes	AZ31B	1.77	221		110	44	16.5	124.9	25.3	1046	25.2	43.6
Sheet	AZ31B-H24	1.77	269	275	199	44	16.5	152.0	25.3	1046	25.2	43.6
Steel PH15-7 MO	RH1050	7.6	1309		1171	200	75.8	172.3	26.3	477	11.0	15.4
4130 Chr. Mdy	1350°F temp	7.83	861		710	200	75.8	110.0	25.5	477	11.3	38.1
Ti6 Al-4 V	Sheet	4.43	1103		999	110	42.7	249	24.9	502	8.8	7.4
	Forgings and bar	4.43	1034		965	110	42.7	233.4	24.9	502	8.8	7.4
Keviar 49 |0|	50%	1.38	1378	29		75	2.1	999.06	54.9		−4.0	1.7
Boron/Al |0|	Monolayer	2.60	1491	137		214	—	573	82	1000	4	—

245

formation. This stage shows a decreasing creep rate which is due to the strain hardening. The second stage shows a constant minimum creep rate caused by the annealing effect. In the third stage, a considerable reduction in the cross-sectional area occurs, resulting in an increase in stress and creep rate which eventually leads to fracture.

Fatigue

In a tensile test, the load is applied gradually to the failure. Such load condition is known as static condition. A spacecraft is subjected to both static and dynamic loads. In a dynamic load, the stresses are repeated. It has been found that when the stresses are repeated a large number of times, the actual maximum stress is below the ultimate strength of the material and quite frequently even below the yield strength. Such failures are called fatigue failures.

A fatigue failure starts with a small crack. Once a crack has developed, the stress concentration effect becomes greater and the crack progresses more rapidly. As the stress area decreases in size, the stress increases in magnitude until the part fails suddenly. The failure is similar to brittle material fracture.

Fatigue Strength

For ferrous materials, the strength under repeated stresses is often referred to as the endurance limit. Endurance limit stress is the stress that can be repeated an infinite number of times without causing the fracture of the material. Nonferrous materials, such as aluminum alloys, do not have an endurance limit, as they continue to weaken when the stress cycles are repeated. Hence the fatigue strength is the maximum stress that can be repeated for a specified number of cycles without producing the failure of the unit. The results of fatigue tests are often plotted in a form which is referred to as the $S-N$ diagram (stress versus number of cycles), as shown in Fig. 4.39. The most convenient method of fatigue testing is to

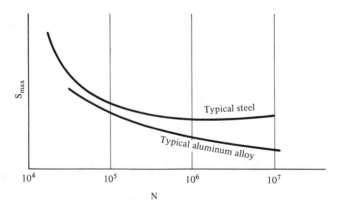

Figure 4.39 S–N curve.

use completely reversed stresses. The common condition, however, is the fluctuation of the stress above some average stress other than zero, as shown in Fig. 4.40.

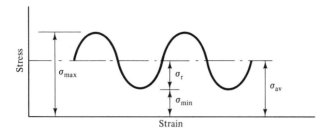

Figure 4.40 Stress cycle.

For a given average stress, σ_{av}, the allowable reversed stress amplitude, σ_r, can be obtained by using the modified Goodman diagram or the more conservative Soderberg diagram, as given in Fig. 4.41. In these diagrams, point C refers to the fatigue strength under reversed stresses (with zero mean) for a specified number of cycles. Point A corresponds to the yield strength and B corresponds to the ultimate strength. If we apply a suitable factor of safety, N, to both the endurance limit and the yield strength, we may draw line DE parallel to the Soderberg line AC as shown in Fig. 4.41. Line DE can be considered as a safe stress line. The allowable mean stress σ_{av} and reversed stress σ_r are related by

$$\frac{1}{N} = \frac{\sigma_{av}}{\sigma_y} + \frac{\sigma_r}{\sigma_E} \tag{4.215}$$

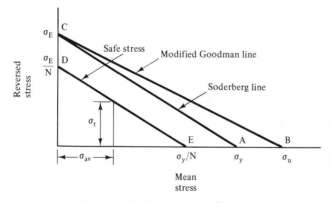

Figure 4.41 Fatigue stress diagram.

Stress Concentration

Stress concentration may be caused by any discontinuity, such as holes, notches, and any abrupt changes in the cross section. A typical example of a stress raiser is shown in Fig. 4.42, where a hole is introduced

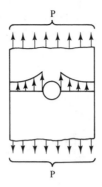

Figure 4.42 Stress concentration.

in a tension member. The value of the maximum stress at the edge of the hole can be considered as a nominal stress multiplied by a stress concentration factor, K.

Under steady loads, the effect of stress concentration is reduced due to the redistribution of the stresses in the region of the stress concentration, resulting from the plastic flow of the material when the maximum stress reaches the yield point. The effect of the stress concentration on the brittle material, under a steady load, may be severe since very little plastic flow occurs. Under repeated loads, however, the endurance strength of even ductile material may be greatly reduced due to stress concentration.

Advanced Materials

To meet the stringent requirements for light weight, high stiffness, and low thermal distortion, advanced materials, such as beryllium, beryllium–lockalloy, and advanced composite materials, are being considered for spacecraft structures. These materials require special design and manufacturing considerations.

Beryllium. Beryllium is a prime candidate for spacecraft structural components in which high stiffness is a prime requirement. Beryllium has low density (1.85×10^3 kg/m^3) and a very high elastic modulus (29.3×10^{10} N/m^2). The specific stiffness of beryllium is over six times greater than that of aluminum alloys. The main concern in the design is its tendency to fracture easily under low-impact loads and small deflection. As expected of a brittle material, beryllium is highly notch sensitive. There is a significant difference in the mechanical properties in the direction of material rolling and transverse to the direction of rolling.

Certain manufacturing precautions are required for beryllium. Beryllium parts can be machined by most conventional techniques such as drilling, cutting, and chemical milling. However, even moderate machining speed produces microcracks on the cut surface. These cracks should be removed by etching to reduce the chance of failure originating from a crack. Mechanical fasteners used on beryllium parts should be limited to those which can be installed without impact loading, or vibration, or without causing interference stresses in the hole. All holes drilled for the joint should be etched to remove any minute surface cracks. Also, beryllium dust particles in the air can cause berylliosis. Therefore, beryllium cannot be cut, sawn, milled, ground, or drilled without special precautions. The complexities of these safety problems have resulted in increased interest in the use of bonded technology. The cost of beryllium material and its manufacturing is significantly higher than that of conventional materials.

Beryllium–Lockalloy. This alloy has been developed for space structures. It is available as sheet, rod, bar, and extrusion and combines the ductile properties of aluminum with the higher strength properties of beryllium. This alloy has high modulus, low density, high formability, and good machining characteristics. It exhibits useful structural properties in the 315 to 425°C service temperature range plus has low notch sensitivity.

Advanced Composite Materials

Advanced composites combine two or more materials to utilize their respective desirable characteristics. They consist of high-strength, high-modulus, low-density filaments embedded in the matrix of essentially homogeneous materials. The filaments most commonly used are boron, boric (silicon carbide–coated boron), and graphite. Boron filaments are generally formed by vapor deposition of boron on a fine tungsten wire. Graphite filaments are made by graphitizing tows or bundles of organic precursor filaments. The matrix materials most commonly used are epoxy in organic matrix resins and aluminum in metals. With composite materials, the structural elements can be formed much closer to the final configuration, thus eliminating a great deal of machining required in conventional materials. Since they are anisotropic materials, composites require a complex analysis and their joint design requires special considerations.

Ultrahigh-modulus graphite/epoxy laminate can surpass beryllium in specific stiffness. Graphite/epoxy is very attractive for those structural elements which require low thermal expansion, such as antenna support structures or antennas. Graphite/epoxy is more versatile than boron/epoxy. It can be contoured into more intricate shapes without filament breakage and can be drilled with existing cutting tools. The major problems encountered in machining boron/epoxy composite result from the extreme hardness of boron filaments and the lack of support that epoxy-resin matrix provides for the filaments. Boron/epoxy composites require diamond or carbide cutters. Graphite/epoxy and boron/epoxy can be joined only by bonding and/or mechanical fasteners.

Boron/aluminum, a metal matrix composite, has been used successfully for structural elements such as thrust cones and tubes. Boron/aluminum has high postbuckling strength compared to graphite/epoxy or boron/epoxy because the epoxy matrix has a lower strain and less strength than aluminum.

Design Considerations

The primary structural components of a spacecraft satellite are the thrust cones, the struts and tubes, the equipment mounting panels, and the rings.

Thrust cones. The thrust cone generally forms the center structure of a spacecraft. It is designed primarily on the basis of its axial compressive loads and bending moments. The primary failure mode involves buckling of the shells. High-modulus materials, such as beryllium and advanced composite materials, provide a lightweight structure. Thrust cones are generally a shell configuration of monocoque, semi-monocoque incorporating longitudinal stiffeners, or sandwich design. The semi-monocoque and sandwich designs result in a structure lighter in weight although more complex in fabrication.

Struts and tubes. The struts and tubes are generally designed for buckling. The mass can be reduced by using beryllium, graphite/epoxy, boron/aluminum, and boron/epoxy. Materials with unidirectional strength/mass ratios are ideal for these components.

Panels. The panels support the subsystem components. They are designed primarily for the minimum natural frequency requirements. A standard design approach is to use a sandwich construction, consisting of an aluminum honeycomb core with bonded aluminum face sheets. Alternative materials for face sheets are magnesium and beryllium. However, the use of beryllium limits component location change. Hence this material should be preferably used on the sandwich panel side where no components are mounted. Graphite/epoxy and boron/epoxy lack the thermal and electrical conductivity of aluminum, although metallic inserts can provide conductivity paths.

Rings. The design criteria for the rings are based on strength and thermal considerations. If the skins of the thrust cones are aluminum, magnesium, or Lockalloy, the same material is generally used for the rings. Beryllium is not a good material for rings due to its brittleness and high notch sensitivity. Graphite/epoxy has poor interlaminar shear capacity. So the high local loads due to fittings can overload normally laid-up graphite/epoxy structures. For thrust cones with beryllium and graphite/epoxy face skins, titanium has been preferred for rings due to its thermal compatibility and high strength.

4.13 STRUCTURAL DESIGN VERIFICATION TESTS

A spacecraft structure is subjected to extensive testing to ensure that it survives the launch loads. A typical verification test plan is given in Fig. 4.43. At first, the primary structure, consisting of the central tube and the

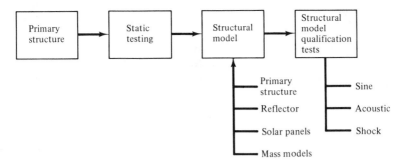

Figure 4.43 Spacecraft structure design verification test plan.

equipment panels, is tested for quasi-static loads as determined by coupled spacecraft/launch vehicle dynamic analysis. After successful testing of the primary structure, the structural test model is assembled by adding reflectors, solar panels, and flight unit mass models to the primary structure. The structural test model of a spacecraft duplicates its material, configuration, and mass distribution. It is subjected to sine, acoustic, and shock tests. These tests are performed primarily to qualify substructures and determine sine, random, and shock test levels for equipment. A brief description of these tests is given in this section.

As we discussed in Section 4.3, a typical launch environment can be divided into 4 distinct types: steady-state accelerations, low-frequency transients in the frequency range 5 to 100 Hz, vibroacoustic accelerations in the frequency range 20 to 2000 Hz, and pyrotechnic shocks in the frequency range 100 to 10,000 Hz. In general, a spacecraft is subjected to a combination of these environments. However, to simplify the testing, these types of launch environments are simulated separately. For each type of environment there is a choice between two tests: static versus centrifugal for steady-state acceleration, swept sinusoidal versus transient for low-frequency transients, random versus acoustic for vibroacoustic accelerations, and firing pyrotechnics versus simulated shocks for shocks. Modal survey tests are also performed on spacecraft structures to verify the accuracy of the analytical structural model which is used in the spacecraft/launch vehicle dynamic analysis to predict launch loads. A brief description of these tests is given in this section.

Steady-State Test

The objective of the steady-state test is to demonstrate the ability of the spacecraft and its components to withstand the quasi-static loads. The alternative approaches for the test are the static testing of the primary structure with detailed stress analysis of the components, or a centrifuge test of the complete spacecraft.

In a static testing, the primary structure of the spacecraft is subjected to quasi-static loads at the qualification (ultimate) level. The structure is loaded at specific locations in specific directions to simulate the load conditions established by the coupled load analysis. These loads are applied through load fixtures, load links, whiffletrees, and cables by hydraulic actuators and deadweights. The test conditions include maximum axial and maximum lateral load cases. Load cells are used to monitor applied loads, strain gauges to determine strain in the structure, and dial gauges to determine deflections of the structure.

The static test configuration for a typical dual-spin-stabilized spacecraft is shown in Fig. 4.44. The primary structure consists of central tubes, spun equipment panels, struts, and the BAPTA. The loads at point A simulate the loads due to the apogee motor. The loads at point B simulate the loads due to the solar array and equipment. The loads from the antennas are simulated at point C. The strain gauges at the adapter are calibrated for notched sinusoidal testing to be discussed later. Deflections are measured by dial gauges.

In a centrifuge test, the spacecraft is placed on a rotating ring. The spin rate and arm length are adjusted to provide the desired centrifugal

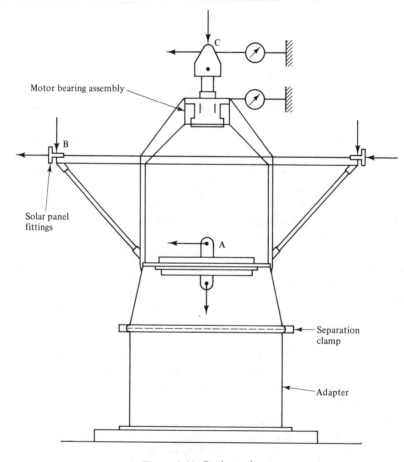

Figure 4.44 Static testing.

acceleration. The orientation of the spacecraft is adjusted to have the acceleration in the desired direction.

Several points should be considered in making the selection between static and centrifuge testing. The main advantages of static testing are an inexpensive testing facility, no constraints on the number and types of transducers because of the absence of rotating parts, and if the critical deformation (i.e., local buckling) starts, the test can be stopped to avoid failure. The main disadvantages are that the concentrated loads are applied at relatively few selected points and only the primary structure is tested.

For centrifuge testing, the main advantage is that every component in the spacecraft experiences acceleration. The disadvantages are the existence of an acceleration gradient, expensive testing facility, and limitations on the number of transducers due to rotation. The failure in the centrifuge test is also more hazardous. Considering the low cost and the simplicity, static testing is generally preferred.

Low-Frequency Transient Test

Swept sinusoidal testing is the most commonly used test to simulate a low-frequency launch environment. In the test, sinusoidal vibrations are applied sequentially at the base of the spacecraft along the three principal axes. The sinusoidal test levels are defined in the launch vehicle user's manual. For Ariane-launched spacecraft, the test levels are given in Table 4.6 for the axial axis and the two lateral axes. The sinusoidal vibration amplitude versus the frequency for the lateral axes are plotted in Fig. 4.45. For acceptance testing, the frequency is increased at the rate of 4 octaves/min. The expression for frequency, f, at time, t, is given by

$$f = f_0 2^{Kt/60} \tag{4.216}$$

where f_0 is the initial frequency at $t = 0$ and K is the sweep rate in octaves/min. Due to the limitations of the shakers, the displacement amplitude is limited to 7.7 mm in the frequency range 5 to 7 Hz.

The schematic diagram of a sinusoidal vibration control system is given in Fig. 4.46. An electrodynamic shaker provides sinusoidal excitation

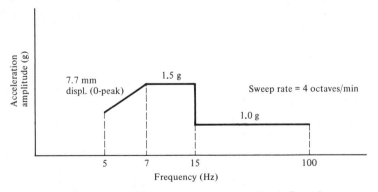

Figure 4.45 Sinusoidal lateral acceptance test levels for Ariane.

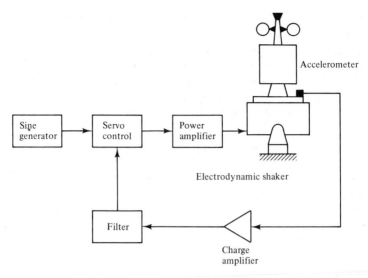

Figure 4.46 Schematic diagram of sinusoidal vibration control.

to the base of the spacecraft. If a voltage having a spectrum of the test specification is used as input to the amplifier of the electrodynamic shaker, the resulting acceleration will be different. Armature resonances of the shaker and the dynamics of the spacecraft react on the system to produce peaks and valleys in the resultant spectrum. If a specified spectrum is to be applied to the spacecraft, the applied voltage spectrum must be altered to correct for the dynamics of the shaker/spacecraft combination. In sinusoidal testing, this function is provided by servo control.

The sine generation produces the signal in accordance with the vibration test parameters such as the frequency range and the sweep rate. The amplitude of the signal is servo controlled. The test specification levels are programmed into the servo control. The acceleration of the shaker table is measured by accelerometers with charge amplifiers. The response signal is fed into a filter which measures the amplitude of the response signals at the exciting frequency. The response signal is compared with the specified test levels and the amplitude of the signal from the sine generator is adjusted accordingly and fed into the power amplifier of the shaker.

Notched Swept Sinusoidal Test

The dynamic response of a large spacecraft will tend to suppress the launch vehicle response at the spacecraft primary resonant frequencies. To understand this phenomenon, let us consider an external force, $P_0 \cos \omega t$, on a large mass, M. To attenuate the motion of the mass, a small spring–mass system of the same natural frequency as the forcing frequency is attached to the larger mass. This causes resonance in the spring–mass system and reduces the response of the larger mass. This is because a significant part of the external energy is transmitted to the spring–mass system. Similarly, the motion of a launch vehicle at the base of the spacecraft will be reduced at the spacecraft resonances. The test specifications, however, do not take this reduction into account, since it is dependent on the characteristics of each spacecraft. Hence test specifications, in general, will result in higher loads than flight loads at the primary spacecraft natural frequencies. To preclude structural failures during sinusoidal testing due to unrealistic loads, the input sinusoidal vibration levels are notched at the primary spacecraft resonant frequencies so that the loads at the critical structural members do not exceed the expected flight loads. The flight loads are determined by coupled launch vehicle/spacecraft dynamic analysis.

A schematic diagram of a notch control system is shown in Fig. 4.47. The structure is subjected to the base excitation in accordance with the test specification with the constraint that the response acceleration does not increase beyond certain levels. The base acceleration and response acceleration signals are fed into the multilevel selector through the normalizer. The multilevel selector is programmed for the maximum allowable values for these signals. The multilevel selector sends the controlling signal and its maximum allowable signal to the servo control. The servo control system compares these signals, modifies the signal from the sine generator, and feeds it to the power amplifier.

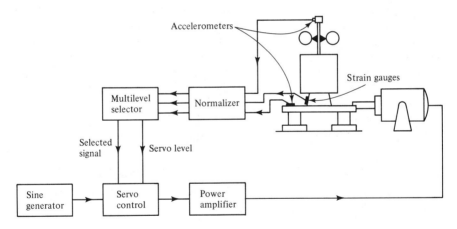

Figure 4.47 Schematic diagram of notched sinusoidal control.

Let us consider a case where the base excitation has a constant amplitude, as shown in Fig. 4.48. In the absence of notching, the base acceleration will be the controlling signal through the test and the response will be given by the dashed lines near resonance. During notched testing, in the beginning the base acceleration will control since the response is less than the maximum allowable value. However, as the response reaches the maximum allowable level, the control shifts from base acceleration to response acceleration and the base excitation is notched. As the resonance passes, the response level goes below the allowable value and the control passes back to the base acceleration signal.

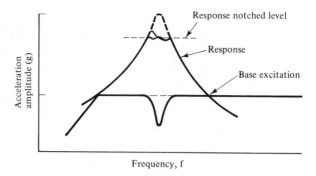

Figure 4.48 Response to notched sinusoidal excitation.

The simulation of the low-frequency transient environment by sinusoidal testing has been most commonly used because of its simplicity in specification and testing. However, it has several deficiencies. Sinusoidal testing results in higher loads at the spacecraft resonant frequencies and they are highly dependent on structural damping in comparison with transient environment. The notching in sinusoidal testing eliminates overtesting of the primary structure, but may result in undertesting of the other structures. With the introduction of digital control systems, it has been possible to generate a specified transient acceleration with a good accuracy at the base of the spacecraft using an electrodynamic shaker. The most important problem related to transient testing is the dependence of the test specification on possible variations of launch vehicle and spacecraft dynamic characteristics.

The schematic diagram of a transient wave control system is shown in Fig. 4.49. The first step is the determination of the system transfer function. The transient calibration input signal, $f(t)$, must be a form such that its Fourier transform is flat in the bandwidth of the system. A pulse of the form

$$f(t)_i = Ke^{-\alpha t} \quad \text{for } t > 0 \qquad f(t)_i = 0 \quad \text{for } t < 0 \qquad (4.217)$$

is used as the calibrating transient. Its Fourier transform is of the same form as a first-order, low-pass filter, with a corner frequency of f_C, where $f_C = \alpha$.

Both the transient calibration input, $f(t)_i$, and the transient calibration output, $f(t)_o$, signals are passed through identical analog-to-digital converters (ADC), and the resultant digital information is stored in the computer memory. From this digital information, the respective Fourier transforms, $F(\omega)_i$ and $F(\omega)_o$, are computed and then divided to produce the test system transfer function

$$H(\omega) = \frac{F(\omega)_o}{F(\omega)_i} \qquad (4.218)$$

The digital description of the required transient waveform, $f(t)_R$, is read into the computer and its Fourier transform, $F(\omega)_R$, is computed. This

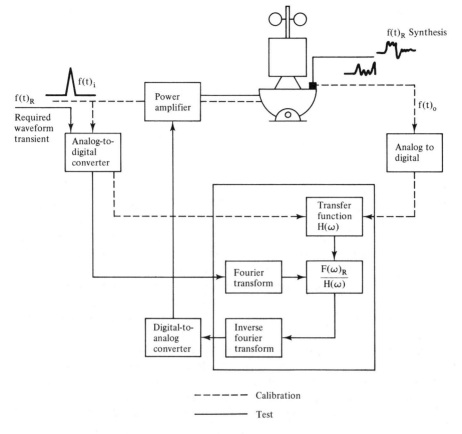

Figure 4.49 Transient wave control schematic diagram.

transform is then divided by the test system transfer function to determine the Fourier transform of the synthesized domain description.

$$F(\omega)_S = \frac{F(\omega)_R}{H(\omega)}$$

$F(\omega)_S$ is then transformed back into the time domain and the digital data are fed into a digital-to-analog converter (DAC), the output signal of which, $f(t)_S$, is fed to the power amplifier of the shaker. This technique is based on the capability of being able to compute accurately and economically and rapidly Fourier transforms and inverse Fourier transforms. This has been made possible with the new algorithm known as the Fast Fourier Transform.

Vibroacoustic Excitation Test

In random vibration testing, the spacecraft is subjected to random vibration through the base by the electrodynamic shaker. A typical spec-

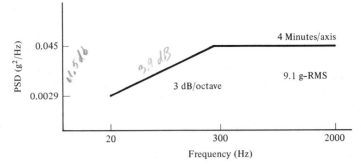

Figure 4.50 Random vibration test specification.

ification for a random vibration test is shown in Fig. 4.50. Random vibration is specified in the terms of a power spectral density. The power spectral density, as defined by Eq. (4.180), is the average of the square of the acceleration in the frequency band f and $f + \Delta f$. The rms value of the acceleration is obtained by integrating the power spectral density over the frequency range and taking the square root of the integrand.

There are two random vibration systems: analog and digital. The early systems were analog but they are now being replaced by digital systems. In an analog system, a noise generator produces a random signal. The random signal is normally divided into 80 narrowband filters. Using attenuators, the amplitudes of these signals are changed, manually or automatically, by comparing the actual random vibration at the base of the spacecraft with the test specification. This process is called equalization and it is needed to correct for the dynamics of the shaker and the spacecraft. Equalization is normally a lengthy process. The attenuated signals are combined in a mixer and fed to the power amplifier of the shaker.

In a digital control system, as shown in Fig. 4.51, the central part of the system is a Fourier analyzer. The shaker acceleration is digitized using

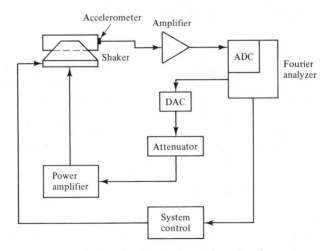

Figure 4.51 Schematic diagram for random vibration system.

an analog-to-digital converter. The Fourier analyzer determines the shaker transmissibility by taking the ratio of the signal transmitted to the power amplifier of the shaker to the acceleration produced at the shaker. Based on this calculated transmissibility and the desired PSD spectrum, a signal is generated for the power amplifier of the shaker. The signal is fed into a digital-to-analog converter (DAC) and then to the power amplifier. In comparison with the analog system, the digital system produces a more accurate PSD spectrum.

In acoustic testing, the spacecraft is subjected to direct acoustic noise in an acoustic chamber. A spacecraft during launch is subjected to random excitation through two paths: direct acoustic noise and random vibration through the interface with the launch vehicle. Random vibration simulates the random excitation at the interface but is unlikely to provide a sufficient excitation to a significant portion of the spacecraft due to transmission characteristics of the structure. The structure exterior to the spacecraft, such as solar panels and the antennas, will not be subjected to adequate excitation. The acoustic test, on the other hand, will not excite adequately the subsystems near the spacecraft interface. Acoustic testing is generally preferred to a random vibration test for most spacecraft because of relatively high acoustic noise levels and a relatively low level of mechanically transmitted vibrations encountered during launch. Smaller spacecraft are more likely to respond adequately to random vibration.

Acoustic tests, as shown in Fig. 4.52, are carried out in a reverberant

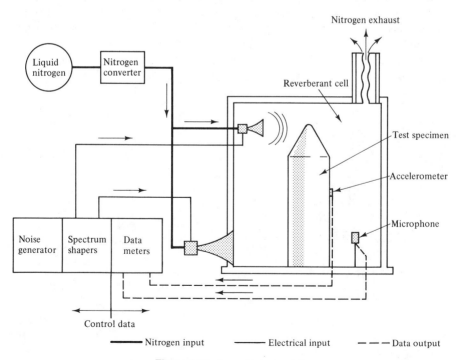

Figure 4.52 Acoustic test setup.

chamber whose shape and volume are configured to maximize the sound pressure level (SPL). Excitation is often provided by both low-frequency horns and high-frequency jets. The fundamental transducer used is the microphone whose voltage output is proportional to the rms acoustic pressure. Typically, five to six microphones are suspended in the chamber to develop an average spectrum. In addition to being able to show an overall SPL, real-time analyzers are available which can display the acoustic spectrum in octave or $\frac{1}{3}$-octave bandwidths. The spectrum of the acoustic noise is shaped in accordance with the test specifications. Acoustic testing can be performed either with or without the shroud. The acoustic levels are adjusted accordingly.

Shock Test

A spacecraft is subjected to shocks due to firing of pyrotechnic bolts for stage separation and spacecraft appendage deployments. Figure 4.53 presents a typical shock acceleration during the separation of a spacecraft from its launch vehicle. The analysis of the shock spectrum is complicated due to its short duration. M. A. Boit developed a concept of shock spectrum which is well suited for shock acceleration. The assumption underlying this concept is that the only significant effect of mechanical shock is the response of the resonant modes of the structure that experiences the shock.

The shock spectrum is defined as the maximum response of a single-degree-of-freedom oscillator for which the natural frequency of the oscillation is variable. The maximum response is calculated for a spring–mass system whose base is excited to the shock acceleration. The natural frequency of the spring–mass system is varied and the shock spectrum is the plot of the amplitude of the maximum response versus natural frequency. The shock spectrum will depend on the assumed damping, which should be specified

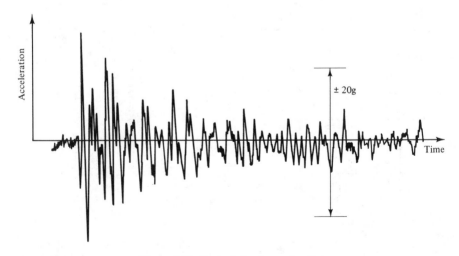

Figure 4.53 Typical shock acceleration.

with the shock spectrum. The common practice is to specify damping in terms of Q, which is defined as

$$Q = \frac{1}{2\xi}$$

where ξ is the damping ratio. The value of Q has intuitive significance since it is the amplification of a resonator when vibrated at its resonant frequency. The most used values of Q are 5 or 10, corresponding to 10% or 5% damping, respectively.

A shock spectrum represents a peak of the response at each frequency. However, there is a distinction between the response observed during the time the shock is acting on the oscillator and the time after the shock is diminished. When a spectrum is plotted from the maximum response attained during the excitation, it is called the primary spectrum. When plotted from responses after the input has subsided, the result is called the residual spectrum. Responses are positive and negative and the maxima in each direction are in general not the same. Hence spectra can be classified as primary positive, primary negative, residual positive, and residual negative. For a typical spacecraft, the separation shock spectrum at the interface is given in Fig. 4.54. The shock spectrum attenuates as a function of distance from the pyrotechnic source.

Electrodynamic vibration shakers can be used to perform shock testing. The shock spectrum does not uniquely define a time waveform. Therefore, many times, waveforms may have the same shock spectrum. One commonly used approach is the summation of sinusoidal wavelets. An individual wavelet with seven half-cycles and a 60-Hz fundamental frequency is shown in Fig.

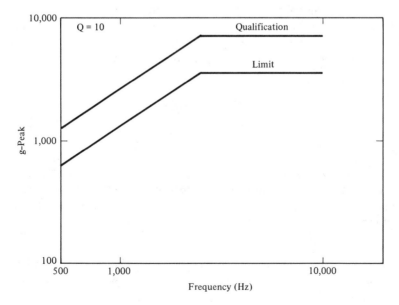

Figure 4.54 Separation shock spectrum.

4.55. This technique was developed by Yang and Saffell for earthquake shock testing. Since the final velocity and displacement values of this waveform are both zero, a waveform of this type is ideal for testing on electrodynamic shakers. In addition, a delay may be specified with each wavelet so that the start time of the wavelet is delayed. This feature permits greater control over the shape of the time waveform. The frequency interval or resolution of the wavelets is generally set automatically to $\frac{1}{3}$- or $\frac{1}{2}$-octave spacing. Initially, the amplitude of each wavelet is approximated by dividing the shock spectrum at the wavelet frequency by the number of half-cycles specified for the wavelet. An iterative procedure is then used which automatically adjusts the wavelet amplitudes until a close fit with the desired spectrum is achieved.

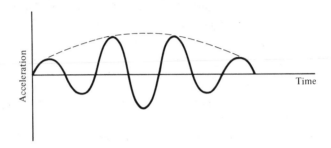

Figure 4.55 Wavelet with seven half-cycles and 60-Hz fundamental frequency.

Once the desired waveform is defined, the next step is to reproduce it on the shaker system. The procedure is similar to that discussed for the transient wave control. It consists of determining the transfer function of the shaker system, taking the Fourier transform of the desired waveform, dividing this by the transfer function, and taking the inverse Fourier transform of the result. The resulting signal is fed to the power amplifier.

PROBLEMS

4.1. The thrust cone of a spacecraft is subjected to a compressive axial load of 3×10^5 N and a bending moment of 2.0×10^4 N·m. The thrust cone dimensions are: height = 0.05 m, top radius = 0.5 m, and bottom radius = 0.75 m. For a monocoque shell and a minimum margin of safety of 10%, calculate the shell thickness and thrust cone mass for aluminum, magnesium, and beryllium.

4.2. The equipment panel in a spacecraft has length 1.5 m and width 1.0 m, and it can be assumed to be simply supported at the edges. The equipment mass is 35 kg and is uniformly distributed over the panel. The fundamental frequency of the panel should be above 35 Hz. The dynamic acceleration is assumed to be $35g$ and uniformly distributed. Determine the structural mass of the panel by assuming it to have a honeycomb cross section with aluminum face skins and aluminum honeycomb.

4.3. A strut of 0.6 m length is subjected to a 4000-N axial compressive force.

Compare the masses of the strut for aluminum and beryllium. The cross section is circular.

4.4. A spacecraft structural model in axial axis is represented by a single spring–mass system. The mass is 1000 kg and the spring stiffness is 4.0×10^4 N/cm. Determine the natural frequency of the system. The spacecraft is subjected to a base acceleration of $1g$ amplitude. Determine the spacecraft accelerations at frequencies of 5, 10, and 15 Hz. Assume the damping to be 10% of the critical damping.

4.5. The maximum flight acoustic levels for STS PAM-D are given in Table 4.3. Calculate the sound pressure power spectral density of the acoustic levels.

4.6. The random vibrations levels (PSD) for a piece of equipment are specified as $+3$ dB/octave for 20 to 300 Hz and 0.045 g^2/Hz for 300 to 2000 Hz. Calculate the g-rms of the level.

REFERENCES

1. W. Glugge, *Handbook of Engineering Mechanics,* McGraw-Hill, New York, 1962.

2. C. Wang, *Applied Elasticity,* McGraw-Hill, New York, 1953.

3. E. F. Bruhn, *Analysis and Design of Flight Vehicle Structures,* S. R. Jacobs & Associates, Inc., Indianapolis, 1973.

4. O. C. Zeinkiewicz, *The Finite Element Method in Structural and Continuum Mechanics,* McGraw-Hill Publishing Company Limited, London, 1967.

5. L. Meirovitch, *Analytical Methods in Vibrations,* Macmillan, New York, 1967.

6. J. S. Prezemieniecki, *Theory of Matrix Structural Analysis,* McGraw-Hill, New York, 1968.

7. R. J. Roark, *Formulas for Stress and Strain,* McGraw-Hill, New York, 1965.

8. H. C. Martin, *Introduction to Matrix Methods of Structural Analysis,* McGraw-Hill, New York, 1966.

9. Anon, *Buckling of Thin-Walled Circular Cylinder,* NASA SP-8007, Aug. 1968.

10. Anon, *Buckling of Thin-Walled Truncated Cones,* NASA SP-8019, Sept. 1968.

11. E. H. Baker, *Shell Analysis Manual,* N68-24802, Apr. 1968.

12. S. Timoshenko and D. H. Young, *Elements of Strength of Materials,* D. Van Nostrand, Princeton, N.J., 1964.

13. R. L. Bisplinghoff, H. Ashely, and R. L. Halfman, *Aeroelasticity,* Addison-Wesley, Reading, Mass., 1957.

14. S. H. Crandall and W. D. Mark, *Random Vibration in Mechanical Systems,* Academic Press, New York, 1963.

15. H. Gerard & G. Backer, *Handbook of Structural Stability,* Part III: *Buckling of Curved Plates and Shells,* NACA TN 3783, Aug. 1957.

16. R. R. Craig and M. C. Bampton, "Coupling of Substructures for Dynamic Analysis," *AIAA Journal,* Vol. 6, pp. 1313–1319, July 1968.

17. R. H. MacNeal, "A Hybrid Method of Component Mode Synthesis," *Computers and Structures,* Vol. 1, pp. 581–601, Dec. 1971.

18. S. Rubin, "Improved Component-Mode Representation for Structural Dynamic Analysis," *AIAA Journal,* Vol. 13, pp. 995–1006, Aug. 1975.

19. *Government/Industry Workshop on Payload Loads Technology,* NASA CP-2075, Marshall Space Flight Center, Ala., Nov. 14–16, 1978.

20. J. P. Barthmair, "Shock Testing under Minicomputer Control," *Proceedings,* 20th Annual Meeting of Institute of Environmental Science, Washington, D.C., 1974.

21. W. L. Day, "The Intelsat IV Spacecraft—Mechanical Design," *COMSAT Technical Review,* Vol. 2, No. 2, pp. 372–376, Fall 1972.

22. *PAM-D User's Requirement Document,* McDonell Douglas Astronautics Company, MDC G6626F, July 1985.

23. S. W. Tsai and H. T. Hahn, *Introduction to Composite Materials,* Technomic Publishing Co., Westport, CT 1980.

24. Ariane User's Manual, European Space Agency, Issue 1980—Rev. 0.

25. C. M. Harris and C. E. Crede, *Shock and Vibration Handbook,* McGraw-Hill, New York, 1976.

5

THERMAL CONTROL

5.1 INTRODUCTION

A spacecraft contains many components which will function properly only if they are maintained within specified temperature ranges. The temperature limits for typical equipment in a spacecraft are given in Table 5.1. The temperatures of these components are influenced by the net thermal energy exchange with the spacecraft thermal environment. The spacecraft thermal environment is determined by the magnitude and distribution of radiation input from the sun and the earth. Component temperatures are established by the heat radiated from external surfaces to the space sink, and internal equipment heat dissipation, together with the characteristics of the conduction and radiation heat-transfer paths between these "sources and sinks." The objective of thermal control design is to provide the proper heat transfer between all spacecraft elements so that the temperature-sensitive components will remain within their specified temperature limits during all mission environmental conditions, including prelaunch, launch, transfer orbit, and synchronous orbit phases.

A typical spacecraft thermal control design process is shown in Fig. 5.1. Initially, system trade-offs are performed to determine the spacecraft configuration. The objective is to meet the requirements of all subsystems within the launch vehicle constraints. The equipment locations are based on a trade-off between structural and thermal requirements. Preliminary analyses are performed to determine the adequacy of the thermal design concept. Next, detailed subsystem designs and trade-offs are performed. In thermal design, the interaction between structural design, equipment power dissipation, equipment location, and the desired component temperature limits are important considerations. A trade-off is performed in terms of

TABLE 5.1 TYPICAL EQUIPMENT TEMPERATURE LIMITS

Subsystem/Equipment	Thermal Design Temperature Limits (°C), Min/Max	
	Nonoperating/Turn-on	Operating
Communications		
Receiver	−30/+55	+10/+45
Input multiplex	−30/+55	−10/+30
Output multiplex	−30/+55	−10/+40
TWTA	−30/+55	−10/+55
Antenna	−170/+90	−170/+90
Electric power		
Solar array wing	−160/+80	−160/+80
Battery	−10/+25	0/+25
Shunt assembly	−45/+65	−45/+65
Attitude control		
Earth/sun, sensor	−30/+55	−30/+50
Angular rate assembly	−30/+55	+1/+55
Momentum wheel	−15/+55	+1/+45
Propulsion		
Solid apogee, motor	+5/+35	—
Propellant tank	+10/+50	+10/+50
Thruster catalyst bed	+10/+120	+10/+120
Structure		
Pyrotechnic mechanism	−170/+55	−115/+55
Separation clamp	−40/+40	−15/+40

thermal control parameters such as thermal coatings, insulation, metal thickness, heat pipes, and louvers, to establish the best thermal design for a spacecraft. After the trade-off studies and preliminary design are completed, an analytical thermal model is developed to predict temperatures. If the equipment temperatures are not within the allowable temperature limits, the thermal design is modifed and the process is repeated.

The accuracy of the analytical thermal model can be improved and verified by performing thermal balance tests on a suitable thermal test model. Equipment heat dissipation is simulated by resistive heaters. Absorbed external flux may be simulated by any of a variety of means, such as lamps which approach the sun in spectral and intensity properties, warm-plate infrared sources, or embedded film-resistive heaters. This test serves as the verification of the spacecraft thermal control subsystem. The flight spacecraft itself is later subjected to thermal vacuum qualification and acceptance tests where test conditions and equipment temperature (expected in orbit) are imposed on the spacecraft. Typically, design margins of 10°C and 5°C beyond those predicted for orbit are used for qualification and acceptance testing, respectively.

This chapter is divided into five sections: heat transfer, thermal analysis, thermal control techniques, spacecraft thermal design, and thermal testing. In the heat-transfer section, the basic principles of heat transfer by conduction

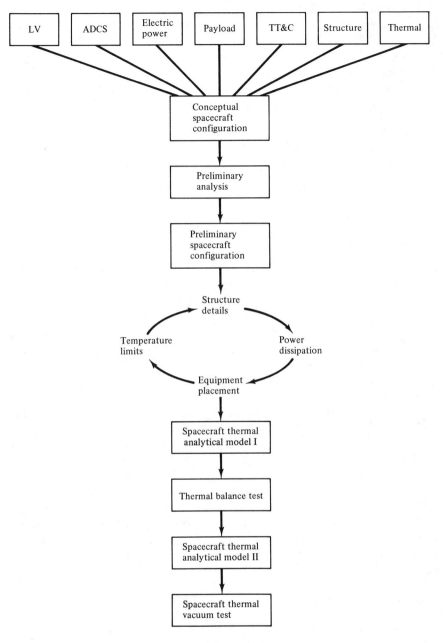

Subsystem Requirements

LV | ADCS | Electric power | Payload | TT&C | Structure | Thermal

Conceptual spacecraft configuration

Preliminary analysis

Preliminary spacecraft configuration

Structure details

Temperature limits

Power dissipation

Equipment placement

Spacecraft thermal analytical model I

Thermal balance test

Spacecraft thermal analytical model II

Spacecraft thermal vacuum test

Figure 5.1 Thermal control design process.

and radiation, the basic modes of heat transfer in spacecraft, are discussed. In the radiative heat-transfer part, basic principles such as Stefan–Boltzmann's law, Kirchhoff's law, view factors, and radiative coupling are reviewed. The major heat fluxes on the spacecraft are also discussed. In the thermal analysis section, the heat-balance equations for the isothermal thermal model and lumped-parameter thermal model are analyzed. In the lumped-parameter model, nodal breakdown techniques, radiative coupling factors, conductive coupling factors, and a typical analysis flowchart are discussed. In the thermal control techniques section, thermal coatings, insulation, heat pipes, louvers, and phase-change material are reviewed. The spacecraft thermal design section reviews various considerations in the thermal design of three-axis-stabilized and spin-stabilized spacecraft. The thermal testing section discusses the thermal balance and thermal vacuum tests for a typical spacecraft program.

5.2 HEAT TRANSFER

The temperature differences inside a body with no heat input and adiabatic external boundaries will reduce in the course of time due to heat flowing from regions of higher temperature to those of lower temperature. There are three modes of heat transfer: conduction, convection, and radiation.

Conduction in a solid is the transfer of heat from one part to another under the influence of a temperature gradient without appreciable displacement of particles. Conduction involves the transfer of kinetic energy from one molecule to an adjacent molecule. The same process takes place within liquids and gases.

Convection involves the transfer of heat by mixing one part of a fluid with another. The fluid motion may be entirely the result of the difference of density resulting from temperature differences (i.e., natural convection), or the motion may be produced by means of mechanical fluid movers (i.e., forced convection).

Solid bodies, as well as liquids and gases, are capable of radiating and absorbing thermal energy in the form of electromagnetic waves. Both conduction and convection rely on a heat-transfer medium, whereas radiation can occur in a vacuum. In a spacecraft, heat is mostly transferred by conduction throughout the solid parts of the spacecraft and radiated across interior volumes and into space from external surfaces.

Conduction

Heat transfer by steady unidirectional conduction is given by the following relation of Fourier proposed in 1822:

$$Q = -KA \frac{dT}{dx} \tag{5.1}$$

where Q = rate of heat conduction along the x axis, W

K = thermal conductivity, a physical property of the body, W/m·K

A = cross sectional area of the path normal to the x axis, m^2

$\dfrac{dT}{dx}$ = temperature gradient along the path, K/m

The thermal conductivities of typical materials are given in Table 5.2.

TABLE 5.2 THERMAL CONDUCTIVITY
OF SPACECRAFT MATERIALS

Materials	Thermal Conductivity, K, at 25°C (W/m · K)
Aluminum	210
Aluminum alloys	117–175
Magnesium	157
Magnesium alloys	52–111
Titanium	21
Stainless steel	16.2
Teflon	0.25

Let us consider a plane wall with thickness X and cross-sectional area A, whose two surfaces are kept in steady state at different but constant temperatures T_1 and T_2. Using Eq. (5.1), the heat-transfer rate from T_1 to T_2, Q, is given by

$$Q = \frac{K}{X} A(T_1 - T_2) \tag{5.2}$$

The amount of heat flow through a unit area of the surface per unit time, the specific heat flux, is given by the equation

$$q = \frac{Q}{A} = \frac{K}{X}(T_1 - T_2) \tag{5.3}$$

Equation (5.3) can be written as

$$Q = \frac{T_1 - T_2}{R_c} \tag{5.4}$$

where

R_c = thermal resistivity, K/w

$$= \frac{X}{KA} \tag{5.5}$$

Fourier's laws for the heat conduction process are similar to Ohm's law for electric current. The heat flow, Q, corresponds to current; the driving force for heat flow; temperature difference $(T_1 - T_2)$ corresponds to voltage E; and thus the thermal resistivity, R_c, is directly analogous to electrical resistivity.

Conductance

When heat is transferred by more than one heat-transfer mode through a structure medium, it is convenient to define the conductance, C, as the gross rate of heat transfer, Q, divided by the overall temperature drop as

$$C = \frac{Q}{T_1 - T_2} \qquad (5.6)$$

It is also convenient to define an overall resistance as the reciprocal of the conductance just defined. In practice it is common to find the terms "conductivity" and "conductance" well defined and distinguishable, but thermal "resistance" is almost always used for both resistivity and resistance.

The electrical analogy is often useful in series and parallel combinations of material. For example, if a wall of three solids of different materials is in series, with thermal resistances R_1, R_2, R_3, and with the temperatures of the two outer surfaces T_1 and T_4, the heat flow, Q, becomes

$$Q = \frac{T_1 - T_4}{R_1 + R_2 + R_3} \qquad (5.7)$$

Thermal contact conductance. In the previous example of different solids in contact, no allowance was made for a temperature drop at their boundary; a perfect contact is assumed. In an actual system, there will be a finite interface resistance and corresponding finite temperature drop. The thermal contact conductance between solids having surface temperatures T_1 and T_2 at the junction is

$$C_j = \frac{Q}{T_1 - T_2} = h_j A \qquad (5.8)$$

The thermal contact conductance per unit area, h_j, is dependent on contacting materials, contact pressure, surface finishes, interstitial medium and is usually determined experimentally for critical joints.

Radiation (Refs. 6 and 7)

Radiation is the transmission of energy by electromagnetic waves. The electromagnetic spectrum and the names given to radiation transmitted in various ranges of wavelengths are shown in Fig. 5.2. The frequency of

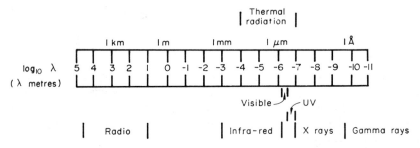

Figure 5.2 Spectrum of electromagnetic waves (Ref. 7).

Thermal Control   Chap. 5

radiation depends on the nature of its source. For example, a metal bombarded by high-energy electrons emits x-rays, high-frequency electric currents generate radio waves, and a body emits thermal radiation by virtue of its temperature. Radiation in the wavelength range 0.1 to 100 μm is called thermal radiation. The radiation within the wavelength band of 0.38 to 0.76 μm is visible.

When radiation falls on a body, a fraction (α) of it is absorbed, a fraction (ρ) is reflected, and the remainder (τ) is transmitted through the body. These fractions are related by

$$\alpha + \rho + \tau = 1 \tag{5.9}$$

where α = absorptivity
  ρ = reflectivity
  τ = transmissivity

Most solid materials absorb practically all radiation within a very thin surface layer, usually less than 1 mm thick. For these opaque materials, $\tau = 0$ and $\alpha + \rho = 1$.

In the analysis of radiative heat transfer, it is useful to introduce the concept of a blackbody. A blackbody can be defined as one that (1) absorbs all radiation incident upon it (and reflects and transmits none) and (2) emits at any particular temperature the maximum possible amount of thermal radiation.

Blackbody = $\alpha = 1$

Stefan–Boltzmann Law

emits max. thermal energy at any given temp.

The rate at which energy is radiated from a blackbody is proportional to the fourth power of its absolute temperature. This law was discovered experimentally by Stefan in 1879 and was deduced theoretically 5 years later by Boltzmann using classical thermodynamics. The law can be stated as

$$E_b = \sigma T^4 \tag{5.10}$$

where E_b = rate at which energy is radiated from a unit area
    of the surface of a blackbody to the hemisphere
    of the space above it, W/m^2
  σ = Stefan–Boltzmann constant = 5.67×10^{-8} W/m^2·K^4
  T = absolute temperature, K

Figure 5.3 shows the radiative flux as a function of wavelength for a blackbody at a number of temperatures. The analytical expression, derived by Planck from quantum theory, is

$$E_{b\lambda} = \frac{C_1 \lambda^{-5}}{e^{(C_2/\lambda T)} - 1} \tag{5.11}$$

where $E_{b\lambda}$ = monochromatic emissive power for a blackbody,
  T = absolute temperature, K
  C_1 = 3.74×10^{-19} kW·m^2
  C_2 = 1.439×10^{-2} m·K
  λ = wavelength, m

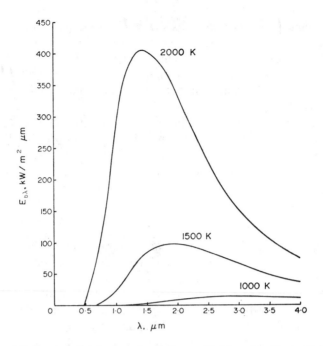

Figure 5.3 Spectral distribution of radiation emitted by a blackbody (Ref. 7).

From Fig. 5.3 it can be seen that in addition to the amplitude increase with temperature, the maximum moves toward the shorter wavelengths. If the temperature of a hot body is less than 500°C (773 K), virtually none of the radiation will fall within the band of the wavelength corresponding to visible light. Even at 2500°C, the temperature of an incandescent lamp, only about 10% of the energy is emitted in the visible range. The wavelength at which maximum flux is emitted (λ_{max}) is inversely proportional to the absolute temperature (T), a relationship known as Wein's displacement law:

$$\lambda_{max}T = 2.9 \times 10^{-3} \text{ m·K} \tag{5.12}$$

Intensity of Radiation

The intensity of radiation in the direction normal to the emitting surface is related to the blackbody emissive power by

$$I_n = \frac{E_b}{\pi} \tag{5.13}$$

When the direction of emission is at an angle ϕ to the normal to the surface, the projected area of emission is $\cos \phi$ and not 1. In this case, the intensity of emission is

$$I_\phi = \frac{E_b \cos \phi}{\pi} \tag{5.14}$$

Equation (5.14) is a statement of the Lambert cosine law.

Kirchhoff's Law

Kirchhoff's law states that for a given wavelength λ, the absorptivity and emissivity are equal:

$$\alpha_\lambda = \varepsilon_\lambda \tag{5.15}$$

For a given range of the spectrum limited by the wavelength λ_1 and λ_2,

$$\alpha(\lambda_1, \lambda_2) = \varepsilon(\lambda_1, \lambda_2) = \frac{\int_{\lambda_1}^{\lambda_2} \alpha_\lambda I_\lambda \, d\lambda}{\int_{\lambda_1}^{\lambda_2} I_\lambda \, d\lambda} \tag{5.16}$$

where I_λ is the monochromatic intensity of the radiation at the wavelength.

For a blackbody, $\alpha_\lambda = \varepsilon_\lambda = 1$ and therefore $\alpha = \varepsilon = 1$. A gray body *black paint* is defined as one whose emissivity is constant and does not vary with wavelength, $\alpha = \varepsilon = $ constant over the entire spectrum range.

For realistic coatings, however, α_λ (or ε_λ) is a function of λ. The absorptivity, α_λ or ε_λ, of white paint, black paint, and second surface mirror surfaces are shown in Fig. 5.4 as a function of wavelength. In practice, the absorptivity, α, and emissivity, ε, are different in most cases, as they depend on the range of the spectrum corresponding to the absorbed and emitted radiation. In spacecraft thermal design, normally α refers to solar absorption α_s and ε to infrared emittance as discussed below.

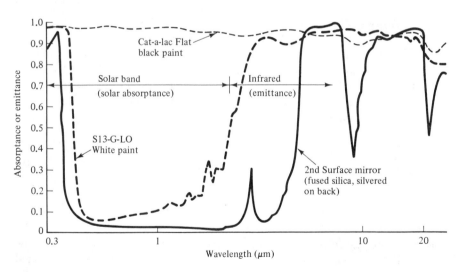

Figure 5.4 Actual representation of three material properties.

Solar Absorptivity

The solar spectral irradiance is given in Fig. 5.5. Since only 2% of the total irradiance corresponds to the range below 0.22 μm wavelength

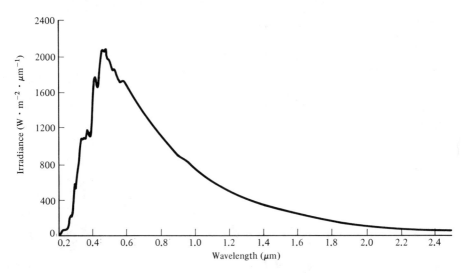

Figure 5.5 Solar spectral irradiance curve.

and 3% above 2.7 μm, a good approximation of α_s is given by

$$\alpha_s = \frac{\int_{0.2}^{2.8} \alpha_\lambda I_\lambda \, d\lambda}{\int_{0.2}^{2.8} I_\lambda \, d_\lambda} \tag{5.17}$$

Infrared Emissivity

The relation between radiation intensity, temperature, and wavelength indicates that emissivity is a function of temperature of the radiating surface. For the range near room temperature (20°C), the effective spectrum may be defined as between 5 and 50 μm. Thus the infrared emissivity, ε, is very closely approximated by

$$\varepsilon = \frac{\int_{5}^{50} \alpha_\lambda I_\lambda \, d\lambda}{\int_{5}^{50} I_\lambda \, d\lambda} \tag{5.18}$$

Solar absorptivity, α_s, and infrared emissivity, ε, for some coatings currently used for spacecraft thermal control are given in Table 5.3. All thermal coatings degrade to a certain extent in the space environment due to an exposure to ultraviolet radiation. Table 5.3 also gives expected degraded values after 7 years in space.

View Factors

The view factor from one surface to another is defined as the fraction of the total radiation emitted by one surface which is directly incident on

TABLE 5.3 THERMAL PROPERTIES OF SURFACES[a]

Surface	Typical Application	Solar Absorptance, α_s		Emittance, ε	
		BOL	EOL	BOL	EOL
Black paint	Interior structure	0.9	0.9	0.9	0.9
White paint	Antenna reflector	0.2	0.6	0.9	0.9
Optical solar reflector	North and south panel radiators	0.08	0.21	0.8	0.8
Graphite/ epoxy	Solar panel and antenna structure	0.84	0.84	0.85	0.85
Aluminized kapton	Thermal insulation	0.35	0.50	0.6	0.6
Tiodized titanium	Apogee motor thermal shield	0.6	0.6	0.6	0.6
Aluminum, aluminum tape, deposited aluminum	Propellant insulation	0.12	0.18	0.06	0.06
Anodized aluminum	Interior structure	0.2	0.6	0.8	0.8
Solar cells	Solar panels	0.65–0.75	0.65–0.75	0.82	0.82
Gold		0.2–0.3		0.03–0.06	

[a] BOL, Beginning of life; EOL, end of life, 7 years.

the other. For a view factor derivation, consider two differential areas as shown in Fig. 5.6. From Eq. (5.14), the radiation, ΔQ, emitted from a small area of a black surface, ΔA_1, at an angle ϕ_1 from the normal to the surface through a solid angle $\Delta \omega$ is

$$\Delta Q = \frac{E_b}{\pi} \cos \phi_1 \, \Delta A_1 \, \Delta \omega \qquad (5.19)$$

For a surface of emissivity, ε_1, Eq. (5.19) may be written as

$$\Delta Q = \varepsilon_1 \frac{E_b}{\pi} \Delta A_1 \cos \phi_1 \, \Delta \omega \qquad (5.20)$$

The area viewed from surface 1 is $\Delta A_2 \cos \phi_2$ and

$$\Delta \omega = \Delta A_2 \frac{\cos \phi_2}{r^2} \qquad (5.21)$$

Substituting Eq. (5.21) into Eq. (5.20) yields

$$\Delta Q = \varepsilon_1 \frac{E_b}{\pi r^2} \Delta A_1 \cos \phi_1 \, \Delta A_2 \cos \phi_2 \qquad (5.22)$$

Expressing Eq. (5.22) in the differential form, the total radiation transmitted

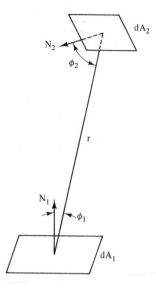

Figure 5.6 View factor geometry for two surfaces of differential area.

directly from a surface A_1 to a surface A_2 is given by

$$Q = \varepsilon_1 E_b \int_{A1} \int_{A2} \frac{dA_1 \cos \phi_1 dA_2 \cos \phi_2}{\pi r^2} \quad (5.23)$$

The view factor, F_{12}, is given by

$$F_{12} = \frac{Q}{\text{total radiation emitted by surface } A_1} \quad (5.24)$$

$$= \frac{Q}{\varepsilon_1 E_b A_1} \quad (5.25)$$

$$= \frac{1}{A_1} \int_{A1} \int_{A2} \frac{dA_1 \cos \phi_1 dA_2 \cos \phi_2}{\pi r^2} \quad (5.26)$$

From the symmetry of the integral equation,

$$A_1 F_{12} = A_2 F_{21} \quad (5.27)$$

The product $A_1 F_{12}$ is called the exchange area between surfaces 1 and 2.

Evaluation of Eq. (5.26) for the determination of view factor is often difficult. The values of view factors for a variety of geometries are, however, available in the literature. These configurations can be extended to other cases by means of the view factor geometry law.

The reciprocity law is given by Eq. (5.27). If a closed system is composed of n surfaces, the following equation for the ith surface applies to all of the n surfaces:

$$\sum_{j=1}^{n} F_{ij} = 1 \quad (5.28)$$

Equation (5.28) is known as a summation law. Hence, when $n - 1$ view factors for a given system are known, the other may be determined from Eq. (5.28). Moreover, using reciprocity equations and if none of the n

surfaces is reentrant (i.e., $F_{11} = F_{22} = \cdots = 0$), then only $\frac{1}{2}(n - 1)(n - 2)$ view factors must be independently determined. The remaining view factors can be determined from Eqs. (5.27) and (5.28).

In determining view factors, it is often convenient to break one surface into many areas. As an example, given two surfaces A_1 and A_2, if surface A_1 is subdivided into A_3 and A_4, the total view factor F_{21} is related to the two subsidiary configuration factors F_{23} and F_{24} according to

$$A_2F_{21} = A_2F_{23} + A_2F_{24} \tag{5.29}$$

Equation (5.29) is known as the decomposition law.

Reflections

The surface of some solids which are highly polished and smooth behave like mirrors to thermal radiation (i.e., the angle of reflection equals the angle of incidence). Such reflection is called specular. Most commercial materials have rough surfaces; that is, their surface irregularities are large compared to the wavelength of radiation. Reflection of thermal radiation from this kind of surface occurs indiscriminately in all directions. Such reflection is called diffuse.

Radiative Coupling

The net heat exchange between two black surfaces i and j is

$$\begin{aligned} Q_{ij} &= A_iF_{ij}E_{bi} - A_jF_{ji}E_{bj} \\ &= A_iF_{ij}(E_{bi} - E_{bj}) \\ &= A_iF_{ij}\,\sigma(T_i^4 - T_j^4) \end{aligned} \tag{5.30}$$

From Eq. (5.30) it can be seen that an analogy may be drawn between radiative heat transfer and the flow of electric current through a resistor. The voltage is analogous to blackbody emissive power, E_b, the current to net heat exchange, Q, and the electrical resistance to the reciprocal exchange area, $1/A_iF_{ij}$. The radiative coupling between two black-bodies i and j is given by

$$R_{ij} = A_iF_{ij} \tag{5.31}$$

In real situations, we are usually concerned with the transfer among more than two nonblack surfaces, which involves reflection.

External Heat Flux

The major heat flux on an earth orbiting spacecraft consists of solar flux, albedo flux, and the thermal radiation of the earth. Other types of incident flux that have an influence on the spacecraft thermal behavior only during short periods are: the infrared flux coming from the internal wall of the fairing before the fairing is jettisoned, the aerodynamic heating after the fairing is jettisoned, which will depend on velocity and altitude, and plume heating coming from the third-stage motor, apogee motor, and so on.

Solar flux.   The yearly average solar flux outside the earth's atmosphere is approximately 1353 W/m^2. The solar constant is defined as the flux existing at a distance of one astronomical unit (AU) from the sun and is closely approximated by this previous quantity. Solar flux varies as the inverse square of the distance from the sun, which causes a variation of 3% during the year due to the eccentricity of the earth's orbit. The maximum (1399 W/m^2) occurs at the perihelion of the earth's orbit around January 3, and minimum (1309 W/m^2) occurs at aphelion around July 4. Solar flux is assumed to be independent of orbit altitude, from near-earth to geosynchronous. Table 6.3 gives a recent estimate of the variation of solar flux. Due to the great distance between the sun and the spacecraft, it is assumed that the solar flux impinges on the spacecraft with parallel rays. Hence the solar intensity (in units of power) on a surface not shadowed by another body is obtained simply as the product of the solar flux times the projected area of the surface on the plane perpendicular to the sun vector. Hence the solar flux incident on a surface is given by

$$Q = S\mu_i A \tag{5.32}$$

where Q = incident solar intensity, W
S = solar flux, W/m^2
A = total area of the surface, m^2
μ_i = ratio of projected (effective) area of the surface/total area
= $\cos \theta$ for a plane surface where θ is
the angle between the sun vector and the plane normal
= $\frac{1}{4}$ for a sphere
= solar aspect coefficient

In a spacecraft, solar panels, antennas, and so on, may shadow the external surface of the spacecraft. The solar aspect coefficients μ_i can include the shadowing effects. The shadow effect may be determined from satellite models by either illuminating the satellite with parallel light (generally the sunlight), or by taking a photograph from a great distance, or by employing an available computer program which performs the numerical integration from inputs of the coordinates of the geometries.

The solar flux variation on the surfaces of a three-axis-stabilized spacecraft in geosynchronous orbit, shown in Fig. 5.7, can be divided into two classes: daily (diurnal) variation and seasonal variation. The spacecraft makes one revolution with respect to the sun in a 24-hour period. The solar flux is a maximum at anti-earth, west, earth, and east spacecraft surfaces at noon, 6 P.M., midnight, and 6 A.M., respectively. There is no daily variation of direct solar flux on the north and south surfaces other than shadowing. The seasonal variation is due to the change of the solar incidence angle through $\pm 23.5°$ and a small change in solar flux intensity (due to the eccentricity of the earth orbit) during a period of a year. Therefore, the north and south surfaces, which have the least solar flux variation and the least projected area to the sun, are prime candidates for providing stable temperatures for high power dissipators, such as TWTs. The spacecraft will

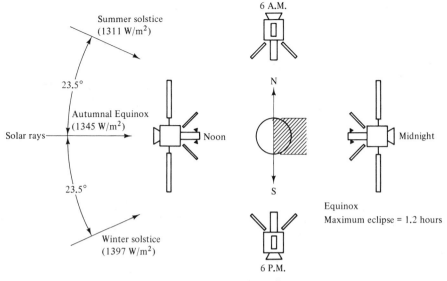

Figure 5.7 Synchronous orbit solar flux variation.

experience two periods of solar eclipses (i.e., zero solar flux), centered around the vernal and autumnal equinoxes, for a total of about 90 days per year. The largest eclipses, of about 72 minutes, occur at equinoxes, around March 21 and September 23. Eclipses lasting longer than 1 hour occur about 50 days per year.

Albedo flux. The albedo flux is the fraction of the total incident solar radiation on the earth which is reflected into space as a result of scattering in the atmosphere and reflection from clouds and the earth surfaces. The albedo flux constant ϕ_a is given by

$$\phi_a = Sa \tag{5.33}$$

where S = solar flux constant
a = albedo coefficient

The recommended annual mean value of a is 0.30 ± 0.02. The value of a varies from 0.1 to 0.8. The variation is related to the increase in the average cloud cover with distance from the equator and the high reflectance of snow- and ice-covered surfaces in high-latitude regions. Low values of mean albedo ($a < 0.3$) exist in the latitude range from 30°N to 30°S.

In the calculation of incident albedo flux on a spacecraft, it is commonly assumed that the albedo coefficient is constant over the earth's surface and that the earth's surface reflects diffusely and obeys Lambert's law. However, the calculation is still complex because the albedo flux depends on the position of the spacecraft, the orientation of the sun, and the spacecraft altitude. For a simple case of a spherical spacecraft and averaging over the spacecraft surface's and sun's orientation, the albedo flux is

$$\phi_A = \frac{Sa}{8}\left(1 - \sqrt{1 - \frac{R_e^2}{\delta^2}}\right) \tag{5.34}$$

where R_e = radius of the earth

δ = distance of the spacecraft from the center of mass of the earth

Thermal radiation of the earth. A portion of the incident solar radiation is absorbed as heat by the earth and its atmosphere and is reemitted as thermal radiation according to the Stefan–Boltzmann law. The mean annual value of thermal radiation near the earth's surface is

$$\phi_r = 237 \pm 7 \text{ W/m}^2 \tag{5.35}$$

It is usually assumed that earth-emitted thermal radiation is constant over the earth surface and that the earth's surface emits diffusely and obeys Lambert's law. Flux incident (ϕ_T) on a spacecraft surface is a function of altitude and for a spherical spacecraft is given by

$$\phi_T = \frac{1}{2}\sigma T_e^4\left(1 - \sqrt{1 - \frac{R_e^2}{\delta^2}}\right)$$
$$= \frac{1}{2}\phi_r\left(1 - \sqrt{1 - \frac{R_e^2}{\delta^2}}\right) \tag{5.36}$$

where T_e = mean earth black-body temperature, K

ϕ_r = earth's thermal radiation, W/m^2

R_e = radius of the earth, km

δ = distance of the spacecraft from the earth center of mass

5.3 THERMAL ANALYSIS

Spacecraft temperatures are computed from solutions of simple heat balance equations of the form

$$\text{heat stored} = \text{heat in} - \text{heat out} + \text{heat dissipated} \tag{5.37}$$

For a spacecraft in orbit above a planetary atmosphere, the heat that is absorbed by the spacecraft includes absorbed sunlight, reflected sunlight (albedo), and planet-emitted radiation. Heat is produced within the spacecraft by power dissipated primarily by electrical and electronic components. Heat is rejected from the spacecraft by infrared radiation from external surfaces. Heat is also exchanged among spacecraft component parts by radiation and conduction.

Let us consider an infinitely conductive, therefore isothermal, spherically shaped spacecraft in geosynchronous orbit where the albedo flux and the earth's radiation are negligible. The spacecraft is assumed to have no dissipative equipment. The temperature of the spacecraft is a direct function of its surface thermo-optical properties: the absorptivity, α_S, and emissivity, ε.

The heat-balance equation is

$$\alpha_s S \pi R^2 = \varepsilon \sigma T^4 4 \pi R^2 \tag{5.38}$$

where R is the radius of the sphere, S is the solar flux intensity (say, 1353 W/m²), α_s and ε are the absorptivity and emissivity of the surface, respectively, σ is the Stefan–Boltzmann constant = 5.67×10^{-8} W/m² · K⁴, and T is the equilibrium temperature.

The equilibrium temperature, T, is therefore

$$T = \left(\frac{\alpha_s S}{4 \varepsilon \sigma} \right)^{1/4} \tag{5.39}$$

If the spacecraft surface is painted white with $\alpha_s = 0.2$ and $\varepsilon = 0.9$ (Table 5.3), the equilibrium temperature is equal to $-83°C$. White paint is sometimes termed a "cold" coating because it absorbs very little solar flux (just like white summer clothing) but has a high emissivity. For a spacecraft surface painted black, $\alpha_s = \varepsilon = 0.9$, the equilibrium temperature is 5°C. Black paint is called a "mean" coating. For a gold surface $\alpha_s = 0.25$ and $\varepsilon = 0.045$, the equilibrium temperature is 154°C. Gold is called a "warm" coating.

Isothermal Radiator

During a spacecraft configuration study, preliminary thermal analyses are performed to estimate the average temperature of the spacecraft and the required radiator size. For such analyses the spacecraft is assumed to be isothermal. The calculation of the average spacecraft temperature is a valuable tool in the configuration study, even though it will be supplemented later by a more detailed calculation of the temperatures at many points in the spacecraft.

The heat-balance equation for an isothermal spacecraft is

$$mc_p \frac{dT}{dt} = \text{external heat flux} + \text{equipment heat dissipation} \tag{5.40}$$
$$- \text{heat radiated in space}$$

where m = total spacecraft mass, kg
c_p = specific heat, W·s/K·kg
T = absolute temperature, K

Neglecting the albedo flux and the thermal radiation of the earth for a geosynchronous orbit satellite, the external heat flux from the sun is given by

$$\text{external heat flux} = \alpha_s S \mu_i a \tag{5.41}$$

where α_s = solar absorptance of the spacecraft surface
S = solar flux intensity, W/m²
a = surface area, m²
μ_i = solar aspect coefficient
= projected (effective) area of the surface/a

$$\text{heat radiated in space} = \varepsilon A \sigma T^4 \tag{5.42}$$

where ε = emissivity

σ = Stefan–Boltzmann constant

$\quad$ = 5.67×10^{-8} W/m$^2 \cdot$ K^4

A = radiator area, m^2

T = absolute temperature, K

Hence the thermal balance equation becomes

$$mc_p \frac{dT}{dt} = \alpha_S S \mu_i a + P - \varepsilon A \sigma T^4 \tag{5.43}$$

where P is the equipment heat dissipation. For a steady-state condition, $dT/dt = 0$. The steady-state temperature is given by

$$T_E = \left(\frac{P + \alpha_S S \mu_i a}{\varepsilon A \sigma} \right)^{1/4} \tag{5.44}$$

In a geosynchronous orbit, a spacecraft is subjected to two eclipse seasons with a maximum eclipse period of 72 minutes per day. During these eclipse periods, $S = 0$ and temperatures drop. To ensure that the equipment is able to withstand eclipse periods, it is necessary to calculate the minimum temperature to be experienced during eclipses. It is also important to know the temperature–time profile during the return to the steady-state conditions.

Equation (5.43) can be rewritten in terms of steady-state temperature T_E [from Eq. (5.44)] as

$$\frac{dT}{dt} = \frac{T_E^4 - T^4}{4\tau T_E^3} \tag{5.45}$$

where the time constant

$$\tau \equiv \frac{mc_p}{4\varepsilon\sigma A T_E^3} \tag{5.46}$$

The solution of Eq. (5.45) depends on whether the temperature is decreasing, such as during an eclipse period, or the temperature is increasing, such as immediately after an eclipse. The steady state temperatures for these two periods, T_E, will be different.

For the case of radiative cooling, no solar flux, and $T > T_E$, the solution can be written as

$$\frac{t}{\tau} + C = 2 \left(\coth^{-1} \frac{T}{T_E} - \cot^{-1} \frac{T}{T_E} \right) \tag{5.47a}$$

For the case of radiative heating, with solar flux and $T < T_E$, the solution is given by

$$\frac{t}{\tau} + C = 2 \left(\tanh^{-1} \frac{T}{T_E} + \tan^{-1} \frac{T}{T_E} \right) \tag{5.47b}$$

where C is the constant of integration and is determined from initial conditions, temperature at $t = 0$, which is known. It should be noted that the temperature cannot be written explicitly as a function of time.

Example 5.1

In a three-axis stabilized spacecraft, the communications equipment is mounted on the north- and south-facing panels. The total thermal dissipation on each panel is 300 W. For the equipment to operate properly, the allowable temperature range for the radiator on the panel exterior is $5°/+37°C$. The mass of the radiator including the mounted equipment is 85 kg and its specific heat c_p is 900 Watts·sec/kg·K. Determine

(a) Radiator size for communications equipment

(b) Equinox/eclipse temperature for the cases: (1) batteries provide full power to communication equipment during eclipse, resulting in no change in thermal dissipation; (2) batteries provide power for only part of the communication equipment such that the dissipation is halved. For the second case, determine the radiator temperature at the end of the eclipse.

(c) Determine the time required for the radiator to reach the minimum operating temperature of 5°C after the eclipse for the second case. The radiator heat dissipation during this period can be assumed to be 375 W by turning on the heaters instead of TWTs.

Solution (a) The (south-facing) radiator is sized for the hottest condition, winter solstice at EOL. The optical solar reflector (OSR) radiator is assumed to be isothermal and to take into account the solar reflection and IR radiation from the solar array and the antennas, the efficiency of the radiator is assumed to be 90% during noneclipse period. The heat-balance equation is

$$\varepsilon \sigma T^4 \eta A = \alpha_S A S \sin \theta + P$$

In this problem, ε = emittance of the radiator = 0.8 from Table 5.3, σ = 5.67×10^{-8} W/m^2 · K^4:η = efficiency = 0.9; A = area of the radiator to be determined; T = radiator temperature in K = 310 K, maximum allowable temperature; α_S = solar absorptance at EOL = 0.21 from Table 5.3; S = solar intensity at winter solstice = 1397 W/m^2; θ = solar aspect angle = 23.5° at winter solstice; and P = 300 W. Substituting these values in the heat-balance equation, the area of the radiator is given by

$$A = \frac{300}{0.8 \times 5.67 \times 10^{-8} \times (310)^4 \times 0.9 - 0.21 \times 1397 \sin 23.5°}$$

$$= \frac{300}{377 - 117.14} \approx 1.16 \text{ m}^2$$

(b) At equinox during the noneclipse period, $\theta = 0$ and the heat-balance equation is

$$\varepsilon \sigma T^4 \eta A = P$$

or

$$T = \left(\frac{P}{\varepsilon \sigma \eta A} \right)^{1/4}$$

$$= \left(\frac{300}{0.8 \times 5.67 \times 10^{-8} \times 0.9 \times 1.16} \right)^{0.25}$$

$$= 282 \text{K} \doteq 9°C$$

radiator or the south panel, no incidence

So the temperature during the equinox noneclipse period is within the prescribed limits. If the temperature was too low, auxiliary heaters would have been necessary. This concept is termed as an augmented passive, or semiactive thermal design.

η - usually .15

F_p - packing factor (.8-.9)

In the first case where the batteries provide full power to the communication equipment during eclipse, the radiator temperature will remain approximately the same during eclipse as during the noneclipse period at equinox.

In the second case, with batteries providing partial power, heat dissipation of 150 W, the equilibrium temperature T_E during eclipse is given by

$$T_E = \left(\frac{P}{\varepsilon \sigma A}\right)^{1/4}$$

$$= \left(\frac{150}{A \left(0.8 \times 5.67 \times 10^{-8} \times 1.16\right)}\right)^{1/4} = 231 \text{ K}$$

The time constant τ from Eq. (5.46) is given by

$$\tau = \frac{mc_p}{4\varepsilon \sigma A T_E^3}$$

Substituting the values of the parameters in the above equation, the time constant τ is given by

$$\tau = \frac{85 \times 900}{4 \times 0.8 \times 1.16 \times 5.67 \times (231)^3 \times 10^{-8}}$$

$$= 2.949 \times 10^4 \text{ s}$$

For the case of radiative cooling, the temperature from Eq. (5.47a) is given by

$$\frac{t}{\tau} + C = 2\left(\coth^{-1}\frac{T}{T_E} - \cot^{-1}\frac{T}{T_E}\right)$$

At $t = 0$, $T = 282$ K, and $T_E = 231$ K. Substituting these values in the equation, the integration constant, C, is equal to 0.936. The equation above is solved by assuming temperature, T, and determining time, t, corresponding to the temperature.

At $T = 273$K $= 0°$C, the time $t = 71$ minutes.

At $T = 272$K $= -1°$C, the time $t = 80$ minutes.

The maximum eclipse period is 72 minutes. Using linear interpolation, the temperature at 72 minutes is 272.9 K.

(c) We want to determine the time required to reach 278 K after eclipse. The radiator heat dissipation is 375 W.

The equilibrium temperature T_E is given by

$$T_E = \left(\frac{375 \times 10^8}{0.8 \times 5.67 \times 0.9 \times 1.16}\right)^{1/4}$$

$$= 298.3 \text{ K}$$

The time constant τ is given by

$$\tau = \frac{85 \times 900}{4 \times 0.8 \times 0.9 \times 1.16 \times 5.67 \times (298.3)^3 \times 10^{-8}}$$

$$= 1.52 \times 10^4 \text{ s}$$

For this case of radiative heating, the temperature is given from Eq. (5.47b) as

$$\frac{t}{\tau} + C = 2\left(\tanh^{-1}\frac{T}{T_E} + \tan^{-1}\frac{T}{T_E}\right)$$

at $t = 0$, $T = 272.9$ K, and $T_E = 298.3$ K. Substituting these parameters in the equation, the integration constant C is determined to be 4.595. Next, we would like to determine the time for the radiator temperature to reach 5°C or 278 K. Substituting the values of τ, C, T and T_E in the equation, t is 64 minutes, the "warm-up period" after maximum period of eclipse.

Solar Array Temperature

In the solar array of a three-axis-stabilized spacecraft, the side facing the sun is covered with solar cells. Therefore, a fraction of the solar flux is converted into the electric power by the solar cells, reducing the heat to the array. The effective solar absorptance, α_{SE}, is

$$\alpha_{SE} = \alpha_S - F_p\eta \tag{5.48a}$$

where α_S = average solar cell array absorptance
F_p = solar cell packing factor (ratio of the total active solar cell area to the total substrate area)
η = solar cell operating efficiency

The steady-state operating temperature of the solar array is given by

$$T_{OP} = \left[\frac{\alpha_{SE}A_F S \cos \alpha}{(\varepsilon_F A_F + \varepsilon_B A_B)\sigma} \right]^{1/4} \tag{5.48b}$$

where A_F = array front-side area
A_B = array back-side area
ε_F = emittance of the array front side
ε_B = emittance of the array back side
S = solar constant
σ = Stefan–Boltzmann constant
α = angle of incidence of the sunlight

Example 5.2
The solar array for a three-axis-stabilized spacecraft has the following characteristics: area = 6 m^2, $\alpha_s = 0.8$, $\varepsilon_F = 0.8$, $\varepsilon_B = 0.7$, $F_p = 0.95$, and $\eta = 0.14$. Calculate the steady-state operating temperatures at vernal equinox and summer solstice.

Solution The effective solar absorptance from Eq. (5.48a) is

$$\alpha_{SE} = \alpha_s - F_p\eta$$
$$= 0.8 - 0.95 \times 0.14 = 0.667$$

The steady-state operating temperature is

$$T_{OP} = \left[\frac{\alpha_{SE} A_F S \cos \alpha}{(\varepsilon_F A_F + \varepsilon_B A_B)\sigma} \right]^{1/4}$$

From Table 6.3, $S \cos \alpha = 1362$ W/m^2 at vernal equinox and 1202 W/m^2 at summer solstice. The temperature at vernal equinox is

$$T_{OP} = \left[\frac{0.667 \times 6 \times 1362}{(0.8 \times 6 + 0.7 \times 6) \times 5.67 \times 10^{-8}} \right]^{1/4}$$
$$= 321.14 \text{ K} = 48.14°C$$

The temperature at summer solstice is

$$T_{OP} = \left[\frac{0.667 \times 6 \times 1202}{(0.8 \times 6 + 0.7 \times 6) \times 5.67 \times 10^{-8}} \right]^{1/4}$$

$$= 311.6 \text{ K} = 38.6°C$$

Lumped-Parameter Analytical Model

In the preceding section, temperatures were calculated for isothermal spacecraft. However, real spacecraft have nonnegligible temperature gradients, representing complex heat conduction and radiative exchanges throughout the spacecraft. The heat-balance calculations of such spacecraft are generally based on a lumped-parameter representation of the physical system known as a thermal model. The spacecraft is divided into n isothermal segments, called nodes. How well this model represents the actual spacecraft is determined largely by how "isothermal" the assigned nodes actually are.

The heat-balance equation for each node, i, is

$$m_i c_i \frac{dT_i}{dt} + \sum_{j=1}^{n} C_{ij}(T_i - T_j) + \sum_{j=1}^{n+1} \sigma R_{ij}(T_i^4 - T_j^4)$$

$$= P_i + \alpha_{si} A_i \mu_i S + \alpha_i A_i \phi_{Ai} + \varepsilon_i A_i \phi_{Ti} \quad (5.49)$$

where $m_i c_i$ = thermal capacity of node, i, JK^{-1}

T_i, T_j = temperatures of nodes i and j, K

$\dfrac{dT_i}{dt}$ = rate of temperature variation of node i, K/s

C_{ij} = conductive coupling between nodes i and j, W K^{-1}

R_{ij} = radiative coupling between nodes i and j, m^2

P_i = internal power dissipation of node i, W

α_{si} = solar absorptivity of node i

ε_i = emissivity of node i

A_i = area of radiation for node i, m^2

μ_i = solar aspect coefficient of node i

S = solar flux incident on node i, W/m^2

ϕ_{Ai} = albedo flux incident on node i, W/m^2

ϕ_{Ti} = earth radiation incident on node i, W/m^2

For a geosynchronous orbit, the albedo and earth radiation may be neglected. The thermal-balance equation becomes

$$m_i c_i \frac{dT_i}{dt} + \sum_{j=1}^{n} C_{ij}(T_i - T_j) + \sum_{j=1}^{n+1} \sigma R_{ij}(T_i^4 - T_j^4) = P_i + \alpha_{si} A_i \mu_i S \quad (5.50)$$

The temperatures at the n node points are determined by solving Eq. (5.50). Since these equations are nonlinear, computer-aided numerical techniques are used to calculate the temperatures.

Radiative Coupling Factors R_{ij}

The fraction of total radiation emitted by node surface i which is directly incident on node surface j is given by the view factor, F_{ij}. In general, however, radiant flux undergoes a number of reflections before it is finally absorbed. In this section a radiation coupling factor R_{ij} will be derived which is a measure of the total amount of radiation absorbed by the jth node from that emitted by the ith node, including all paths, direct and reflected. Uniform incident radiation on each nodal surface is assumed.

Let us consider radiation exchange on surface i as shown in Fig. 5.8. Q_i is the thermal flux incident on the surface i. It should be noted that for infra-red flux, the absorptivity and emissivity for surface would be equal. A fraction, $\varepsilon_i Q_i$, will be absorbed and the remainder will be reflected. The surface will also emit W_i due to its temperature, T_i. The total heat flux leaving the ith surface is equal to

$$W_i + r_i Q_i \tag{5.51}$$

where $W_i = \varepsilon_i A_i \sigma T_i^4$

$\quad r_i = \text{reflection coefficient} = 1 - \varepsilon_i$

$\quad Q_i = \text{thermal flux incident on the surface } i$

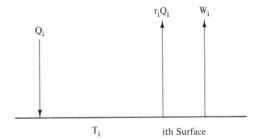

Figure 5.8 Incident, reflected, and emitted radiation.

The total heat flux incident on the jth surface will be

$$Q_j = \sum_{i=1}^{n} (W_i + r_i Q_i) F_{ij} \tag{5.52}$$

where F_{ij} is the view factor from the ith surface to jth surface. In Eq. (5.52) it is assumed that surfaces radiate and reflect diffusely (i.e., according to the cosine law). A completely nondiffuse surface is a specular one, where incident and reflected rays have equal angles with the surface normal. The specular surface must be treated quite differently. Equations similar to Eq. (5.52) can be written for each of the n surfaces. These equations can be written in matrix form as follows:

$$[M]\{Q\} = [\bar{F}]\{W\} \tag{5.53}$$

where

$$[M] = \begin{bmatrix} 1 - r_1 F_{11} & \cdots & -r_1 F_{i1} & \cdots & -r_1 F_{n1} \\ & \ddots & & & \\ -r_i F_{1i} & \cdots & 1 - r_i F_{ii} & \cdots & -r_i F_{ni} \\ & & & \ddots & \\ -r_n F_{1n} & \cdots & -r_n F_{in} & \cdots & 1 - r_n F_{nn} \end{bmatrix} \qquad (5.54)$$

$$[\overline{F}] = \begin{bmatrix} F_{11} & \cdots & F_{i1} & \cdots & F_{n1} \\ & & & & \\ F_{1i} & \cdots & F_{ii} & \cdots & F_{ni} \\ & & & & \\ F_{1n} & \cdots & F_{in} & \cdots & F_{nn} \end{bmatrix} \qquad (5.55)$$

$$\{Q\} = \begin{Bmatrix} Q_1 \\ \vdots \\ Q_i \\ \vdots \\ Q_n \end{Bmatrix} \quad \{W\} = \begin{Bmatrix} W_1 \\ \vdots \\ W_i \\ \vdots \\ W_n \end{Bmatrix} \qquad (5.56)$$

Multiplying Eq. (5.53) by $[M]^{-1}$ on the both sides, the result is

$$\begin{aligned} \{Q\} &= [M]^{-1} [\overline{F}] \{W\} \\ &= [\overline{M}] [\overline{F}] \{W\} \qquad (5.57) \\ &= [C] \{W\} \end{aligned}$$

where $[\overline{M}] = $ inverse of $[M] = [M]^{-1}$
$[C] = [\overline{M}] [\overline{F}]$
From Eq. (5.57), Q_i is given by

$$Q_i = \sum_{j=1}^{n} C_{ij} W_j \qquad (5.58)$$

where C_{ij} are elements of matrix $[C]$ and are given by

$$C_{ij} = \sum_{k=1}^{n} \overline{M}_{ik} \overline{F}_{kj} = \sum_{k=1}^{n} \overline{M}_{ik} F_{jk} \qquad (5.59)$$

where $\overline{M}_{ik}$ and $\overline{F}_{kj}$ are the elements of matrices $[M]^{-1}$ and $[F]$, respectively, and F_{jk} is the view factor. Substituting Eq. (5.59) into Eq. (5.58), the result is

$$Q_i = \sum_{j=1}^{n} W_j \sum_{k=1}^{n} \overline{M}_{ik} F_{jk} = \sum_{j=1}^{n} \varepsilon_j A_j \sigma T_j^4 \sum_{k=1}^{n} \overline{M}_{ik} F_{jk} \qquad (5.60)$$

The total absorbed power on the ith surface, Q_{ai}, is given by

$$Q_{ai} = \varepsilon_i Q_i = \sum_{j=1}^{n} R_{ji} \sigma T_j^4 \qquad (5.61)$$

where R_{ji} is the radiative coupling between surfaces j and i. Comparing Eqs. (5.60) and (5.61), we have

$$R_{ji} = \varepsilon_j A_j \varepsilon_i \sum_{k=1}^{n} \overline{M}_{ik} F_{jk} \qquad (5.62)$$

where the ε terms are the emissivities, F_{jk} the view factors, A_j the surface area, and $\overline{M}_{ik}$ are the elements of matrix $[M]^{-1}$. The heat flux lost by surface i, Q_{li}, can similarly be written as

$$Q_{li} = \sum_{j=1}^{n} R_{ij} \sigma T_i^4 \qquad (5.63)$$

The radiative coupling, R_{ij}, is equal to R_{ji}. The net heat balance of the surface i will be given by

$$Q_{ai} - Q_{li} = \sum_{j=1}^{n} R_{ij} \sigma (T_j^4 - T_i^4) \qquad (5.64)$$

Therefore, the radiative coupling factors can be calculated from Eq. (5.62) by knowing surface radiative properties, emissivities, areas, and view factors. Several computer programs are available to compute the radiative coupling factors. Some of these programs employ Monte Carlo techniques whereas others use the contour integration method.

Conductive Coupling C_{ij}

The conductive coupling C_{ij} is usually calculated on the basis of two assumptions:

1. The contact surfaces of the nodes are isothermal.
2. The heat flux Q_{ij} transmitted by conduction from node i to node j is given by

$$Q_{ij} = C_{ic}(T_i - T_{ic}) = h_c A_c (T_{ic} - T_{jc}) \qquad (5.65)$$
$$= C_{jc}(T_{jc} - T_j)$$

where C_{ic} = conductance between node i and its contact surface with node j, W/K

C_{jc} = conductance between node j and its contact surface with node i, W/K

T_i, T_j = temperature of nodes i and j, K

T_{ic}, T_{jc} = temperature of contact surfaces, K

h_c = heat-transfer coefficient between the surfaces in contact, W/m²K

A_c = total area of contact surface, m²

From Eq. (5.65), we get

$$Q_{ij} = C_{ij}(T_i - T_j) \qquad (5.66)$$

where

$$\frac{1}{C_{ij}} = \frac{1}{C_{ic}} + \frac{1}{h_c A_c} + \frac{1}{C_{jc}} \qquad (5.67)$$

If nodes i and j are parts of the same homogeneous material, as frequently occurs when subdividing a spacecraft equipment panel, then $1/h_c A_c = 0$ and Eq. (5.67) is reduced to simply one-dimensional conduction heat transfer.

Solar Aspect Coefficient

External spacecraft surfaces may be subjected to a solar input flux which is partially blocked as a function of time by certain parts of the spacecraft, such as solar arrays and antennas. The solar aspect coefficients are calculated to take these shadowing effects into account. The following methods may be used for their determination.

1. *Photographic determination:* A mock-up of the satellite is photographed from an adequately long distance with a long-focal-length lens (to decrease the parallax effects) and solar-exposed areas of the shades are then determined directly from prints with a surface integrator.

2. *Numerical determination:* The solar aspect coefficients may be determined by numerical integration with the aid of a computer. Existing programs need as input such parameters as the coordinates of surface boundaries, spin axis, solar vectors, and surface properties.

Nodal Breakdown

To build a thermal mathematical model of a spacecraft, the thermal engineer must have all the drawings of the spacecraft structure to determine the dimensions, the arrangement of the equipment, the types of assembling, the nature of the materials which are used, and so on. The nodes in the model should be chosen to be as isothermal as possible. However, the number of conductive and radiative exchange coefficients one has to calculate increases rapidly with the number of nodes. Hence modeling is a compromise between the inaccuracies inherent in a limited number of nodes and the cost (labor and computer time) of a big model. The nodes must be chosen judiciously in accordance with the physical problem being studied. If local temperatures are being sought, nodes must at least fall on these points; nodes must fall at the endpoints when a temperature gradient is sought; lumping may be coarse in high-thermal-conductivity areas but must be fairly fine if isotherms are sought. Typically, a bulk (coarse) thermal model of a spacecraft is developed first; then detailed thermal models of sections of the spacecraft are developed, as required, using the results of the bulk model as boundary temperatures.

Thermal Analysis of a Spacecraft

Several general-purpose computer programs are available for thermal analysis. A typical analysis flowchart is shown in Fig. 5.9. The first step is to break the spacecraft into *n* nodes. As discussed earlier, to perform thermal analysis, we need to calculate radiative coupling factors, conductive coupling factors, and external heat flux and should know spacecraft equipment heat dissipation. Radiative coupling factors and orbital heat fluxes are calculated, as shown in the figure, by using the LOHARP computer program (other programs are available to perform this task). Input data for this program consist of spacecraft structural parameters, thermo-optical properties

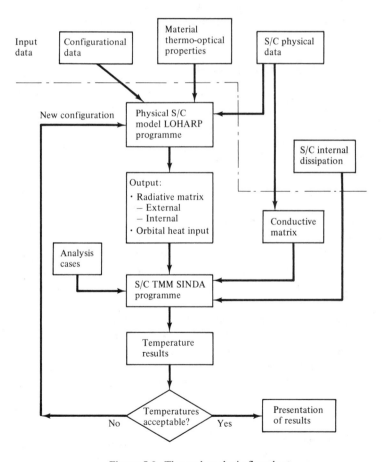

Figure 5.9 Thermal analysis flowchart.

of the coatings, and orbit parameters. The conductive coupling factors are calculated directly from physical data. The radiative coupling factors, orbital heat flux, conductive coupling factors, and spacecraft equipment heat dissipation are the input data for the SINDA program, which calculates the temperatures of the nodes. Other thermal analyzer programs are also available. If the temperatures are not acceptable, the thermal design is modified and the process is repeated.

For thermal analysis of the Intelsat V main body, a 107-node bulk thermal model was used for a geosynchronous orbit. The nodal breakdown of the model is shown in Fig. 5.10. A detailed description of nodes for north and south panels and antenna deck is given in Fig. 5.11. A modified 117-node model, including 10 additional nodes to account for the stowed solar panels, was used to analyze transfer orbit phases. The thermal analyses of antenna module were performed using a 33-node model for the transfer orbit condition and 178-node model for the geosynchronous orbit condition. In addition, very detailed thermal models (up to 300 nodes) of sections of the north-south panels were developed to predict temperature gradient (results of the bulk model were used for boundary temperatures).

5.4 THERMAL CONTROL TECHNIQUES

The typical evolution of a spacecraft thermal design can be considered in three stages: the conceptual design, the preliminary design, and the detailed design. The main task of the thermal designer during the conceptual design stage is to influence the spacecraft configuration in such a way as to achieve effective thermal control. Feasibility studies, mission planning, and gross design trade-offs also occur at this stage.

After the finalization of the conceptual design, the preliminary design begins. System-level trade-offs are performed on the basis of spacecraft or mission requirements. A more detailed configuring of the spacecraft, such as the overall electronics packaging layout, begins. Preliminary definitions of subsystem thermal requirements and characteristics are obtained and power and mass budgets established. Analyses are performed to verify the adequacy of the conceptual thermal design. If warranted, modifications to the thermal design are made.

The foregoing process is continued in greater depth in the detailed design stage. To assist a detailed thermal analysis of spacecraft design, extensive use is made of digital computers.

The thermal control techniques can be divided into two classes: passive thermal control and active thermal control.

Passive Thermal Control

A passive thermal control system maintains the component temperature within the desired temperature range by control of the conductive and the radiative heat paths through the selection of the geometrical configurations

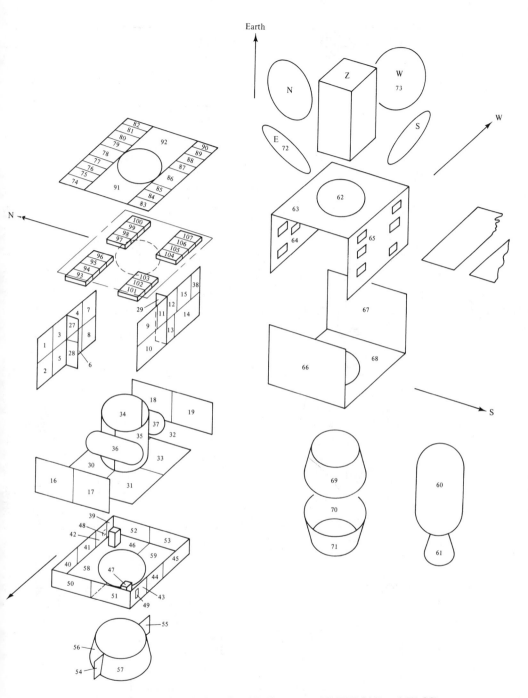

Figure 5.10 Thermal model of Intelsat V. (Courtesy of INTELSAT and FACC)

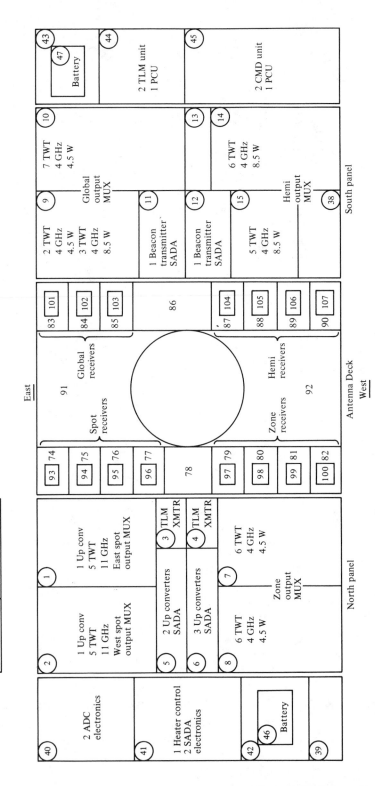

Figure 5.11 Nodal breakdown of the north and south panels and of the antenna deck of Intelsat V. (Courtesy of INTELSAT and FACC)

and thermo-optical properties of the surfaces. Such a system does not have moving parts, moving fluid, or electric power input other than the power dissipation of spacecraft functional equipment. Passive thermal control techniques include thermal coatings, thermal insulations, heat sinks, and phase-change materials.

Thermal coating. External surfaces of a spacecraft radiatively couple the spacecraft to space, the only heat sink available. Because these surfaces are also exposed to external sources of energy, their radiative properties must be selected to achieve a balance at the desired temperature between internally dissipated and external sources of power and the heat rejected into space. The two properties of primary importance are the emittance of the surface, ε, and the solar absorptance, α_s. Table 5.3 gives the properties of different thermal coatings. Two or more coatings can be combined in an appropriate pattern to obtain a desired average value of α_S and ε (e.g., a checkerboard pattern of white paint and polished metal).

For a radiator, low α_S and high $\varepsilon(\alpha_S/\varepsilon \ll 1)$ are desirable to minimize solar input and maximize heat rejection to space. For a radiator coating, both the initial values of α_S and ε and any changes in these values during a mission lifetime are important. For long-life missions, 7 to 10 years, degradation can be large for all white paints. For this reason, the use of a second surface mirror coating system is preferred. Such a coating is vapor-deposited silver on typically 0.2-mm-thick fused silica, called an optical solar reflector.

The values of the radiative properties are subject to uncertainties arising from four sources: (1) property measurement errors; (2) manufacturing reproducibility; (3) contamination before, during, and after launch; and (4) space environment degradation.

Degradation of the thermal coating in the space environment results from the combined effects of high vacuum, charged particles, and ultraviolet radiation from the sun. The last two factors vary with the mission trajectory, and orbit degradation data are obtained continually from flight tests and laboratory measurements. The Intelsat IV spacecraft series provided an excellent statistical sampling of in-flight data on solar absorptance degradation of silver quartz mirrors located in the north-facing end plane thermal radiator. The changes in solar absorptance, $\Delta\alpha_s$, versus time in geosynchronous orbit, shown in Fig. 5.12, were inferred from the temperature measurements of the mirrored radiator. Table 5.3 gives the degraded values of α_s at the end of 7 years in synchronous orbit for typical thermal coatings.

Thermal insulation. Thermal insulation is designed to reduce the rate of heat flow per unit area between two boundary surfaces at specified temperatures. Insulation may be a single homogeneous material, such as low-thermal-conductivity foam, or an evacuated multilayer insulation in which each layer acts as a low-emittance radiation shield and is separated by low-conductance spacers.

Multilayer insulations are widely used in the thermal control of spacecraft

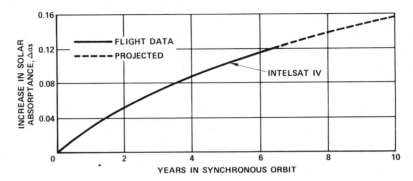

Figure 5.12 Intelsat IV increase in solar absorptance of silvered-fused-silica mirror radiators versus time in synchronous orbit (Ref. 14).

and components to (1) minimize heat flow to or from the component, (2) reduce the amplitude of temperature fluctuations in components due to time-varying external radiative heat flux, and (3) minimize the temperature gradients in components caused by varying directions of incoming external radiative heat.

Multilayer insulation (MLI) consists of several layers of closely spaced radiation-reflecting shields which are placed perpendicular to the heat-flow direction. The aim of radiation shields is to reflect a large percentage of the radiation the layer receives from warmer surfaces. To avoid direct contact between shields, low-conductivity spacers are commonly used. Sometimes embossing or crinkling of the shields produces small contact areas whose thermal joint conductance is low. The space between shields is evacuated to decrease gas conduction. For space applications, proper venting of the interior insulation is provided to avoid undue pressure loads on the shields during ascent and to pass outgassing products from the insulation material for long-term on-orbit missions. An evacuated MLI provides, for a given mass, insulation which is orders of magnitude greater than that furnished by more conventional materials such as foams and fiberglass batting.

For typical applications, the multilayer insulations (MLI) are shown in Fig. 5.13. For normal temperatures, the outer skin is 25-μm aluminized kapton and the remaining layers are made of aluminized Mylar separated by Dacron mesh to provide low thermal conductivity. The outer aluminized kapton layer provides an outer covering for handling a moderate α_S/ε ratio and blanket protection from the space environment for a 7- to 10-year mission. The Kapton provides high-temperature protection since its useful maximum temperature is 343°C compared to 121°C for Mylar. The insulation blanket that is subjected to apogee motor plume heating has an outer layer of titanium or stainless material. These materials can withstand the high temperature (1400°C) of the engine exhaust gases while protecting the remaining blanket assembly. The outer layer for such an insulation blanket is painted black to provide a high emittance in order to radiate the absorbed apogee motor heating.

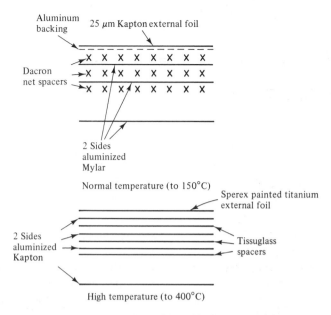

Figure 5.13 Typical multilayer blanket composition.

Heat sinks/thermal doublers. Heat sinks are materials of large thermal capacity which are placed in thermal contact with the component whose temperature is to be controlled. When heat is generated by the component, the temperature rise is restricted because the heat is conducted into the sink. The sink will then dispose of this heat to adjacent locations through conduction or radiation. The heat sink can serve the same function in reverse; that is, heat sinks can prevent severe cooling during the periods of low heat absorption or generation. Heat sinks are commonly used to control the temperature of those items of electronic equipment which have high dissipation or a cyclical variation in power dissipation. The equipment and structure of the spacecraft itself usually provides a heat sink. Thermal doublers, sometimes referred to as heat sinks, are used for TWTAs to spread the heat over the OSR radiator area, resulting in a reduction in maximum temperature and temperature gradient. The thermal doublers are optimized to minimize mass by employing high-thermal-conductivity aluminum, such as type 1050 tempered aluminum, which offers a 70% increase in conductivity over conventional aluminum, and where possible an optimum profile and mounting area. A typical configuration of a TWTA mounted with a thermal doubler on the OSR radiator is shown in Fig. 5.14.

Phase-change materials. Solid–liquid phase-change materials (PCMs) present an attractive approach to spacecraft passive thermal control when the incident orbital heat fluxes or on-board equipment dissipation changes widely for short periods.

The PCM thermal control system consists primarily of a container which is filled with a material capable of undergoing a chemical phase

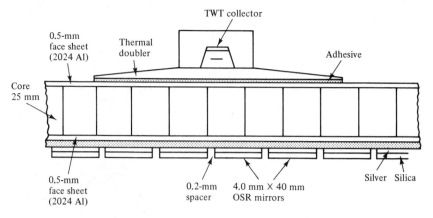

Figure 5.14 TWTA mounted with heat sink on OSR radiator configuration.

change. When the temperature of the spacecraft surface increases, the PCM will absorb the excess heat through melting. When the temperature decreases, the PCM solidifies. Phase-change materials used for temperature control are those whose melting point is close to the desired temperature of the equipment. Then the latent heat associated with the phase change provides a large thermal inertia when the temperature of the equipment is passing through the melting point. However, the phase-change material cannot prevent a further temperature rise when all the material is melted.

Active Thermal Control

Passive thermal control may not be adequate and efficient in terms of spacecraft added mass for the applications where the equipment has a narrow specified temperature range and there is a great variation in equipment power dissipation, surface thermo-optical properties due to space degradation, and solar flux during the mission. In such cases, temperature sensors may be placed at the critical equipment location. When critical temperatures are reached, mechanical devices are actuated to modify the thermo-optical properties of surfaces or electrical power heaters come on or off to compensate for the variation in the equipment power dissipation. For spacecraft with high-power-dissipation equipment, such as high-power TWTA, it may be more efficient in terms of added mass to use heat pipes to increase thermal conductivity in place of heat sinks.

This section provides a brief review of active control elements, such as heat pipes, louvers, and electric heaters.

Heat pipes (refs. 4 and 5). A heat pipe is a thermal device that provides efficient transfer of a large amount of thermal energy between two terminals with a small temperature difference. Heat pipes provide an almost isothermal condition and can be considered an extra high-thermal-conductivity device.

A heat pipe, as shown in Fig. 5.15, consists of a closed tube whose

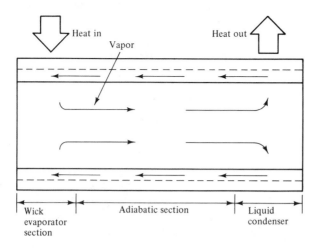

Figure 5.15 Schematic diagram of heat pipe.

inner surfaces are lined with a capillary wick. Heat at the evaporator portion of the heat pipe vaporizes the working fluid. The resulting difference in pressure drives vapor from the evaporator to the condenser, where it condenses and release the latent heat of vaporization. The loss of liquid by evaporation results in the liquid–vapor interface in the evaporator entering the wick surface, and a capillary pressure is developed there. This capillary pressure pumps the condensed liquid back to the evaporator for reevaporation. So the heat pipe transports the latent heat of vaporization continuously from the evaporator section to the condenser without drying out the wick, provided that the flow of working fluid is not blocked and a sufficient capillary pressure is maintained. The amount of heat transported as latent heat of vaporization is usually several orders of magnitude higher than that which can be transmitted as sensible heat in a conventional convective system.

Figure 5.16 shows a typical pressure distribution in a heat pipe. The flow of the vapor from the evaporator to the condenser creates a pressure gradient along the path. The flow of the liquid from the condenser to the evaporator also causes a pressure gradient in the liquid. So, along the length

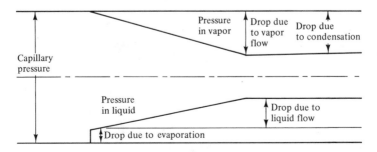

Figure 5.16 Pressure distribution in heat pipe.

of the pipe, the pressure at the liquid side of the liquid–vapor interface is different from that at the vapor side. This pressure difference between the two sides of the liquid–vapor interface is balanced by capillary pressure, which is established by the meniscus that forms at the interface by forcing the interface back into the wick structure.

For the heat pipe to work, the maximum capillary pressure must be greater than the total pressure drop in the pipe. Neglecting the pressure drops in the evaporator and the condenser, the condition can be stated as

$$(P_c)_{max} \geqslant \Delta P_l + \Delta P_v + \Delta P_g \tag{5.68}$$

where $(P_c)_{max}$ = maximum capillary pressure
 ΔP_l = pressure drop required for the liquid to flow from the condenser to the evaporator
 ΔP_v = pressure drop required for the vapor to flow from the evaporator to the condenser
 ΔP_g = pressure head due to gravity which is near zero in space

If Eq. (5.68) is not satisfied, the wick will dry out in the evaporator region and the pipe will not operate.

Maximum Capillary Pressure. At the liquid–vapor interface when a meniscus is formed, the capillary pressure, which is the pressure difference between vapor pressure and liquid pressure $(P_v - P_l)$, can be calculated by the equation

$$P_c = \sigma \left(\frac{1}{R_1} + \frac{1}{R_2} \right) \tag{5.69}$$

where P_c = capillary pressure
 σ = surface tension coefficient of the liquid
 R_1, R_2 = principal radii of curvature of the meniscus

The maximum value of $(1/R_1 + 1/R_2)$ will depend on the type of wick. Defining $(1/R_1 + 1/R_2)$ as equal to $2/r_c$, where r_c is the effective capillary radius, Eq. (5.69) can be written as

$$P_c = \frac{2\sigma}{r_c} \tag{5.70}$$

The effective capillary radius for the commonly used wick of rectangular groove can be determined analytically. For the rectangular groove, one of the radii of curvature is infinity and the other radius for maximum capillary pressure, zero wetting angle, is equal to half of the groove width (assuming that the groove path is greater than half of the groove width). Therefore, for a rectangular groove,

$$r_c = W \tag{5.71}$$

where W is the width.

The function of a wick is threefold: (1) to provide surface pores at the vapor–liquid interface for capillary pressure, (2) to provide the necessary flow passage for the return of the condensed liquid, and (3) to provide the heat flow path between the inner wall of the pipe and the vapor–liquid interface. An axially grooved wick structure is a popular structure. It provides highly conductive metal fins which provide a low resistance for heat flow across the wick. A compilation of the major wick designs is given in Fig. 5.17. Wicks 1 to 3, screen, sintered structure, and standard axial groove, are homogeneous wicks. Wick 4 is a slab design, wick 5 is a special combination of groove and slab design, and wick 6 is a multiwrap or filled artery design. Wicks 7 to 13 are examples of composite or arterial wicks, which in general offer a higher heat pipe performance but are more sensitive to the presence of gas within the heat pipe.

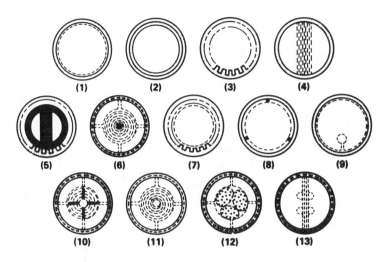

Figure 5.17 Types of heat pipe wicks (Ref. 15).

An increase in heat input to a conventional heat pipe will result in an increase of operating temperature if the sink temperature is unchanged. Therefore, the heat pipe acts like a heat-transport element of constant thermal conductance. The heat flow can be expressed as

$$Q = C_{HP}\Delta T_{HP} \tag{5.72}$$

where Q = heat flow, w
$\quad C_{HP}$ = heat pipe thermal conductance, w/k
$\quad \Delta T_{HP} = (T_e - T_V) + (T_V - T_C)$, k
$\quad T_e$ = evaporator surface temperature, k
$\quad T_V$ = vapor temperature, k
$\quad T_C$ = condenser temperature, k

When the external heat source and heat sink are interchanged, the heat flow in the pipe is reversed. An important parameter used in specifying a heat pipe is the transport capability (or pump capability) of the pipe. The transport capability is defined as the product of the heat to be pumped and the effective length. The effective length is normally defined as the distance from the midpoint of the condenser to the midpoint of the evaporator. The transport capability decreases with an increase in the operating temperature. Constant-conductance heat pipes have a disadvantage in applications where equipment heat dissipation and/or the outside thermal input is variable. For such applications, variable-conductance heat pipes have been developed.

Variable-conductance Heat Pipes. The purpose of a variable-conductance heat pipe (VCHP) is to control the operating temperature of given equipment against variations in the heat dissipations of the equipment and/or in the thermal environment.

The change in the conductance of the heat pipe may be exerted by controlling the mass flow using either of the following techniques:

1. Interruption of the fluid flow, either the liquid flow in the wick or the vapor flow in the core, by means of thermostatically controlled valves
2. Reduction of the condensation rate

The second technique is most widely used. It is normally achieved by the introduction of a noncondensible gas in the condenser, partially displacing the condensible gas.

The operation principle of a simple VCHP using a nonheated gas reservoir which is lined with a capillary wick is shown in Fig. 5.18. The temperature profile along the heat pipe is shown for three cases: maximum heat transfer, Q_{max}; nominal heat transfer, Q; and minimum heat transfer, Q_{min}. The sink temperature is kept constant. For maximum heat transfer, Q_{max}, the total condenser length is active and available for heat transfer to the sink. For minimum heat transfer, Q_{min}, the condenser is essentially blocked for heat transfer (i.e., full of gas). The indicated location of the vapor/gas front refers to a nominal heat transfer which requires part of the condenser to be active for heat transfer.

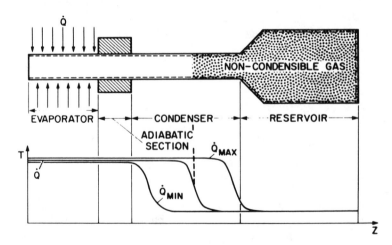

Figure 5.18 Schematic of variable conductance heat pipe with cold gas reservoir (Ref. 15).

Heat pipe radiators are used in spacecraft which use high-power TWTAs for direct broadcast television. The traditional thermal control technique employing a thick heat sink to distribute the TWT's heat over the radiator area results in excessive weight. A study was performed at Comsat Labs (Ref. 12) for the design and analysis of heat pipe north-south radiators for a three-axis-stabilized spacecraft for such applications. The heat pipes were assumed to be ATS-F-type ammonia-filled axially grooved heat pipes. The north-south equipment panels of the spacecraft were standard aluminum honeycomb panels with embedded heat pipes. Optical solar reflectors (OSRs) on the panel's exterior radiate the heat efficiently into space. The heat pipes are used as structural members to provide additional stiffness to the panel.

The primary parameters governing a radiator's heat rejection capability are heat pipe spacing, face sheet thickness, and heat pipe vapor temperature. In the Comsat Labs study, the masses of the radiator elements are assumed to be: face sheet, 2.77 g/cm^3; core, 0.05 g/cm^3; adhesive, 0.029 g/cm^2; OSR, 0.049 g/cm^2; and heat pipe, 1.49 g/linear centimeter. The thermal parameters were: OSR solar absorptance, 0.1 (nondegraded) and 0.21 (degraded); aluminum conductivity, 1.68 $W/cm\cdot°k$; bonded heat pipe; joint conductance, $0.85w \ cm^2\cdot k$; and heat pipe evaporator film coefficient, 0.7 $W/cm^2\cdot K$. The variation of the radiator area with weight per watt of dissipation as a function of heat pipe spacing, face sheet thickness, and vapor temperature is given in Fig. 5.19. According to Fig. 5.19, the minimum radiator mass is obtained with a heat pipe separation of approximately 25 cm and a face sheet thickness of 0.5 mm. In terms of mass, it is advantageous to design the radiator to operate at the highest possible temperature. For the study, the allowable temperature for the TWT baseplate was 50°C. The temperature difference between the TWT baseplate and heat pipe vapor is 10°C. Therefore, the radiator was sized for an average temperature of 40°C under the worst environmental and operational conditions.

Louvers (ref. 13). For a spacecraft in which the changes in internal power dissipation or external heat fluxes are severe, it is not possible to maintain the spacecraft equipment temperatures within the allowable design temperature limits unless the α_S/ε ratio can be varied. A very popular and reliable method which effectively gives a variable α_S/ε ratio is through the use of thermal louvers. When the louver blades are open, the effective α_S/ε ratio is low (low α_S, high ε) and when the blades are closed, the effective α_S/ε ratio is high (low α_S, low ε). The louvers also reduce the dependence of spacecraft temperatures on the variation of the thermo-optical properties of the radiators.

Louvers consist of five main components: baseplate, blades, actuators, sensing elements, and structural elements. The base plate is a surface of low absorptance-to-emittance ratio which covers the critical set of equipment whose temperature is being controlled. The blades, driven by actuators, are the elements of the louvers which give variable radiation characteristics at the baseplate. When the blades are closed, they shield the baseplate from its surroundings. When they are fully open, the coupling by radiation

5.5 SPACECRAFT THERMAL DESIGN

A spacecraft thermal design is highly dependent on the mission and the type of attitude stabilization system. In a dual-spin-stabilized spacecraft, the spinning solar array drum equalizes the solar flux and provides a comfortable temperature to the internal equipment. The north face can radiate freely from the spacecraft except for the obstruction caused by antennas. The south face normally contains the apogee motor nozzle and is usually covered by the shield or the multilayer insulation blanket. Recent spacecraft, such as SBS and Intelsat VI, have radiators on the solar array drum. Only the antennas on the despun section experience diurnal variation of the solar flux (except for eclipse during equinox).

In three-axis-stabilized spacecraft, the main body is basically non-spinning. It rotates about N/S axis at one revolution per day in order that the antennas continually orient toward the desired area on the earth surface. This results in a diurnal variation of solar flux on the east, antiearth, west, and earth surfaces of the spacecraft. These surfaces are normally covered with multilayer insulation to avoid the extreme daily temperature variation. The north and south surfaces are affected only by seasonal variation of solar incidence angle of $\pm 23.5°$ and shadowing by spacecraft appendages. The thermal control design of a three-axis-stabilized spacecraft is normally more difficult than that for a dual-spin-stabilized spacecraft.

As examples, this section provides a review of the thermal designs of a dual-spin-stabilized spacecraft, Intelsat IV, and a three-axis-stabilized spacecraft, Intelsat V and ATS-F, which used active thermal control techniques.

Dual-Spin-Stabilized Spacecraft

Figure 5.20 shows the thermal control elements of Intelsat IV, which is a dual-spin-stabilized spacecraft. The despun section consists of an antenna/communications subsystem. The remaining subsystems, propulsion, attitude control, electric power, and so on, are in the spun section.

Since the antenna/mast assembly is despun, the sun makes one complete revolution relative to it in a 24-hour period. Although the antenna assembly is not generally sensitive to temperature extremes, the temperature gradients cause structural distortion, which results in antenna pointing error. To reduce temperature gradients, multilayer insulation blankets are used extensively.

The TWTs and their power supplies provide the majority of the internal power dissipation of the spacecraft. Hence it is necessary to control the local temperature and the bulk temperature in the despun compartment. The bulk temperature control requires rejecting the power into space, necessitating radiators with low α_s/ε ratio. The forward-end sunshield and spinning solar array perform this function. The forward sun shield has two sections. The in-board section is covered on the space side with aluminized Teflon ($\alpha_s = 0.16$, $\varepsilon = 0.66$). The outboard section (spinning) is covered

cooling. To improve the contact surface conduction, TWTs and all electronic boxes are mounted with RTV filler. The communication receiver and certain other electronics boxes are gold plated ($\varepsilon = 0.03$) for required radiation isolation during eclipse. Output multiplexers are covered with MLI to decouple them from sunshield cooling during eclipses. The remaining boxes are painted with high-emittance white paint ($\varepsilon = 0.85$).

Positioning and orientation subsystem (P&O). The temperature control in the P&O subsystem is designed to prevent hydrazine from freezing in orbit. The propellant tanks are covered with MLI. The lines and valves are conductively decoupled from the spacecraft structure with low-conductivity spacer materials and wrapped with low-emittance aluminum-foil tape ($\varepsilon = 0.05$) to reduce radiation cooling during eclipse. Heaters are used around the lines and valves. The thruster chamber is enclosed in a canister consisting of two concentric shells of stainless steel separated by refrasil batting to shield the spacecraft from very high thruster firing temperatures (800°C).

The temperature of the apogee motor, propellant, and nozzle throat area, is kept in the temperature range 4 to 32°C throughout the transfer orbit in order that the apogee motor may ignite properly. The apogee motor case is covered with multilayer insulation and the heaters heat the nozzle throat on command. Post-apogee motor firing heat soakback to the spacecraft is minimized by multilayer insulation and the conduction isolation at the mounting ring.

The solar panel provides a very suitable environment for the spun section except during eclipse. The aft thermal barrier is used to minimize heat loss out of the aft end and shields the spacecraft from apogee motor exhaust plume heating and contamination. The thermal barrier is composed of a stainless sheet with black paint on the outside and gold coating on the inside, and nickel-foil layers. The low-power-dissipating equipment, the spacecraft housekeeping, are mounted on the spacecraft structure in the spun section.

Three-Axis-Stabilized Spacecraft

As discussed earlier, in a three-axis-stabilized spacecraft the main body is nonspinning. This results in the sun making a complete revolution with respect to the east/west surfaces in a 24-hour period and the north/south surfaces see the variation of solar incidence angle of only $\pm 23.5°$ with a period of a year. Therefore, the solar flux variation is significantly smaller on the north/south surfaces in comparison to the east/west surfaces. Considering the solar flux variation on the spacecraft surfaces, the thermal control design for a typical three-axis-stabilized spacecraft consists of locating high-power-dissipation equipment, such as TWTs, on the north/south surfaces where the heat can be radiated directly to space. The east/west surfaces are covered with multilayer insulation to minimize the effect of solar flux variation on the equipment.

Intelsat V.   The thermal control configuration for Intelsat V, a three-axis-stabilized spacecraft, is shown in Figs. 5.21 and 5.22 for a transfer orbit and a synchronous orbit, respectively. The thermal control of Intelsat V spacecraft is accomplished using passive techniques, including a selective location of the power-dissipating equipment, a selective use of surface finishes, and the regulation of thermal paths. The passive design is augmented with heater elements for the batteries, hydrazine-propellant lines and tanks, and the apogee motor to keep the temperatures above the minimum allowable levels. Heater elements are employed in the catalytic thrust chambers and electrothermal thrusters to optimize performance and useful lifetime. The thermal control for three major parts of the spacecraft—main body, antenna module, and solar arrays—is achieved essentially individually with heat transfer between these parts minimized.

Main Body Thermal Control.   The overall thermal control of the main body is achieved by (1) the heat dissipation of components in the communications and support subsystem modules, (2) absorption of solar energy by the OSR radiator on the north/south panels, and (3) the re-emission to space of infrared energy by the OSR radiator. High thermal dissipators,

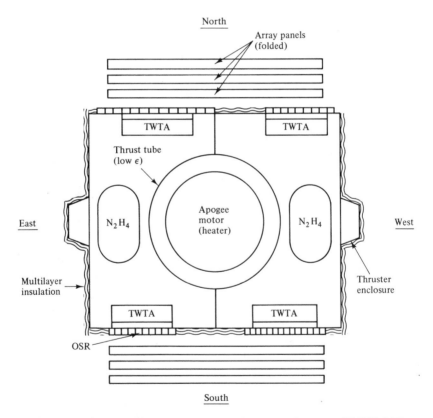

Figure 5.21 Preoperational-phase thermal configuration. (Courtesy of INTELSAT and FACC)

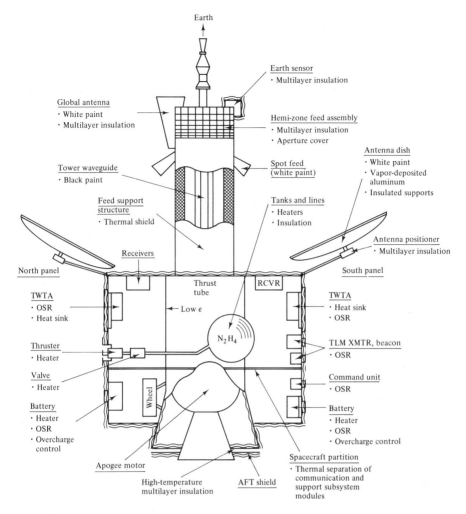

Figure 5.22 Thermal subsystem operational configuration features (solar array and shunt limiters not shown). (Courtesy of INTELSAT and FACC)

such as TWTAs, are located on the north and south panels so that they may efficiently radiate their energy to space via the heat sinks and the OSR radiators. The east and west panels, antenna deck, and antiearth surfaces are covered with multilayer insulation to minimize the effect of solar incidence on equipment temperature control during a diurnal cycle. Most of the equipment is directly attached to the interior surface of the north and south radiator panels, thereby maximizing the thermal efficiency of transporting heat from the source to the radiator. The requirements for some equipment are best satisfied, however, by utilizing other locations. As an example, receivers are mounted on the earthward panel to minimize waveguide lengths and to achieve a more stable temperature level than would occur in locations near to TWTAs. The propulsion subsystem components are mounted on or near to the east and west panels to accommodate thruster location requirements and to minimize propellant line lengths. The concentrated heat load of the TWTA collectors is distributed by means of high-conductivity ASTM 1050 tempered aluminum heat sinks between the TWTA units and their mounting panels.

The conductive heat transfer with the antenna module has been minimized primarily to limit the large diurnal temperature effect of the antenna module on the main body. This is achieved by using multilayer insulation between the antenna module and the main body, graphite/epoxy for the tower legs, and low-thermal-conductance spacers.

The propellant tanks, lines, and valves are covered with insulation, and heaters are provided to keep the temperature above the freezing temperature of hydrazine. To protect against plume heating during the apogee motor firing, high-temperature insulation blankets, which consist of a titanium outer layer to permit a temperature of up to 500°C, are used on the aft side of the spacecraft. The heat-soak-back effects of the apogee motor are minimized by using a low-emittance surface inside the thrust tube and on the motor casing ($\varepsilon < 0.25$) and by conductively decoupling the apogee motor. The high-temperature insulation blankets minimize the plume heating effects.

Antenna Module. For spacecraft in a geosynchronous orbit, the sun appears to travel completely around the antenna module once each day. The various parts of the module are subjected to full sun and complete shadowing, the shadowing is caused by the main body, the module elements, and by the earth during the equinox. Therefore, the antenna module is subjected to a rather severe environment, considering that minimal heat is dissipated within the module itself.

The thermal control of the tower structure and tower waveguide is achieved mainly by the use of a three-layer thermal shield around the tower structure. The shield is painted black on the outside to prevent solar reflections from concentrating on antenna reflectors. The interior surface of the shield is bare Kapton, enhancing radiative heat transfer within the tower. The shield serves to minimize the diurnal temperature variation within the tower. The thermal control of reflectors is normally achieved by the use of white paint on the concave surfaces of the reflectors and a multilayer insulation blanket on the convex surface.

Solar Array Thermal Control. The primary elements of a solar array are the solar array panels, the shunts, and the yoke. Thermal control of the solar array panels in geosynchronous orbit is achieved by the absorption of solar fluxes on the solar cells on the front surfaces of the panels and reemission of infrared energy from the front and graphite/epoxy back surface of the panels.

Shunts with beryllium heat sinks are located on the antisun surfaces of each of the two in-board panels. The front side of the solar panel opposite the shunt is painted white. The shunts and beryllium heat sinks are covered with aluminum tape on the anti-sun surface. The white paint reduces absorbed solar energy. The aluminum tape lowers the shunt's radiative coupling to space, raising the minimum shunt temperature experienced at the end of the equinox eclipse. Thermal control of the yoke is achieved by painting the yoke white.

Application technology satellite (ATS-F) (ref. 10). The ATS-F was a three-axis-stabilized spacecraft for advanced communications and scientific experiments. The spacecraft consisted of two major elements, a precision (30-ft-diameter) deployable antenna reflector and an earth-viewing module (EVM) which housed the electronic components and scientific packages. The thermal design of the EVM employ active thermal control elements such as louvers and heat pipes.

The EVM, shown in Fig. 5.23, was divided into three distinct modules: experimental module, service module, and communications module. The experimental module (EM), closest to the earth viewing side, contained earth sensors and experiments. Thermal control of this bay was critical due to the alignment requirements of the sensors and the tight temperature specification (5 to 25°C) for the batteries. The service module (SM) contained most of the housekeeping units, resulting in nearly constant internal power dissipation and temperature. The communications module (CM) contained some of the highest power dissipators, such as the C-band, S-band, and L-band transponders and the ion engine electronics and thrusters. The thermal control of this bay was complicated by large fluctuations of heat dissipation.

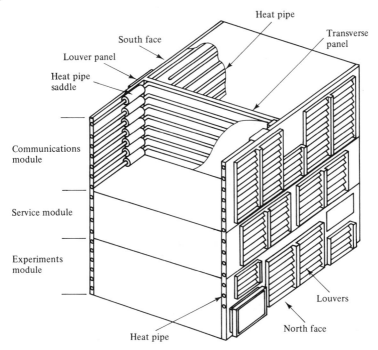

Figure 5.23 Thermal control system of ATS-F earth viewing module (Ref. 10).

The equipment mounting surfaces of the EVM were fabricated from 12.7-mm (0.5-in.) thick honeycomb with straight heat pipes bonded between the fact sheets of the north and south panels. The north and south panels were connected by a similar transverse honeycomb panel with heat pipes in the shape of a "C" or a "Z," which served to equalize the heat loads between the two panels. All heat pipes were in the same plane and defined a series of "H" segments, as shown in Fig. 5.23. The outboard surfaces on the north and south panels were covered with OSR, which, together with thermal louvers, provided the primary heat rejection system for the spacecraft.

The thermal louvers consisted of a support frame, individually activated blades, and bimetallic actuator springs located in a central enclosure. The frames and the blades were highly polished. The OSRs, 0.2 mm (8 mils) thick with vapor-deposited silver, were used as the thermal control surfaces behind the louvers. The effective emittance of the louver radiator, ε_{eff}, is a function of louver blade angle and the effective absorptance, α_{eff}, is a function of louver blade angle, θ, and solar aspect angle, ϕ. The net heat rejection of the louver radiator to space is given by

$$Q = A_b[\varepsilon_{eff}(\theta)\sigma T_b^4 - \alpha_{eff}(\theta, \phi)S] \qquad (5.73)$$
$$= A_b F_{net}\sigma T_b^4$$

where $F_{net} = \varepsilon_{eff} - \alpha_{eff}S/\sigma T_b^4$
- = net effective emittance of louver
- Q = net heat leaving the louver radiator
- A_b = radiator area
- ε_{eff} = effective emittance of the louver radiator
- α_{eff} = effective absorptance of the louver radiator
- S = solar constant
- σ = Stefan–Boltzmann constant

The nominal actuation range of the ATS louvers was 16.6°C (30°F) from the fully closed to fully open blade positions. Figure 5.24 shows the net effective emittance of the ATS louver with the solar aspect angle fixed at 23.5°. The temperature set point for the louver blade opening and closing defines the net effective emittance curve for each louver assembly.

The ATS heat pipe was made of internally grooved aluminum tubing with ammonia as the working fluid. The 11.4-mm (0.45-in.)-internal-diameter tubing has 27 axial grooves for capillarity, with average groove dimensions of 0.76 mm (0.03 in.) wide and 1 mm (0.04 in.) deep. The groove geometry and the cross section of the heat pipe are shown in Fig. 5.25. For the straight heat pipes, 305 mm (12 in.) of the length of the heat pipe was evaporator and the remainder was condenser, maximum heat transport capability was 1500 W-in., and the heat load was 60 W. The temperature differences were less than 5.5°C between maximum evaporator and minimum condenser temperatures and maximum 1.66°C (3°F) within evaporator or

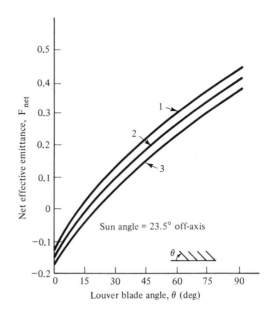

Louver Actuator
Set-point Temperatures

	Closed		Open
1	18.8°C	~	35.5°C
2	13.3°C	~	30°C
3	7.7°C	~	24.4°C

Sun angle = 23.5° off-axis

Figure 5.24 Net effective emittance of ATS louvers with OSR radiator versus louver blade angle (Ref. 10).

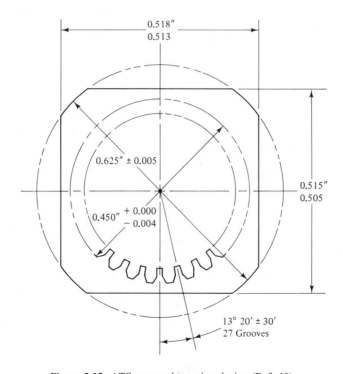

Figure 5.25 ATS groove heat pipe design (Ref. 10).

condenser. The pressure was 1700 psia at 132°C and the operating temperature was 5 to 40°C.

5.6 THERMAL TESTING

Thermal testing is performed at various stages of spacecraft development. A typical thermal test program consists of (1) a thermal design verification test, also known as a thermal balance test, on the thermal model; (2) qualification thermal vacuum tests on the components, subsystems, and prototype spacecraft; and (3) acceptance thermal vacuum tests on the components, subsystems, and flight spacecraft. The objectives of the thermal balance test are to evaluate the ability of the thermal control subsystem to maintain equipment temperatures within the allowable limits and to verify the analytical model. The qualification thermal vacuum test, where the spacecraft equipment temperatures are maintained 10°C beyond the expected extreme temperatures, has the main purpose of identifying whether any adverse effects on spacecraft performance result from any weaknesses in the thermal design. The acceptance thermal vacuum test, where the spacecraft equipment temperatures are maintained 5°C beyond the expected extreme temperatures, has the objective of identifying any adverse effects that result from defects in materials or workmanship related to the fabrication of the unit. Both qualification and acceptance tests involve the collection and analysis of spacecraft subsystem performance data. This section provides further details of these tests.

Thermal Balance Test

The thermal balance test is performed to evaluate the spacecraft thermal design by accurately simulating the worst-case environment, to verify the thermal mathematical model, and to establish qualification and acceptance test temperature levels.

In this thermal test, the test chamber must simulate vacuum and the effective space temperatures. At geosynchronous altitude, the vacuum is generally less than 10^{-6} torr, which eliminates heat exchanges by convection or gaseous conduction. The vacuum reproduced in the test chamber must be such that these types of heat exchanges are negligible and a pressure below 10^{-5} torr is sufficient. The spatial environment temperature experienced by an orbiting satellite is in the region of 4 K. Fortunately, this temperature does not need to be reproduced in the thermal vacuum chamber via a cold shroud. Higher chamber temperatures are more readily achievable, more cost-effective, and have a negligible effect on the test results, with the optimum cold shroud temperature of 100 K (LN_2 temperature) the most often used. An ideal cold shroud which simulates the space heat sink should not reflect either solar radiation or thermal radiation emitted by the spacecraft back toward the spacecraft. It should also emit only negligible thermal radiation toward the spacecraft.

In the thermal balance test, the most critical thermal balance conditions are simulated. Analytical models of the spacecraft thermal design are developed to predict the performance of the thermal design in a known test chamber environment. The correlation of the results between the chamber thermal balance tests and the analytical model can thus provide a means of validating the thermal design and improving the analytical model accuracy in predicting performance.

Thermal balance tests can be performed by utilizing three different techniques for simulating the thermal environment: solar simulation, infrared simulation, and skin heaters. In the solar simulation, the solar beam is simulated by lamps which closely match the sun intensity and spectrum. In infrared test, the absorbed solar irradiation predicted by analysis is simulated by IR lamps, IR plates, and the appropriate chamber shroud temperature. With skin heaters, the absorbed solar energy is applied to heaters attached to the spacecraft exterior surfaces.

A thermal balance test most often includes steady-state phases corresponding to different environmental conditions and transient phases to simulate eclipses. The steady-state phases allow a check and eventually the correction of conductive, radiative, and thermo-optic characteristics of the mathematical model. The thermal capacities are checked by a transient phase.

The goal in a thermal balance test is not necessarily to subject the spacecraft to a thermal environment identical to in-orbit conditions but to simulate this environment as economically as possible. As discussed earlier, the thermal design verification is an iterative process which includes the prediction of on-orbit temperatures based on the analytical model, prediction of thermal control design performance for the test chamber environment, correlation of test results with predictions, and the use of the verified analytical model for final in-orbit temperature predictions. Therefore, the imperfection of any thermal balance test must be justified with respect to the correct qualification of the analytical model and with respect to the influence of these imperfections on the absolute temperature values.

Solar Simulation Test

The Jet Propulsion Laboratory (JPL) 8.2-m solar beam facility (5.6-m beam) is shown in Fig. 5.26. The minimum operating pressure of the chamber is 10^{-6} torr. During the Intelsat V spacecraft testing, chamber pressure was nominally held at 2×10^{-6} torr. The walls and floor are lined with thermally opaque aluminum cyrogenic shrouds controlled over the temperature range -185 to $95°C$ using liquid and gaseous nitrogen. An off-axis solar simulation system is provided by an array of 37 xenon 20/30-kW compact arc lamps. This array provides a simulated solar beam that is reflected down into the test volume by a collimating mirror, which is temperature controlled with gaseous nitrogen through the range -75 to $95°C$. The nominal solar beam is 5.6 m (18.5 ft) with 10% beam uniformity and a 2% nominal intensity control. Beam collimation accuracy is $\pm 2°$ and

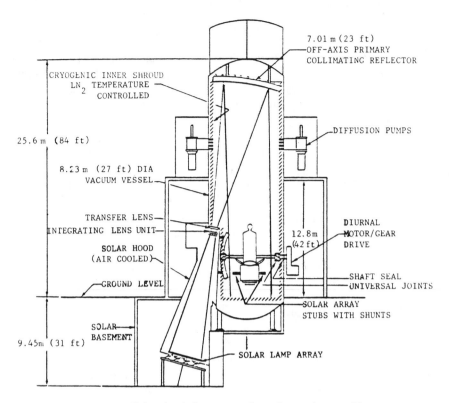

Figure 5.26 Solar simulation test configuration-equinox position.

the solar spectrum is that of xenon lamps as modified by the simulator optics.

Since the solar beam is fixed, the spacecraft is rotated to simulate the different solar aspect angles, namely, the diurnal variation of the solar aspect angles of 360°/day and the annual variation of the solar aspect angle of ±23.5°. Solar simulation test is an expensive test. However, for complex shapes, such as complex antenna structures, with multiple reflection and shadowing effects, solar simulation is the only reliable thermal balance test technique. Such effects are difficult to take into account by the analytical predictions.

As an example of what a thermal test model consists of, Intelsat V, a three-axis-stabilized spacecraft, is described below. In the main body, the structure consists of flight design equipment panels, including thrust tube, ADCS deck, and OSR on the north and south panels. Electronic equipment is simulated by dummy boxes designed to the same mass, center of gravity, form factor, foot print, conductance, thermal dissipation (simulated by resistors), mounting arrangement, and surface finish as the flight units. The antenna system is prototype hardware as far as is possible. The internal and external insulation and thermal finishes are identical to the flight design. The deployed solar array configuration is represented by two stub solar array wing assemblies on Intelsat V due to limitations of the JPL chamber size. The assembly includes accurate thermal simulation of the solar array drive attach bracket, the yoke, and the in-board deployed panel and shunt assembly. For large spacecraft, due to the limitations of the solar beam chamber size, it will be necessary to test the spacecraft in parts, which should be taken into consideration during the thermal design of the spacecraft, to make the test meaningful.

A typical test profile for a solar thermal vacuum test is shown in Fig. 5.27. It consists of basically the following five cases:

1. Summer solstice, BOL, minimum dissipation
2. Equinox, BOL, minimum dissipation, coldest design condition
3. Winter solstice, BOL, minimum dissipation
4. Summer solstice, EOL, maximum dissipation, hottest design condition–north panel
5. Winter solstice, EOL, maximum dissipation, hottest design condition–south panel

To simulate EOL condition for Intelsat V, solar absorptances of the OSR surfaces and white painted surfaces of hemi-zone feed are artificially increased by using black-painted tapes.

It should be noted that the spectral distribution and the collimation angle are generally different for a solar simulator than for the real solar radiation, and one has to account for these differences when correcting the mathematical model. If the spectral distribution is too different from that of the sun, the absorptances of the coatings have to be recalculated for this spectrum.

Infrared Simulation

The basic principle of IR techniques consists of analytically transforming the solar heat flux to the external surface of the satellite to an equivalent IR source such as lamps or heated wires and rods. This method requires detailed knowledge of the external heat balance; the total incident energy on the spacecraft; the ratios of reflected, emitted, and absorbed energy; and the external surface properties. Therefore, it relies more on analytical methods and assumptions than on solar simulation. Ideally, the radiation source must meet the following requirements:

1. The IR source must be sufficiently dense to provide a homogeneous flux.
2. The IR radiation must be sufficiently directional and collimated not to illuminate areas normally shadowed.
3. The spectrally irradiance of the IR source must be perfectly known or measured.

Another technique for IR simulation is to regulate the chamber shroud temperature so that the radiative interchanges between the spacecraft external

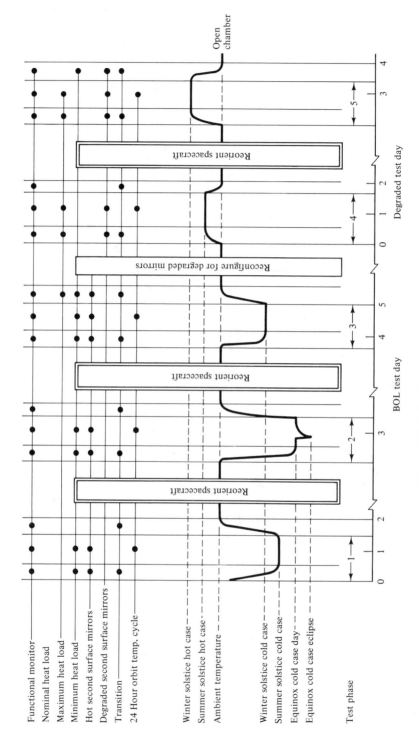

Figure 5.27 Solar thermal vacuum test profile.

coatings and the shrouds are identical to those in flight. However, this method has a shortcoming in that the surfaces with different coatings have a different heat-exchange coefficient with the shrouds, resulting in an inaccurate heat balance. This is obvious for the differing characteristics of the OSR radiator and multilayer insulation-covered surfaces. It is also not possible to simulate the local difference between solar illuminated and shadowed areas due to the global radiative interchanges.

Skin Heaters

In this method, the heaters are attached directly to the spacecraft body. The heater power is adjusted according to the absorbed heat calculated by thermal analysis. The advantages of this method are uniformity in heating; variation of heat input and gradients by electrical power control; no calibration is required for heaters; and surface emission to the cold shroud is not blocked by any test heater assembly. The disadvantages are that the absorbed heat is determined by analysis only; direct contact to the flight unit is not always suitable; and heater failure could damage the skins and these cannot be repaired during tests.

Spacecraft Thermal Vacuum Test

The objective of the spacecraft thermal vacuum test is to verify proper operation of the spacecraft under conditions of simulated space vacuum and imposed temperatures that are 10°C beyond the mission predicted temperatures for the qualification test and 5°C for the acceptance test. The specified temperatures are based on the thermal balance test results. The specified temperatures are forced at subsystem locations by modifying the operational modes of the spacecraft and by adjusting local temperature boundary conditions to provide additional heating, or cooling, or both heating and cooling.

The Intelsat V qualification thermal vacuum test configuration is shown in Fig. 5.28 for a geosynchronous orbit. All the black chamber shroud surfaces facing the spacecraft were flooded with liquid nitrogen to simulate the space thermal sink. A liquid-nitrogen-cooled fixture with Calrod element heaters was used to simulate the solar heating of the spacecraft. The principal features of the test fixture included:

1. Liquid-nitrogen enclosures and polished aluminum reflectors with Calrod heater elements for the north and south panels of the spacecraft
2. Calrod heater arrays with polished aluminum reflectors for east and west panels
3. Open frame Calrod heater arrays with horizontal polished aluminum reflectors for antenna module temperature control
4. Adiabatic mounting pedestal provided by conductive isolation and guard heaters

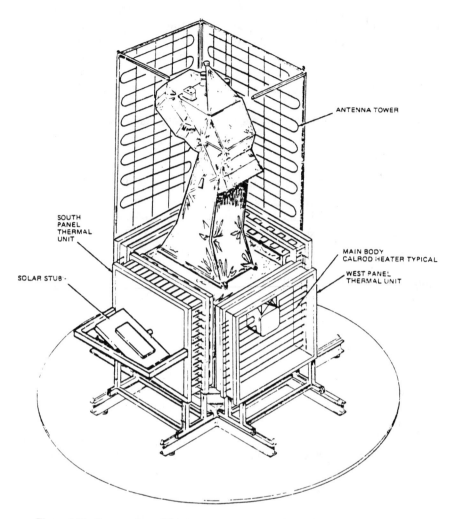

ANTENNA TOWER

SOUTH
PANEL
THERMAL
UNIT

MAIN BODY
CALROD HEATER TYPICAL

WEST PANEL
THERMAL UNIT

SOLAR STUB

Figure 5.28 Spacecraft qualification thermal vacuum test configuration. (Courtesy of INTELSAT and FACC)

For a transfer orbit, solar panels are kept in the stowed position. For synchronous orbit testing, however, solar simulation stubs were employed to support shunts on the north and south sides of the spacecraft. The antenna tower did not have reflectors. A simulated apogee motor casing was used. In the test, approximately 550 copper–constantan thermocouples were employed. Approximately 80 black absorbed-flux calorimeters were utilized to provide an infrared flux distribution, to provide a feedback control signal for automatically controlling power levels within zones of Calrod elements, and to measure the background energy levels when no IR solar simulator was provided.

Thermal qualification for the Intelsat V, a three-axis-stabilized spacecraft, consisted of the following test conditions:

1. Transfer orbit/eclipse, coldest case, nominal, minimum internal electrical heat load
2. Transfer orbit, coldest case, $-11°C$ margin, minimum heat load
3. Transfer orbit, hottest case, nominal, maximum heat load
4. Equinox/eclipse, nominal, BOL, minimum heat load
5. Winter solstice, hottest south panel, nominal, EOL, maximum heat load
6. Winter solstice, hottest south panel, 11°C margin, EOL, maximum heat load
7. Equinox/eclipse, coldest case, $-11°C$ margin, BOL, minimum heat load
8. Summer solstice, hottest north panel, nominal, EOL, maximum heat load
9. Summer solstice, hottest north panel, 11°C margin, EOL, maximum heat load

The spacecraft equipment performance testing is performed after stabilization at each test condition. The test conditions, sequence, and test period vary in spacecraft programs depending on the complexity of the spacecraft design and the time required to perform functional tests. Nominally, about 168 hours of thermal vacuum testing is performed. For acceptance tests, the temperature margin of 5°C is used.

PROBLEMS

5.1. A spherical satellite has a diameter of 1 m. The surface, which is covered with solar cells, has an effective solar absorptance 0.55 and emittance of 0.8. The internal electrical power dissipation is 200 W. Determine the average temperature of the surface of the satellite at vernal equinox.

5.2. The solar array for a three-axis-stabilized spacecraft has the following characteristics: area $= 4$ m^2, $\alpha_S = 0.75$, $\varepsilon_F = 0.75$, $\varepsilon_B = 0.75$, $F_p = 0.95$, $\eta = 0.14$. Calculate steady-state operating temperatures at summer solstice and vernal equinox.

5.3. A direct broadcast satellite has TWTAs. The RF power output of each TWTA is 200 W at 48% efficiency. The maximum allowable temperature for the TWTA baseplate is 50°C. The heat dissipation by the other equipment is 250 W. TWTAs and other equipment are equally divided between the north and south heat pipe radiators. Determine the mass and area for each radiator by using Fig. 5.19.

5.4. The radiator of a three-axis-stabilized spacecraft is designed to radiate 400 W of equipment thermal dissipation. It is assumed to be isothermal and is covered with OSR. The solar absorptivity at EOL $= 0.21$ and emissivity $= 0.78$. To take into account the reflection from the solar array and antennas, the radiating efficiency is assumed to be 90% during a noneclipse period. The allowable temperature range for the radiator is 0 to 40°C. Determine (a) the area of the radiator, and (b) the temperature at equinox during a noneclipse period.

5.5. In Problem 5.4 it is assumed that the battery provides power only to the housekeeping equipment and, after an eclipse, the spacecraft is warmed up by heaters to bring the radiator temperature up to the minimum operating temperature. The mass of the radiator is 100 kg and the specific heat is 900 W · s/kg K. The heat dissipation during an eclipse is assumed to be 100 W and 600 W during warm-up. Determine (a) the temperature of the spacecraft at the end of an eclipse of 72 minutes, and (b) the time required to warm up the radiator.

5.6. Equipment in a spacecraft has a thermal dissipation of 15 W. The equipment radiates heat to the radiator which radiates it to space. Both the equipment surfaces and the radiator inside surface are painted black. The effective exchange area, AF, between the equipment and the radiator is 0.16 m^2. The steady-state temperature of the radiator is 40°C. Determine the temperature of the equipment surfaces.

REFERENCES

1. R. M. Van Villet, *Passive Temperature Control in the Space Environment*, Macmillan, New York, 1965.

2. J. A. Wiebelt, *Engineering Radiation Heat Transfer*, Holt, Rinehart and Winston, New York, 1966.

3. W. H. McAdams, *Heat Transmission*, McGraw-Hill, New York, 1954.

4. P. D. Dunn and D. A. Reay, *Heat Pipes*, Pergamon Press, Elmsford, New York, 1976.

5. S. W. Chi, *Heat Pipe Theory and Practice*, McGraw-Hill, New York, 1976.

6. J. R. Howell and R. Siegal, *Thermal Radiation Heat Transfer*, NASA SP-164, 1969.

7. W. A. Gray and R. Muller, *Engineering Calculations in Radiative Heat Transfer*, Pergamon Press, Elmsford, New York, 1972.

8. *Spacecraft Thermal Control*, NASA SP-8105, May 1973.

9. *Comsat Technical Review*, Vol. 2, No. 2, 1972.

10. H. Hwangbo, J. H. Hunter, and W. H. Kelly, "Analytical Modelling of Spacecraft with Active Thermal Control System," AIAA 8th Thermophysics Conference, Palm Springs, Calif., July 16–18, 1973, AIAA Paper No. 73-773.

11. J. A. Robinson, "The Intelsat IV Spacecraft—Thermal Control," *COMSAT Technical Review*, Vol. 2, No. 2, Fall 1972.

12. W. H. Kelly and J. H. Reisenweber, Jr., "Optimization of a Heat Pipe Radiator for Spacecraft High-Power TWTAs," IV International Heat Pipe Conference, Sept. 7–10, 1981, The Royal Aeronautical Society, London.

13. W. H. Kelly, J. H. Reisenweber, Jr., and H. W. Flieger, "High Performance Thermal Louver Development," AIAA 11th Thermophysics Conference, San Diego, Calif., July 14–16, 1976, AIAA Paper No. 76-460.

14. Hughes Aircraft Company Geosynchronous Spacecraft Case Histories, Jan. 1981.

15. M. Groll, "Heat Pipe Technology for Spacecraft Thermal Control," *Spacecraft Thermal and Environmental Control Systems*, ESA SP-139, Nov. 1978.

6

ELECTRIC POWER

6.1 INTRODUCTION

The electric power subsystem provides power to the spacecraft during all mission phases, beginning with the switchover to internal battery during lift-off on through the on-orbit phase. The electric power system for a communications satellite is designed primarily for synchronous orbit load requirements, where, as shown in Fig. 6.1, typically 87% of the total power is allocated to the communications subsystem, while only 13% is required by housekeeping subsystems. In the communications subsystem, a majority of the power (82% of the total power) goes directly to the high-voltage power supplies for the traveling-wave tubes or power supplies for solid-state RF amplifiers.

The primary source of electric power is from the solar arrays, in which the solar cells convert solar energy into electric power by photovoltaic conversion. During a solar eclipse, when the power from the solar arrays is not available, power storage devices such as rechargeable batteries provide the power. These batteries are discharged during eclipses and are charged slowly during sunlight. For geosynchronous orbits, there are two periods of solar eclipse, each lasting 45 days centered around the equinoxes, with a maximum period of 1.2 hours per day. The solar arrays and the batteries have highly unregulated power characteristics. Hence power control electronics are used to regulate spacecraft bus voltage and the charge rates for the batteries. Therefore, an electric power subsystem consists of three basic elements: solar array; electric power storage devices, generally batteries; and power control electronics.

A simplified block diagram for a conventional electric power subsystem is shown in Fig. 6.2. The power from the main solar array is supplied to

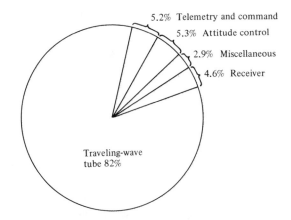

Figure 6.1 Power distribution in a typical communications satellite.

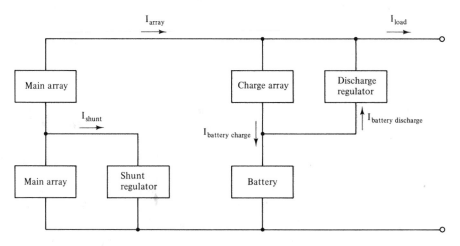

Figure 6.2 Electric power subsystem block diagram.

the spacecraft subsystems through the primary power distribution bus. The voltage of the solar array can vary extensively as a function of several factors, such as load and temperature. After eclipse exit, when the array is cold, its output voltage can double. To prevent the bus voltage from exceeding the voltage design range of the subsystems, regulators, such as a shunt regulator in this case, are used. The regulators bypass some of the output current from the solar panel to control the bus voltage when the solar power exceeds the load. A battery discharge regulator is designed to turn on when a predetermined decrease in the bus voltage is sensed. Such a case would be represented by a decreasing solar panel output during entry into a solar eclipse. The regulator then provides a minimum regulated output voltage on the bus from the battery. The on-charge voltage level of the battery must be higher than the bus voltage in order to achieve full recharge. To meet this need, an additional small-area solar array is often connected in series with the main array to boost the voltage for battery charging. This array can be sized to provide constant-current charging of the battery.

This chapter provides an overview of the basic elements of an electric power subsystem. In the solar array section, performance characteristics of different solar cells, radiation degradation, solar cell parameters, solar array electric configuration, and different deployable solar array designs are discussed. The section on batteries provides an overview of electrical characteristics of Ni-Cd and Ni-H$_2$ batteries. Different types of power regulators, such as series, shunt, and pulse modulated, are discussed in the section on control electronics. Typical electric power subsystem configurations are also discussed in this chapter.

6.2 SOLAR ARRAY

A solar array, sometimes called a solar generator (in Europe), consists of solar cells which convert solar energy into electric power by photovoltaic conversion, interconnectors to connect the solar cells, panels on which the solar cells are mounted, and deployment mechanisms to deploy the panels in orbit. This section presents the basic characteristics of the solar array elements.

Solar Cells

The photovoltaic (PV) effect is the basis of the conversion of light to electricity in solar cells. Solar cells are made with semiconductor materials. To explain PV effects, a review of some of the basic atomic characteristics will be necessary.

All materials are made from atoms. Atoms are composed of three kinds of particles: protons, neutrons, and electrons. The positively charged protons and neutral neutrons reside in a nucleus. The negatively charged electrons, much lighter than protons and neutrons, orbit the nucleus. Although an atom contains charged particles, overall it is electrically neutral. The electrons can exist only at discrete energy levels.

The distribution of electrons in the highest energy band determines most of the thermal and electrical properties of the materials. In an atom, the highest band that is occupied is called the valence band and the electrons in this band are called valence electrons. In an insulator, the valence band is full and the energy gap between the valence band and the next allowed band, called the conduction band, is so large that under ordinary circumstances the application of an external field is insufficient to raise a valence electron into the conduction band. In a conductor, the conduction band is partially filled. Hence the electrons in the conduction band will be able to accept energy, resulting in the conduction of the electric current. Semiconductors are similar to insulators except that in semiconductors the forbidden gap is much smaller. For example, alumina (Al$_2$O$_3$) at room temperature has an energy gap (E_g) of 10 eV, while a semiconductor, silicon, has an energy gap of only 1.1 eV.

The silicon atom has four valence electrons. When atoms are brought close together, as in a crystal, their valence electrons bond them together. As a solid, each silicon atom usually shares each of its four valence electrons with another silicon atom. When light strikes a silicon crystal, if the energy of light is greater than the energy gap, it will separate electrons from their atomic bond, producing an electron–hole pair. The electrons are free to move through the crystal. However, if no other mechanism is involved, the light-generated electrons and holes would meander about the crystal randomly for a time and then lose their energy thermally as they return to their valence position. For the electrons and holes to produce an electric current, another mechanism, called the potential barrier, is neeeded. The potential barrier selectively separates light-generated electrons and holes, sending more electrons to one side of the cell and more holes to the other. Thus separated, the electrons and holes are less likely to rejoin each other and lose their electric energy. This charge separation sets up a voltage difference between either terminal of the cell which can be used to drive an electric current in an external circuit.

There are several ways to form a potential barrier in a solar cell. The common method is to form a *p-n* junction, which is explained here briefly. In pure silicon, each of the four valence electrons associated with each atom are part of bonds in the crystal. Therefore, there are no free electrons. Let us assume that some of the silicon atoms are replaced by phosphorus atoms, which have five valence electrons. In phosphorus atoms, four of its valence electrons will assume positions similar to those assumed by the four valence electrons it replaces. The fifth electron of the phosphorus atom is not part of the bond and is relatively free. The impurity atom will have positive charge because of giving up an electron. Such a crystal is known as *n*-type because it has free negative charges. Similarly, suppose that some of the silicon atoms in a pure silicon crystal are replaced by boron atoms, which have three valence electrons. In this case, one of the bonds of the boron atom with the adjacent silicon atom would be missing an electron. The missing electron acts in many ways as though it were a free positive electron and is called a hole. Such a crystal is called *p*-type because of the presence of free positive charges in it.

Figure 6.3(a) shows a diagram of the *n*-type and the *p*-type silicon crystals. The charges with circles represent the positive phosphorus and negative boron ions, which are fixed in the crystal as they are bonded to the silicon atoms. The uncircled charges are free to wander through the crystal. When *n*-type and *p*-type materials are joined, as shown in Fig. 6.3(b), the diffusion of the charges across the junction occurs. The free electrons in the *n*-type material will see a region to their left where there are no free electrons, and hence there will be a flow of these electrons to the left. The free holes in the *p*-type material will see a region to their right where there are no holes, so there will be a flow of holes, positive charges, to the right. As time goes on, on the left-hand side there is now an excess of negative charges and on the right side an excess of positive

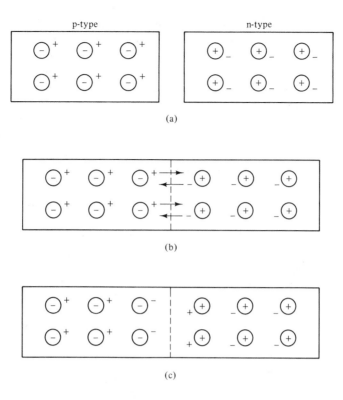

(a)

(b)

(c)

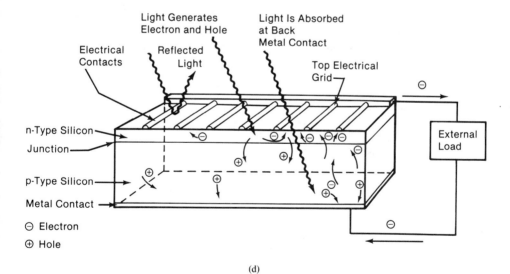

(d)

Figure 6.3 Solar cell operation.

charges. Thus an electrical field has been established in the region of interface of the two materials. This process continues until the electrical potential becomes of such a magnitude that further diffusion of holes and electrons is prohibited. When this equilibrium is established, the construction of the p-n junction has been completed, resulting in a permanent electric field without the aid of an external electric field, as shown in Fig. 6.3(c).

The operation of a solar cell is illustrated in Fig. 6.3(d). Let us assume that the light striking the solar cell has enough energy to free an electron from a bond in the silicon crystal. This creates an electron–hole pair. Let us also assume that the electron–hole pair is generated on the p-type silicon side of the junction. The electron has only a short time during which it is free because it is likely to combine with one of numerous holes on the p-type side. The solar cells are designed such that the electron will encounter a junction before it has the chance to combine with a hole. Once the free electron is in the neighborhood of a junction, it is accelerated across the barrier into the n-type silicon. The hole partner of this electron–hole pair, however, remains on the p-type side of the junction because it is repelled by the barrier at the junction and it is not in danger of recombining because there is already a predominance of holes on the p-type side. A similar situation occurs when the electron–hole pairs are generated by light on the n-type side of the junction. The free electron remains on the n-type side of the junction and the hole crosses the junction into the p-type side. If we connect the n-type side to the p-type side of the cell by means of an external electric circuit, current flows through the circuit. The amount of light incident on the cell creates a nearly proportional amount of current. The voltage depends on the potential barrier of the junction and is always less than the energy gap, for example, for the particular semiconductor.

Solar cell operation can also be explained by the energy-level diagram of the n-p junction shown in Fig. 6.4. At the n-p junction, the potential

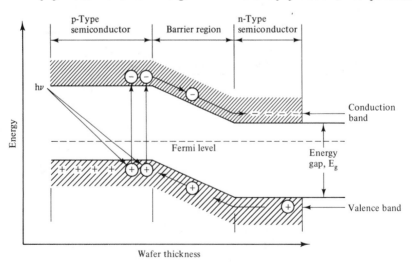

Figure 6.4 Energy-level diagram of solar cell.

barrier is formed because thermodynamics require that the Fermi level be the same in the two materials. The Fermi level is by definition the energy at which the probability of a state being filled is exactly one-half and it is near the bottom of the energy gap for p-type silicon and near the top of the energy gap for n-type silicon. Because of the potential barrier, the excess electrons flow to the right and the excess holes to the left. As explained earlier, if due to light photons, electron–hole pairs are generated on the p-type silicon side of the junction, the electrons flow to the right into the n-type silicon. Similarly, if electron–hole pairs are generated on the n-type silicon side, holes will move to the left to the p-type silicon. If suitable electrical connections are made, electric current will flow, resulting in conversion of solar energy into electricity.

Figure 6.5 shows the I–V curves of an unilluminated and an illuminated solar cell. Quadrants 1 and 3 correspond to a power sink and quadrants 2 and 4 correspond to a power generator. A nonilluminated silicon cell is a power sink and is similar to a semiconductor diode. Its resistance is low in the forward direction and high with current flow in the reverse direction. When the solar cell is illuminated, its I–V curve is displaced in the $-I$ direction by a distance proportional to the illumination intensity. Now the I–V curve passes through quadrant 4 and the solar cell becomes a power source. If we turn the quadrant 4 diagram upside down, we get the conventional characteristic I–V output curve for a solar cell. The maximum power delivered, P_{max}, is the largest I–V product, which is represented by the largest rectangle that can be fitted in this area. Therefore, the squarer the current–voltage characteristics, the higher the efficiency. The power, the product of I and V, as a function of voltage is shown in Fig. 6.6. The maximum power occurs in the "knee" area of the I–V curve. A solar cell I–V curve, as shown in Fig 6.6, has the following three significant points: I_{sc}, short-circuit current;

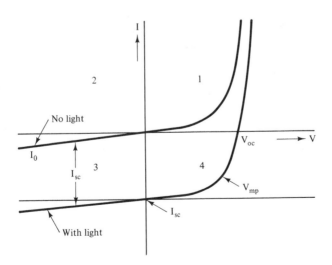

Figure 6.5 Solar cell characteristic curve.

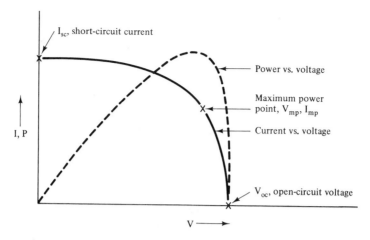

Figure 6.6 Solar cell current/power versus voltage.

P_{mp}, maximum power output point; and V_{oc}, open-circuit voltage. Corresponding to P_{mp}, there is a maximum power current, I_{mp}, and a maximum power voltage, V_{mp}. The output voltage of the cell decreases as the current increases from no load. The rate of change is indicative of a low source impedance ($<1.0 \ \Omega$). As the maximum current capability of the solar cell is approached, the solar cell then acts as a current-limited source with a high source impedance ($>1000 \ \Omega$).

Solar cell efficiency. The efficiency, η, of a solar cell is defined as

$$\eta = \frac{P_{out}}{P_{in}}$$

where P_{out} is the electrical power output of the cell and P_{in} is the solar energy input to the cell. The spectral radiance of the sun at the distance of 1 AU in the absence of the earth's atmosphere is given in Fig. 5.5. The sun emits radiation in wavelength ranges between 1×10^{-10} m (x-rays) and 30 m (radio frequency). The peak of the spectrally emitted energy occurs at 0.48 μm and approximately 77% of the emitted energy lies in the band from 0.3 to 1.2 μm, which is of interest to the present silicon solar cells, and 22% is above 1.2 μm. Radiation is propagated in small units called photons, each photon containing one quantum of energy, which is given by Planck's equation,

$$E = h\nu \tag{6.1}$$

where h = Planck's constant = 6.6262×10^{-34} J · s

$\nu = c/\lambda$ = radiation frequency

λ = wave length of the radiance

c = velocity of light = 3×10^8 m/s

Solar cells produce power by the absorptance of light. The absorptance coefficient, a, is defined as follows:

$$N_{ph} = N_{pho}e^{-at} \qquad (6.2)$$

where N_{ph} is the photon density of monochromatic light at a distance t from the front surface. At the front surface, the density of the entered photon density is N_{pho}.

Figure 6.7 shows a plot of the absorption coefficient a versus $h\nu$ for a number of semiconductors. It can be seen that for $h\nu < E_g$, the value of a is very small while for $h\nu > E_g$ the value of a rises rapidly. Since more than 50% of the energy in sunlight is carried by photons whose energy lies between 1.0 and 2.5 eV, for the materials shown in Fig. 6.7, most of the solar photons will be absorbed in a thickness of material between 10^{-3} and 10^{-5} cm. This means that from the point of view of nearly total absorptance of light, photovoltaic cells need not be thicker than a few times 10^{-3} cm. From Fig. 6.7 it can be seen that from the thickness point of view, a silicon cell is not the most desirable. Presently, the cells are made thicker than the optimum value mainly for manufacturing and handling considerations.

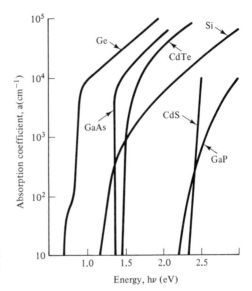

Figure 6.7 Optical absorption coefficient as a function of photon energy ($h\nu$) in electron volts.

The light absorbed in a solar cell will generate heat in the form of atomic vibration if $h\nu < E_g$, will separate an electron–hole pair if $h\nu = E_g$, and will produce an electron–hole but have an excess of energy which becomes heat if $h\nu > E_g$. The nearly perfect means of transforming sunlight into electricity is when photon energy is equal to bandgap energy for the solar cell material E_g. Where monochromatic light with energy equal to the bandgap is used to illuminate a solar cell, the efficiency can be well over 75%. The solar energy, however, has a spectrum of a wide variety of energies and intensities.

The theoretical maximum efficiency under the solar spectrum is plotted against the energy gap, E_g, for a few photovoltaic materials in Fig. 6.8. The efficiency of GaAs nearly coincides with the peak efficiency associated with

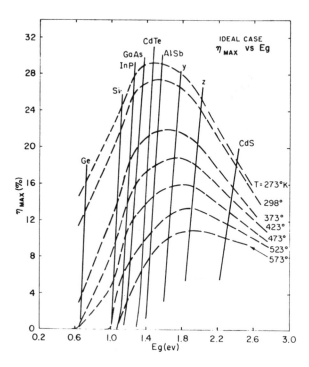

Figure 6.8 Temperature-dependent maximum efficiency as a function of energy gap for a few photovolatic materials (Ref. 3).

the spectrum. The output of every material monotonically decreases with an increasing temperature.

Silicon solar cells. The first practical solar cell was produced in 1954 by the Bell Telephone Laboratories. While these cells were initially considered for terrestrial use only, the success of solar cells in space began in 1958 when rectangular 0.5 × 2 cm cells were selected for Vanguard, the first U.S. satellite. Later, cell sizes were increased to 1 × 2 cm, 2 × 4 cm, and larger. Contact gridlines over the active cell area were used to reduce cell internal resistance. The antireflective coatings of silicon monoxide and improved processes increased mean cell efficiency to around 10 to 11%. These cells are called conventional cells.

Between 1961 and 1971, silicon solar cell technology was relatively static. In 1972, a significant improvement in solar cell efficiency was announced. This improvement was based on increasing the cell blue response, decreasing the internal resistance to about 0.05 Ω, and improving the charge carrier collection processes within the cell. These solar cells, known as violet cells, have an efficiency in the neighborhood of 14%. Different surface treatments to minimize light reflections led to the development of the black cell, also known as the textured surface cell. The drawback with this cell is that it operates at a higher temperature due to the increase in the solar absorptance. In flat, sun-oriented arrays, the increased temperature nearly counteracts any efficiency improvement due to the nonreflective cell treatment. Another noteworthy achievement has been in the ultrathin silicon solar cells. Starting in the mid-1970s, 0.05-mm-thick solar cells of 2 × 2 cm size have been fabricated.

Modern n-on-p solar cells are fabricated by, first, growing single-crystal p-type silicon ingots, and cutting and slicing them into thin wafers. An n-type impurity is then diffused at high temperature into the wafer surface, thereby forming a junction less than 1 μm from the surface. The diffused layer is subsequently removed from all but one large surface, which is then referred to as the cell's active, or light-sensitive, area. Next, metallic contacts are applied to both the diffused n-layer and the p-base layer. In a final step, an antireflective coating is deposited over the active area.

Conventional solar cells. These cells utilize pre-1972 fabrication technology. The junction depth is held between 0.3 and 0.5 μm. The ohmic collector bar is between 0.9 and 1.25 mm in width. There are three gridlines per centimeter and each gridline is 0.15 to 0.20 mm in width. Antireflective coating is silicon monoxide (SiO). The cell thickness ranges from 0.20 to 0.35 mm nominal. The series resistance is 0.2 to 0.25 Ω for a 2 × 2 cm cell. The maximum power output under AMO sunlight for a 2 × 2 cm cell is about 55 mW. The maximum efficiency for a conventional cell is about 10.6%. The energy losses for a conventional n/p solar cell are given in Table 6.1.

Violet cells. Since the solar radiance is abundant in the blue and ultraviolet region, the efficiency can be increased by increasing the spectral response in these regions. A blue-rich solar spectrum creates a heavier carrier concentration near the front surface of the cell than in the bulk.

TABLE 6.1 ENERGY LOSSES FOR A CONVENTIONAL n/p, 10 ohm-cm, AMO

Loss Source	Fraction of Available Energy Used in Conversion	Percentage of Input Used in Conversion
Long wave length photons not absorbed	0.76	76
Excess photon energy not utilized in conversion	0.57	43.4
Voltage factor $\dfrac{V_{oc}}{E_g}$	0.49	21.3
Curve factor $\dfrac{I_{max} V_{max}}{I_{sc} V_{oc}}$	0.81	17.2
Collection efficiency	0.72	12.4
Additional curve factor	0.91	11.3
Series resistance loss	0.97	11.0
Reflection loss	0.97	10.6
Useable power		10.6

Therefore, for a shallow junction, near the surface, more carriers will be collected before recombination than in the cell with a deep junction. Since a shallow junction introduces a high sheet resistance in the diffused layer, resulting in a higher potential drop, the sheet resistance is compensated for by increasing the number of current collecting gridlines. Increasing the number of gridlines, however, also increases the shadowing effect on the cell. The gridlines can be made thinner to reduce shadowing, but this, in turn, increases the voltage drop in the grid. Hence a careful optimization of all the parameters is required.

The violet cell, which was developed at COMSAT, uses an improved diffusion technology which eliminates the so-called "dead layer" from the top surface layer. Violet cells contain 10 to 30 grids/cm. The series resistance is lower than that in the conventional cells and is about 0.05 Ω for a 2 $\times$ 2 cm cell. Tantalum pentoxide (Ta_2O_5) is used as an antireflective coating because of a high refractive index and less absorption in the short-wavelength region. With these improvements, the spectral response in the blue and ultraviolet region is greatly increased. The maximum efficiency of violet cells is reported to be about 14 to 14.5%. The maximum power output under AMO sunlight for a 2 $\times$ 2 cm cell is 76 mW.

Black cells. The black cell, also known as textured surface cell, is based on chemically etching the surface geometry of the violet cell to reduce reflectivity. The textured surface consists of randomly spaced, four-sided, pointed tetrahedra of varying heights, ranging from approximately 1 μm to as high as 15 μm and with a spacing of 3 to 10 μm. The incident light undergoes two reflections before complete escape, causing reduction in the reflection losses. The black cell surface not only results in reduced reflection, but also greatly enhances the response at each end of the spectrum as compared to flat surface cells. Under AMO sunlight, the maximum output is 84 mW for a 2 $\times$ 2 cm cell. This corresponds to 15.5% efficiency.

Radiation Degradation

The major types of radiation damage in solar cells are ionization and atomic displacement. Both the voltage and current output are affected. At synchronous altitude, the radiation environments which are considered for solar cell damage are: solar flare protons on station, trapped electrons on station, and trapped electrons and protons during transfer orbits. The electrons trapped in the earth's magnetic field are the principal cause of solar cell degradation. During maximum solar activity, solar flare protons cause some additional solar cell damage. The total radiation environment causes two radiation components to enter the solar cell: one through the coverglass and the other through the substrate.

Solar flares are eruptions of the sun with the emission of energetic particles. The frequency of eruptions increases to a maximum and decreases again during approximately 11-year-long "solar cycles." The duration of a "solar maximum" is approximately 7 years.

Damage Equivalent 1-MeV Fluence

The wide range of electron and proton energies present in the space environment necessitates some method of describing the effects of various types of radiation in terms of a radiation environment which can be produced under laboratory conditions. For analytical and test convenience, 1-MeV fluence has been used as a standard radiation environment for comparison. The actual damage produced in solar cells by electrons of various energies is related to the damage produced by 1-MeV electrons by damage coefficients. Similarly, the damage caused by the protons of various energies is related to the damage of 10-MeV protons. Experimental studies indicate that a given fluence of normally incident 10-MeV protons induces the same amount of damage as 1-MeV electrons at 3000 times the fluence of the 10-MeV protons. The total 1-MeV fluence is calculated from the equation

$$\phi_e(1\text{MeV}) = \int_{E_c} K_e(E) \frac{d\phi_e(E)}{dE} \cdot dE + 3000 \int_{E_c} K_p(E) \frac{d\phi_p(E)}{dE} \cdot dE$$

where E_c = cutoff energy for the coverglass shield

$K_e(E)$ = relative damage coefficient for normally incident 1-MeV electrons

$K_p(E)$ = relative damage coefficient for normally incident 10-MeV protons

$\dfrac{d\phi_e(E)}{dE}$ = electron differential space fluence

$\dfrac{d\phi_p(E)}{dE}$ = proton differential space fluence

$\phi_e(1\text{MeV})$ = equivalent 1-MeV electrons

The damage coefficient for a solar cell is dependent on the shield thickness. The radiation shield is provided at the front by the coverglass and the adhesive, and at the back by the substrate and the adhesive. To a first-order approximation, the shielding effectiveness is proportional to mass per unit area. Therefore, the shield thickness is usually expressed in units of mass density times shield thickness. The thicknesses for the front and back shields are calculated by adding the thicknesses of all the front elements and the back elements, respectively. Having calculated the equivalent 1-MeV electron fluence, the degradation of a cell power is determined from its radiation degradation characteristics. For communications satellites operating in geosynchronous orbit, the amount and effects of radiation fluence varies principally with the degree of radiation protection provided on solar cells. For a spin-stabilized spacecraft in which the back side of the solar cell is inherently protected, the radiation dosage is equivalent to typically 3×10^{14} electrons/cm^2 during 7 years of synchronous orbit operation. Flat sun-oriented solar arrays in three-axis-stabilized spacecraft receive radiation from both sides of the cells, so it is considerably higher in comparison to spin-stabilized solar arrays. For a flat sun-oriented solar array, the typical

1-MeV electron fluence for 7 years synchronous orbit operation is 10^{15} electrons/cm^2 for a rigid panel array and about 2×10^{15} electrons/cm^2 for a semirigid flexible blanket solar array. Figure 6.9 illustrates the reduction of cell maximum power as a function of electron irradiation fluence for three types of cells: conventional, violet, and black cells. Also marked on this figure are the typical amount of radiation fluences for 7- and 10-year operation in geosynchronous orbit. The optical (noncell) losses, which consist mostly of coverglass adhesive darkening, are relatively large, 4 to 10%, during the first year in orbit, and increases very little thereafter. The solar cell damage is about 3% during the first year and approximately 15% for 7 years in operation. Therefore, the total solar cell degradation is in the range 19 to 25% in 7 years. The degradation rate decreases with time. Approximately one-half of the total degradation occurs in the first two years of a 7-year orbit life. In extending the spacecraft lifetime from 7 years to 10 years, the incremental array degradation results in an additional loss of only 5% power.

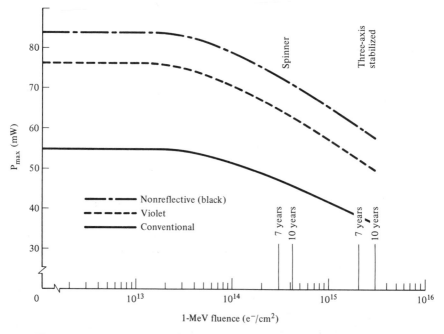

Figure 6.9 Maximum power output of conventional, violet, and nonreflective (black) cells as a function of 1-MeV electron irradiation fluence for a 2×2 cm area.

Solar Cell Parameters

Base resistivity. The base resistivity of a solar cell affects both the preirradiation energy conversion efficiency and the radiation damage. Solar cells with high base resistivity will have lower efficiency and higher radiation damage resistance. Hence, for low-fluence missions, lower-base-resistivity cells (1 to 3 $\Omega \cdot$ cm) provide a higher power output, and for high-fluence missions, high-resistivity cells (7 to 14 $\Omega \cdot$ cm) provide the higher power output.

Thickness. If solar cells are made thin enough, less than 0.30 mm, so that long-wavelength light (approximately 1 μm) cannot be completely absorbed, the long-wavelength (red) response will be reduced, as shown in Fig. 6.10(a), and efficiency will decrease. This will result in making short-wavelength (blue) response a more important factor in determining thin-cell efficiency. Electron irradiation will have less effect on these cells since

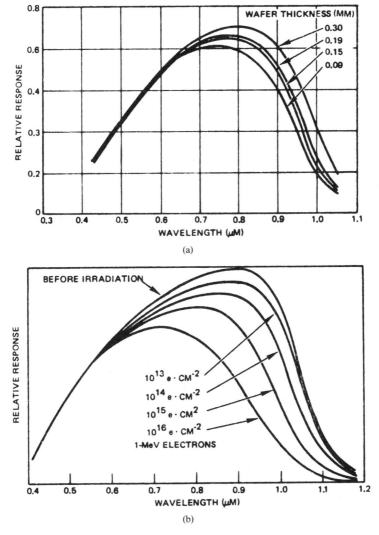

(a)

(b)

Figure 6.10 Effects of cell thickness: (a) changes in spectral response with cell thickness; (b) changes in spectral response with cell thickness; (c) effects of cell thickness and radiation on electric power.

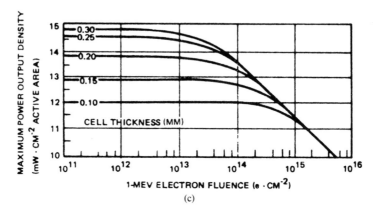

Figure 6.10 (*cont.*)

electron damage manifests itself mainly as a loss in red response, as shown in Fig. 6.10(b). On the basis of Fig. 6.10(a) and (b), we can see in Fig. 6.10(c) that the effect of thickness on electric power output decreases as the radiation dosage is increased.

Solar Cell Optical Design

The optical interface between the cell and the sunlight has the following components:

1. Antireflective layer on the cover
2. Cover bulk material
3. Ultraviolet-reflective layer on the surface
4. Cover adhesive
5. Antireflective layer on the active solar cell surface
6. Solar cell bulk material

The primary objective of the optical design is to minimize absorption losses in the wavelength region in which the solar cell is responsive. The solar cell covers with reflective filters are typically coated on both sides. The outer side has a single-layer antireflective coating which is designed to enhance the transmission of light energy through the cover into the solar cell. This coating is subjected to the following significant environments: ultraviolet, charged particle, and corpuscular radiation. Therefore, the coating must be resistant to these environmental exposures. The commonly used coating that meets these requirements is magnesium fluoride, MgF_2. The commonly used material for the cover is fused silica Corning glass 7940. The inner side of the cover has an ultraviolet (UV) energy-reflective coating which has two functions: to protect the adhesive from the damage by the UV radiation and to reflect the wavelength (short) region of the solar spectrum which is not converted by the solar cell into the electrical energy. Alternatively, a cerium-doped microsheet glass which inherently provides ultraviolet filtering can be used. The cover is normally bonded to the front side of the solar cell with a transparent silicone adhesive. Dow-Corning DC 93-500 is widely used.

The purpose of antireflective coating on the front active surface of solar cells is to minimize reflection losses at the interface between the cover adhesive and the silicon wafer. For conventional cells, silicon monoxide (SiO) has been used as an antireflective coating. However, for high-efficiency cells, violet, and black cells, tantalum pentoxide (Ta_2O_5) is used as the antireflective coating. It has a high refraction index with less absorption in the short-wavelength region.

The installation of a cover on a solar cell decreases or increases the amount of light energy reaching the solar cell. The properties and characteristics of antireflective solar cell coating can be adjusted in such a way that the glassed cell output is maximized. With Ta_2O_5 coatings, the coating can be adjusted to provide approximately 0 to 6% output gain due to glassing.

Effect of the Angle of Incidence

The power from the solar cell will be maximum when the angle of incidence of the illumination is zero (i.e., it is normal to the solar cell surface). The power decreases as the angle of incidence deviates from zero. Theoretically, the power should be proportional to the cosine of the angle of incidence. However, measured data show that the solar cell maximum available power will fall off faster than the cosine of the angle of incidence indicates. The principal factor that causes this deviation is the change in reflection coefficients at large angles.

Effect of Temperature

The electric power characteristics of the solar cells are usually obtained at 25 to 28°C. A change in cell temperature will change its $I-V$ curve. An increase in the cell operational temperature causes a slight increase in the cell current and a significant decrease in the voltage. The temperature coefficient for current is typically less than 0.03%/°C and is dependent on the cell thickness, junction depth, antireflective coating, and state of radiation damage. The temperature coefficient for voltage is negative and is typically 2.2 to 2.3 mV/°C. The voltage coefficient is essentially independent of radiation damage.

The solar cell characteristics for Intelsat IV, Intelsat V, and Intelsat VI are given in Table 6.2.

Electrical Elements

The electrical contacts on the solar cell p- and n-type semiconductor surfaces aid in electrical power collection from the illuminated active cell area. The contact geometries differ from type to type and from one cell

TABLE 6.2 SOLAR CELL CHARACTERISTICS

Characteristics	Intelsat IV	Intelsat V	Intelsat VI	
			K4-3/4	K7
Power BOL (28°C) (mW)	55.6	133.5	177.5	307.8
POWER EOL (28°C) (mW)	45.4	116.6	156.3	230.8
BOL				
I_{mp} (A)	0.125	0.2966	0.391	0.644
V_{mp} (V)	0.445	0.450	0.454	0.478
I_{sc} (A)	0.141	0.315	0.4187	0.6887
V_{oc} (V)	0.560	0.548	0.545	0.590
Size (cm)	2×2	2.1×4.04	1.8×6.2	2.5×6.2
Thickness (cm)	0.033	0.025	0.02	0.02
Material	Si	Si	Si	Si
Base resistivity $\Omega \cdot$ cm/type	10/N/P	10/N/P	10/N/P	10/N/P
Front junction depth (μm)	>0.3	0.2	0.2	0.2
Back surface field	No	No	No	Yes
Back surface reflector	No	No	Yes	Yes
Contact metallization	TiPdAg	TiPdAg	TiPdAg	TiPdAg
Front contact width (cm)			0.06	0.06
Antireflective coating	SiO_2	Ti_2O_5	TiO_x	$Ti_xAl_2O_3$
Cover type	Fused silica	Ceria-doped microsheet	cmx microsheet with antireflective coating	cmx microsheet with antireflective coating
Cover thickness (cm)	0.030	0.015	0.021	0.021
Cover adhesive	DC 634-89	DC 93-500	DC 93-500	DC 93-500
Cover front surface	Polished	Polished	Polished	Textured

manufacturer to another. Normally, the front surfaces have gridlines to reduce the shadow and the back surfaces have full contact surfaces.

The purpose of a solar cell interconnector is to conduct electrical energy from the individual solar cells to the power collection wire harness. One of the most severe problems with interconnectors in sun-oriented deployed solar arrays is the thermal stress cycles. In geostationary orbit, the solar array experiences 700 to 1000 eclipse cycles over the spacecraft life. Because of the low thermal capacity of the array, the arrays will cycle from approximately +55 to −196°C for a flexible substrate. Furthermore, as shown in Fig. 6.11, the maximum rate of temperature change, thermal shock, for a lightweight flexible array is increased to more than six times that of a drum spinner array. To reduce thermal stresses, several alternative interconnector configurations have been developed. These configurations use new interconnecting techniques, such as parallel gap resistance welding, ultrasonic bonding, and thermal compression bonding; and new interconnector material, such as silver-plated molybdenum or silver-plated Invar. The advantages of these techniques are that they join the same metals (i.e., the silver contact on the solar cell to the silver-plated interconnect) without requiring an intermediate layer such as solder, which has poor fatigue characteristics at low temperature and can crack with repeated cycling.

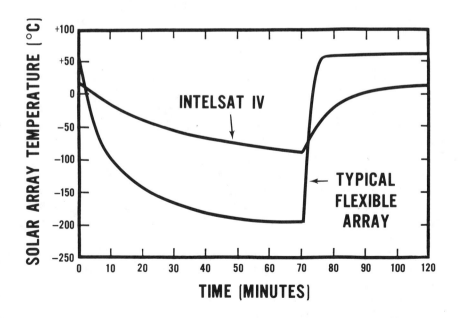

Figure 6.11 Solar array temperature profiles (longest eclipse) (Ref. 16).

Gallium Arsenide Cell

Gallium arsenide solar cells have slightly higher efficiency than silicon solar cells. However, due to their higher cost and nonavailability in production quantities, these cells have not been used on communications satellites. In future, these cells may become cost competitive with silicon cells.

Currently, GaAs cells of 17% efficiency have been produced, which compares to 15% efficiency for the best available silicon solar cells. The difference between their efficiencies becomes significant at higher temperatures. A typical silicon cell degrades approximately 0.5% for every degree Celsius rise in temperature, mostly due to voltage effects. The degradation due to thermal environment of GaAs is approximately one-half that of silicon. As an example, at EOL (1×10^{15} electrons/cm^2), the power densities for K7 silicon cells are 14 mW/cm^2 and 7 mW/cm^2 for 28°C and 128°C, respectively, a degradation of 50% due to higher temperature. For a typical GaAs cell, the corresponding power densities are 19.08 mW/cm^2 and 13.8 mW/cm^2, a degradation of merely 27%. Hence GaAs cells are particularly attractive at high temperatures, such as in a solar concentrator system. Tests have shown GaAs cells to have considerably better ability to anneal than that of Si cells. As a result of annealing, the EOL efficiency of GaAs may be very close to that of BOL efficiency. The GaAs solar cell mass for a similar-size cell is approximately twice that of a silicon cell. However, fewer GaAs cells are needed for a given power, resulting in a smaller solar array. As an example, a GaAs solar cell spinning array having a BOL 1-kW power level in synchronous orbit will result in the reduction of 27% in area and 7% in mass over the most efficient silicon solar array.

Solar Array

The power output of a single cell being quite low, the individual cells are arranged in series to provide the desired voltage and in parallel to provide the desired current requirements. In addition, solar array modules are often constructed with several strings in parallel which are connected together in a series parallel "ladder" network as shown in Fig. 6.12. This is done to minimize power loss with a single cell failure. If each string were used independently, the loss of a single cell would open circuit the entire string, and the output from that string would be totally lost. With the ladder network arrangement, the remaining cells in any string with a failed cell have an output current path, and the output is degraded to a lesser degree.

A typical solar array construction is shown in Fig. 6.13. The basic elements for a solar array are: solar cells, coverglasses, interconnectors, insulator, adhesives, antireflective coating, and panel substrate.

In order to meet the increasing power requirements of satellites, deployable solar arrays are required. These solar arrays are stowed into a compact package during launch and deployed in the orbit. The total array

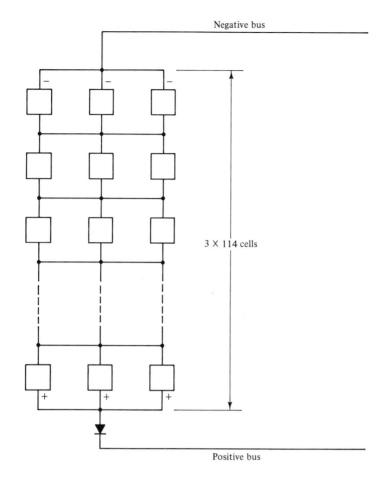

Figure 6.12 Solar cell array matrix.

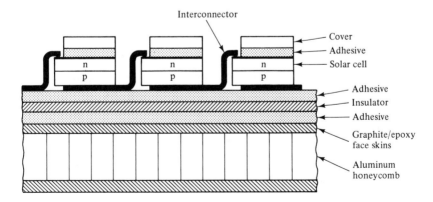

Figure 6.13 Typical solar array construction.

mass consists of the masses of the structural elements, the solar cells, and the coverglasses. The structural elements constitute the major part of the array mass. A variety of design concepts have been developed for the structural elements, which exhibit low mass, high deployment reliability, and high natural frequencies. The specific performance of a solar array is expressed in W/kg. Most of the recently developed arrays are designed for a certain range of power output. Therefore, it is not possible for such arrays to have the same specific W/kg performance at all power levels within the design range.

For spin-stabilized spacecraft, the solar panels are normally rigid honeycomb cylindrical panels. Early designs for Intelsat I to IV used undeployed panels which imposed area limitations on the available power. The current designs, SBS and Intelsat VI, however, use two telescoping cylindrical panels, the larger panel positioned over the fixed panel during launch and deployed in orbit. For three-axis-stabilized spacecraft, several deployment design concepts have been developed. Most of the current solar array designs can be divided into the following groups:

- Rigid honeycomb panels
- Honeycomb panels with stiffeners
- Flexible substrates with rigid frames
- Flexible fold-up blankets
- Flexible roll-up blankets

The deployment design concepts of these solar arrays are shown in Fig. 6.14.

Honeycomb panels. The common type of deployed solar array is one that employs spring activation with some form of pantograph mechanism to deploy several panels of rigid substrates covered with solar cells. In the past, these arrays have typically used aluminum or glass/epoxy face skins. The recent solar arrays use less dense aluminum honeycomb with graphite/epoxy or Kevlar/epoxy face skins.

A spin-stabilized spacecraft normally uses the honeycomb cylindrical panel. For Intelsat VI, it consists of Kevlar epoxy face skins with a 13-mm thick aluminum honeycomb core. The outer face skins are covered by Kapton for ultraviolet protection and an insulated bonding surface for cell laydown. The effective illumination area for a spinning cylindrical panel is $1/\pi$ times the total area because only half of the panel is illuminated at one time and the angle between the surface normal and solar rays varies from 0 to 90° on the illuminated surface. Hence the specific performances of solar arrays in spin-stabilized spacecraft are significantly lower than those for sun-tracking flat panels in three-axis-stabilized spacecraft.

Honeycomb panels with stiffeners. In this design the solar array panels are made efficient by using lightweight honeycomb panels reinforced

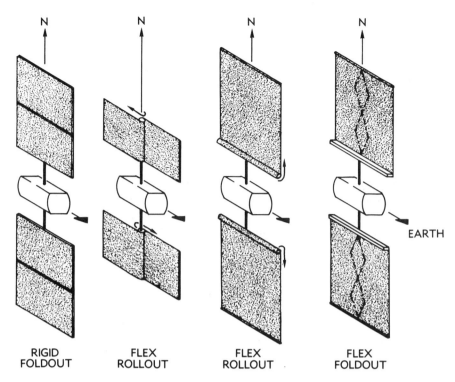

Figure 6.14 Deployed solar array concepts (Ref. 16).

by tubular frames. The solar arrays of Fltsatcom, TDRSS, and Intelsat V were based on this design concept.

The solar array for Intelsat V, a three-axis-stabilized spacecraft, consisted of three panels and an open-frame yoke hinged together and folded for stowage during the launch. Each panel measured 1.6 m in width by 2 m in length. The panel substrates were made from open-weave graphite/epoxy face skins and aluminum honeycombs. The substrates were reinforced at their perimeters with graphite/epoxy beams. The solar cells were insulated from the substrate by a Kapton sheet. A closed cable loop and pulley system synchronized the panel deployment on each wing. The natural frequencies in stowed and deployed configurations were 34 Hz and 0.36 Hz, respectively. The total array mass of 64.1 kg consisted of 22.8 kg electrical, 37.4 kg mechanical, and 3.9 kg hold-down components. A total of 17,380 solar cells were mounted to the 18.12-m^2 substrate area. The characteristics of the solar cells are given in Table 6.2. The calculated solar array power at the end of 7 years in synchronous orbit was 1358 W at equinox and 1288 W at summer solstice. The specific performance values for equinox were 26.6 W/kg and 21.2 W/kg at the beginning of life and the end of life, respectively.

Flexible substrates with rigid frames. This type of design consists of a pretensioned flexible substrate mounted within a rigid frame. Some very early work along these lines was performed for the NASA/JPL-sponsored LASA project at Boeing. A recent version of this concept, known as ultra-lightweight panel (ULP), was developed at MBB.

The ULP array consisted of two identical wings, each made up of a yoke and an empty panel frame, and between 2 to 25 panels per wing for array power levels of 1 to 13 kW. Each panel measured approximately 3.3 m (in the wing-width direction) × 1.15 m. Solar cells are mounted on a flexible graphite/epoxy cloth substrate, pretensioned in the direction parallel to the 3-m width only. Kapton was used as insulation under the solar cells. The panel frame consisted of rectangular-cross-section graphite/epoxy tubes. The natural frequencies of the array were about 24 Hz stowed and 0.3 Hz deployed for a 1.58-kW array and 0.048 Hz deployed for a 7.12-kW array. The deployment mechanism is similar to that for Intelsat V. Each panel carried AEG Telefunken 10-Ω · cm solar cells of 20 × 40 × 0.2 mm size, covered by 0.10-mm-thick ceria-stabilized microsheet. The cell output at 25°C after irradiation with 2×10^{15} 1 MeV electrons/cm^2 was 95.7 mW. Each panel produced 268 W at 58°C during equinox after 7 years in synchronous orbit. The specific end-of-life performance was 30.4 W/kg for the 1.58-kW system and 39.3 W/kg for the 7.12-kW system.

Flexible fold-up blankets. In this solar array design, the blankets are folded up, accordion fashion, during launch and unfurled in orbit by a deployable boom or mast. The solar array on the Canadian Technology Satellite (CTS), also known as Hermes, was the first operational solar array of this design. This satellite launched in 1976 also carried a fixed body-mounted array. This body-mounted array provided about 100 W during transfer orbit and was jettisoned thereafter. The deployable array consisted of two wings, each 6.53 m long by 1.30 m wide, comprised of 30 foldable panels, 26 of which were solar cell covered. Deployment was achieved with a motorized Bi-Stem boom of stainless steel, 34 mm in diameter and 0.18 mm thick, and located behind the blanket.

The blanket consisted of a laminate of 25-μm Kapton and 36-μm fiberglass. The solar cells were 20 × 20 × 0.2 mm bonded directly to the substrate. The coverglasses were 0.1-mm-thick ceria-doped microsheet covers. The array produced a total power of 1330 W at the beginning of life. The total mass of each wing was 30 kg, which included 6.97 kg for the blanket mass. The specific beginning-of-life performance was 22.2 W/kg.

Flexible roll-up blankets. Flexible roll-up arrays consist of one or two solar array blankets that are stowed on a cylindrical drum. For deployment, a deployable boom or mast draws the blanket from the drum against a retracting force that keeps the blanket stretched flat.

Hughes Aircraft Corporation, under U.S. Air Force contract, developed a flexible roll-up solar cell array (FRUSA). It was launched as a flight experiment in 1971 aboard an Agena spacecraft into a low earth orbit. It consisted of two solar cell blankets 4.9 m long by 1.7 m wide, rolled up on a single 20-cm-diameter drum. A 50-μm-thick Kapton cushion protected the solar cells in the stowed configuration during launch. The blankets were deployed from the common drum by a pair of extensible Bi-Stem steel booms, 22 mm in diameter and held under a nominal 14 N tension. The deployed fundamental natural frequency was 0.25 Hz.

The solar array blankets were continuous laminates of Kapton and fiberglass. The beginning-of-life power was approximately 1500 W. The total system mass consisted of 16.2 kg for the drum mechanism, 13.1 kg for the solar cells and covers, and 2.4 kg for blankets, resulting in the total mass of 31.7 kg. The specific performance at the beginning of life was 47.4 W/kg.

Solar Array Power

The power of a solar array varies with time due to (1) the variation in solar intensity, (2) the variation in the angle between the solar array surface normal and solar rays, and (3) the radiation degradation in solar cell power characteristics. The variation of solar intensity is given in Table 6.3. The solar declination angle varies during the year with a maximum $+23.4°$ during summer solstice (June 21), $-23.4°$ during winter solstice (December 22), and 0 during vernal equinox (March 21) and autumnal equinox (September 23). The power from a solar cell is proportional to the solar intensity, F, and the cosine of the angle, θ, between the solar array surface normal and the solar rays. As given in Table 6.3, the effective solar intensity factor ($F \cos \theta$) will be maximum at vernal equinox and minimum at summer solstice.

The solar array power characteristics during the satellite lifetime are obtained by superimposing the seasonal variation in power output on the time-varying radiation degradation characteristic. A typical solar array maximum power output versus time in geosynchronous orbit is shown in Fig. 6.15. As shown in Fig. 6.15, two maxima representing equinoxes and two minima representing solstices occur every year. The solar intensity variation is slightly out of phase with the change in declination. Hence, the solar intensity value at vernal equinox will be slightly higher than that for autumnal equinox. Similarly, the minimum for summer solstice will be lower than that for winter solstice. The design requirement for a solar array is generally to produce the demanded current and voltage output at the EOL summer solstice condition.

6.3 BATTERIES

For a spacecraft, rechargeable batteries are used during the periods when the electric power from the solar array is not available or is not large enough. The batteries have to meet the following operational conditions.

During launch, the housekeeping functions of the spacecraft are maintained from the batteries until the fairing is jettisoned or the spacecraft is deployed, in case of Space Shuttle launch. Even then, in many spacecraft

TABLE 6.3 VARIATION OF SOLAR INTENSITY WITH EARTH–SUN DISTANCE (Ref. 2)

Date	Solar Intensity, S ($mW \cdot cm^{-2}$)	Relative $F = \dfrac{S}{135.3}$	Inclination Angle, θ	$Cos\ \theta$	$F\ cos\ \theta$
Jan. 3 (perihelion)	139.9	1.034	−23.11	0.919	0.95
Feb. 1	139.3	1.0296	−17.21	0.955	0.983
Mar. 1	137.8	1.0185	−7.73	0.99	1.008
Apr. 1	135.5	1.0015	−4.23	0.997	0.998
May 1	133.2	0.9845	−14.95	0.966	0.951
June 1	131.6	0.9727	22.0	0.927	0.902
July 4 (aphelion)	130.9	0.9675	23.13	0.919	0.889
Aug. 1	131.3	0.9704	18.13	0.95	0.922
Sept. 1	132.9	0.9823	8.43	0.989	0.971
Oct. 1	135.0	0.9978	−3.03	0.998	0.996
Nov. 1	137.4	1.0155	−14.3	0.969	0.984
Dec. 1	139.2	1.0288	21.75	0.928	0.955
Mar. 21 (vernal equinox)	136.2	1.0066	0	1.0	1.0066
June 21 (summer solstice)	131.1	0.96896	+23.44	0.917	0.8885
Sept. 23 (autumnal equinox)	134.5	0.99408	0	1.0	0.99408
Dec. 22 (winter solstice)	139.7	1.0325	−23.44	0.917	0.94680

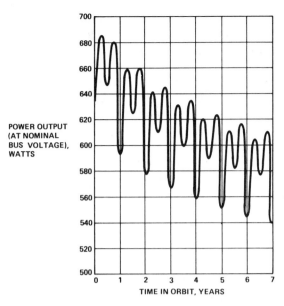

Figure 6.15 Typical solar array power output versus time in geosynchronous orbit.

missions the solar array in transfer orbit will deliver power only intermittently, requiring a battery to supplement it during eclipses or power load peaks. Another vital function of the battery prior to final acquisition on station is to deliver the electric power for various pyrotechnic actuators, such as for the ignition of the apogee boost motor or the deployment of the solar array.

During synchronous orbit, the battery has to deliver electrical energy during two solar eclipse seasons per year. These eclipse seasons are 45 days long and center around the vernal and autumnal equinoxes. There is one eclipse period per 24 hours with the maximum period of 72 minutes. The batteries discharge during an eclipse and are charged during the sunlight period. Between the two eclipse periods the battery is basically on storage for about 4.5 months, during which time a low battery temperature is desirable. Hence, for a synchronous satellite, the battery must be sealed to prevent loss of electrolyte, the energy density and charge/discharge efficiency of the battery must be high, and it must survive approximately 700 charge–discharge cycles for a 7-year application. Silver–zinc (Ag-Zn), nickel–cadmium (Ni-Cd), and nickel–hydrogen (Ni-H$_2$) batteries are the primary candidates for these space applications.

The most attractive advantage of an Ag-Zn battery is its high energy density (110 to 132 W-h/kg). It also possesses good temperature-discharge performance. The major disadvantage is that it has low-cycle-life (20 to 200) capability, which limits its application to short-life satellites, such as scientific satellites. In addition, the control of overcharge is critical with Ag-Zn batteries because oxygen evolved during overcharge does not readily recombine.

Nickel–cadmium batteries have significantly higher cycle capabilities than those of Ag-Zn and have an increased tolerance to overcharge exposure because of better oxygen-recombination capabilities. The disadvantages are lower energy density (22 to 26 W-h/kg) and degraded high-temperature performance. However, due to their higher life-cycle capability, Ni-Cd batteries have been used exclusively for communications satellites in the 1960s and 1970s for 7-year life-applications. In spite of their higher life-cycle capability compared to other batteries, Ni-Cd batteries are often a life-limiting factor for satellites.

Significant improvements in the lifetime of batteries have been achieved with the recent development of nickel–hydrogen batteries. Nickel–hydrogen batteries have fewer inherent failure mechanisms, resulting in considerably longer lifetimes than Ni-Cd batteries when used for the same depth of discharge, or they could operate for a higher depth of discharge for a similar life requirement. Therefore, the current trend is to use Ni-H$_2$ batteries for satellites that have a 7- to 10-year design life. This section provides a brief overview of Ni-Cd and Ni-H$_2$ batteries.

Nickel–Cadmium Batteries

The cell is a basic unit of a battery. It has four main components: the cadmium negative electrode, which supplies electrons to the external circuit when it is oxidized during discharge; the nickel positive electrode, which accepts the electrons from the external circuit; the aqueous electrolyte, 35% KOH, which completes the circuit internally, and a separator made of nylon or perhaps of polypropylene, which holds the electrolyte in place and isolates the positive and negative plates. Figure 6.16 shows a typical satellite Ni-Cd cell. The electrode plate structure consists of a porous sintered nickel plaque which is impregnated with the active materials.

The chemical reactions are reversible, that is, chemical energy which is converted into electrical energy during discharge can be restored by charging. The chemical reactions are as follows:

Negative electrode:

$$Cd + 2OH^- \overset{discharge}{\underset{charge}{\rightleftharpoons}} Cd(OH)_2 + 2e$$

During discharge, cadmium is oxidized to cadmium hydroxide and releases electrons to the external circuit.

Positive electrode:

$$2\,NiOOH + 2\,H_2O + 2e \overset{discharge}{\underset{charge}{\rightleftharpoons}} 2Ni(OH)_2 + 2(OH)^-$$

During discharge, the charged nickel hydroxide, NiOOH, goes to a lower valence state, Ni(OH)$_2$, by accepting electrons from the external circuit. During charging the reactions shown above are reversed. The net reaction is

$$Cd + 2\,NiOOH + 2\,H_2O \overset{discharge}{\underset{charge}{\rightleftharpoons}} Cd(OH)_2 + 2Ni(OH)_2$$

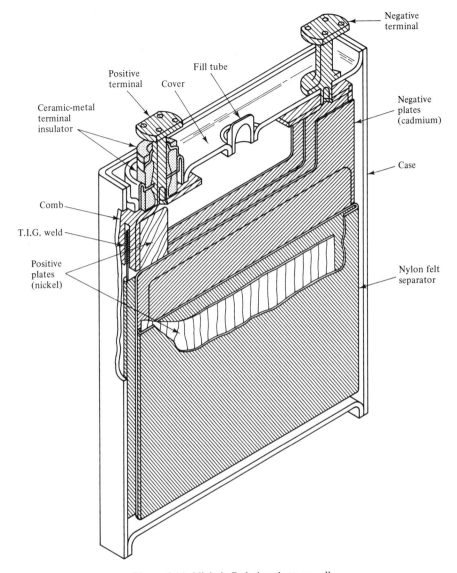

Figure 6.16 Nickel–Cadmium battery cell.

Electrical characteristics. Batteries are rated according to their capacities in units of ampere-hours. Battery charge and discharge rates are expressed in multiples of the "C" (capacity) rate. A battery discharging at the C rate will expend its nominal capacity in 1 hour. For example, for a 10-Ah-capacity battery, the C rate will be 10 A. At the 0.25C rate, the rated capacity will be delivered in about 4 hours and at the 4C rate in about $\frac{1}{4}$ hour.

Discharging characteristics. The nickel–cadmium cell's voltage remains relatively constant until all its capacity is very nearly discharged. At this point the voltage drops off sharply. Figure 6.17 shows the discharge voltage and capacity for different discharge rates. The nominal capacity is based on the discharge voltage down to 1.0 V at its 1-hour discharge rate. As shown in Fig. 6.17, a reduction in this rated capacity of nickel–cadmium cells occurs at higher discharge rates and an improvement in the capacity as seen at lower discharge rates. The optimum operating temperature is from −10 to +20°C.

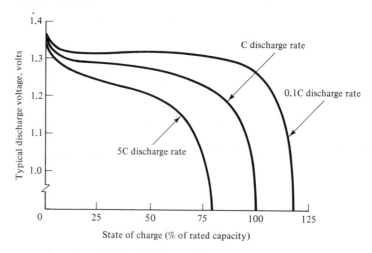

Figure 6.17 Ni-Cd battery discharge characteristics.

Charging characteristics.   Cell voltage rises substantially above the characteristics discharge voltage when a depleted cell is put on charge. During the majority of the charge period, the terminal voltage remains steady. As it reaches the full state of charge, the voltage rises rapidly. It reaches a level of 1.47 or 1.48 V when full charge is reached. The required charging voltage, a product of 1.5 V times the number of batteries in series, is usually higher than the primary distribution bus. Therefore, it is necessary to provide a boost-voltage capability for battery charging. Constant current control is used on many spacecraft to provide recharging and to limit battery heating due to overcharging.

Battery-charging efficiency varies significantly as a function of charge rate and temperature. A charge rate lower than 0.01C results in more current being used to generate oxygen than is used to convert the active material. As the charge rate is increased, the proportion of the current utilized to charge the cell increases. It is for this reason that the minimum charge rates used in practice are in the range 0.025C to 0.05C. Figure 6.18 shows the effects of charge rates and temperature on charging efficiency and charged capacity. As shown in Fig. 6.18, if the charging temperature is raised to 45°C, the available capacity is reduced because the battery never gets fully charged. At temperatures much below 5°C, the efficiency of oxygen recombination reaction is reduced, resulting in higher pressure or nonacceptance of long overcharging.

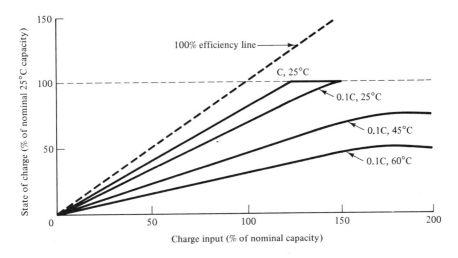

Figure 6.18 Ni-Cd battery charging characteristic.

Battery life. The important factors affecting the useful life of a nickel–cadmium cell are battery temperature, depth of discharge, and excessive overcharge. The most important effect of high battery temperatures is the reduction in the separator life. Prolonged exposure of a nickel–cadmium battery to high temperature will hasten the decomposition of the separator material. Very low temperature will also reduce life, because at low temperatures the efficiency of oxygen recombination reaction is reduced. So the repeated overcharging at a low temperature can result in pressure buildup. Therefore, battery temperature is an extremely critical parameter in the battery life design. The battery temperature has to be kept within a narrow temperature range. It is common practice to use a thermal radiator to keep battery temperature below 24°C and to use heaters to keep it above 4°C. A typical temperature profile during the eclipse season is shown in Fig. 6.19. Repeated cycling to a deep depth of discharge tends to degrade the cell plate structures, causing cracking. These microcracks absorb electrolyte, resulting in gradual separator dry-out. Ni-Cd battery cycle life as a function of depth of discharge and temperature is shown in Fig. 6.20. For a synchronous orbit application for 7 to 10 years, a battery will encounter approximately 1000 charge–discharge cycles. For this number of cycles, Ni-Cd battery depth of discharge is generally limited to 50 to 60%.

Reconditioning. The batteries exhibit a gradual decay of terminal voltage during successive discharge periods. This effect is most pronounced when the charge–discharge cycle is repetitive, and is referred to as the "memory effect." When the battery is cycled to a fixed depth of discharge, the active material that is not being used gradually becomes unavailable, resulting in an effective increase in the depth of discharge. In addition to the gradual decay of discharge voltage, the batteries will also exhibit a tendency toward divergence of the individual cell voltages during charge

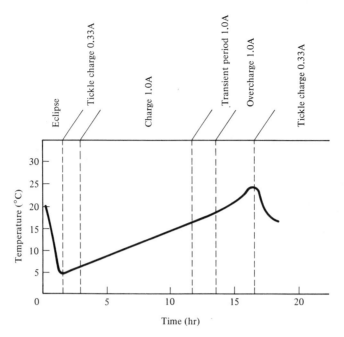

Figure 6.19 Battery temperature profile.

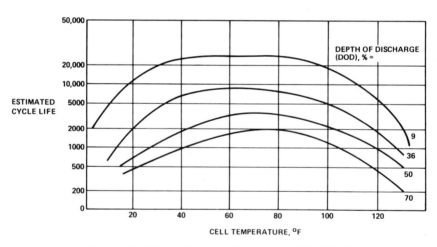

Figure 6.20 Nickel–Cadmium battery cycle life (Ref. 10).

and discharge. Battery performance can be restored to a certain extent by reconditioning. A typical reconditioning process for a rechargeable battery consists of effecting a deep discharge by connecting individual resistors across the terminal of each cell and then recharging at a high rate. A few weeks before the eclipse season, the batteries on many spacecraft are reconditioned to restore their performance.

To enhance battery life, the battery is kept within a small temperature range, it is reconditioned before each eclipse season, and during what is essentially a storage period between eclipse seasons it is trickle-charged (C/45) to prevent cadmium migration from negative electrode to positive electrodes.

Nickel–Hydrogen Batteries

Nickel–cadmium batteries were used in communication satellites in the 1960s and 1970s. For long-term applications, however, Ni-Cd batteries have demonstrated certain limitations. To achieve the required lifetime, the depth of discharge (DOD) must be controlled, and careful thermal and recharge management is required. These limitations led to the development of nickel–hydrogen batteries which are expected to have a considerably longer lifetime than Ni-Cd batteries when used for same DOD, and could operate at higher DOD for a similar life requirement. The nickel–hydrogen battery combines the most stable and reliable electrodes from both the hydrogen–oxygen fuel cell and nickel–cadmium cell. The current trend in communications satellites, such as Intelsat VI, is to use nickel–hydrogen batteries in place of nickel–cadmium batteries.

The electrochemical reactions for normal operation of a nickel–hydrogen cell are as follows:

At the positive electrode, nickel, the reaction is

$$NiOOH + H_2O + e \underset{charge}{\overset{discharge}{\rightleftharpoons}} Ni(OH)_2 + OH^-$$

At the negative electrode, which is the gas electrode, the reaction is

$$\frac{1}{2} H_2 + OH^- \underset{charge}{\overset{discharge}{\rightleftharpoons}} H_2O + e$$

The net reaction is

$$\frac{1}{2} H_2 + NiOOH \underset{charge}{\overset{discharge}{\rightleftharpoons}} Ni(OH)_2$$

Since hydrogen is needed for discharge, and evolves during charging, it must be contained. Figure 6.21 shows an Intelsat V Ni-H$_2$ battery cell. The cell contains 24 positive electrodes assembled in 12 back-to-back pairs and 24 negative electrodes. The electrodes are stacked on a center rod between two end plates. The busbars are welded to the positive and negative electrode tabs projecting from the opposite sides of the stack. The nickel electrodes used are similar to standard aerospace electrodes. Since hydrogen is a gas, the negative electrode contains a catalyst material, platinum, and the gas diffuses to the electrode for the necessary reaction. The separator is fuel cell asbestos. This material is a nonwoven mat of chrysotile asbestos fibers. Aqueous KOH solution is used as an electrolyte for the cells. The center rod is made of stainless steel covered with Teflon tubing to eliminate electrical contacts with plates. The busbars and weld tabs are made of nickel.

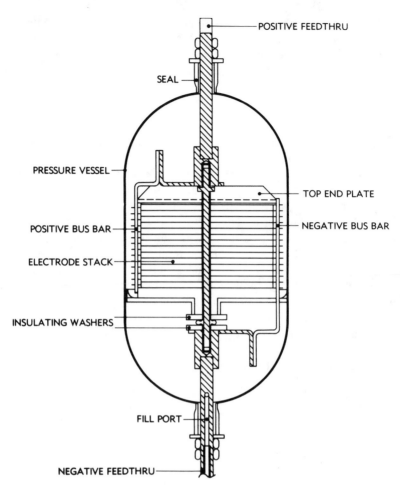

Figure 6.21 Nickel–Hydrogen battery cell (Ref. 7).

The salient electrical characteristics of Ni-H$_2$ battery performance as obtained from NTS-2 spacecraft data are given in Figs. 6.22 and 6.23. The dependence of the Intelsat V Ni-H$_2$ battery capacity on the operational temperature variations are given in Fig. 6.24. It is shown that the maximum battery capacity occurs approximately between 10 and 15°C and that at either side of this optimum temperature, the capacity decreases at the rate of 1 Ah per °C of temperature variation. Figure 6.25 shows the laboratory life test results of Ni-H$_2$ batteries as a function of battery DOD for geosynchronous orbit operations. The results indicate that for 10-year operations, which correspond to approximately 900 cycles, about 65% DOD is acceptable.

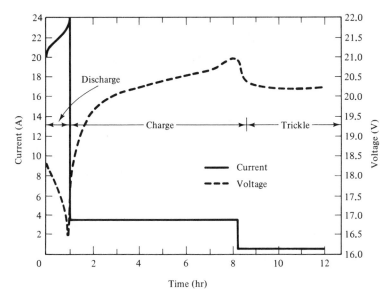

Figure 6.22 Ni-H₂ battery current and voltage.

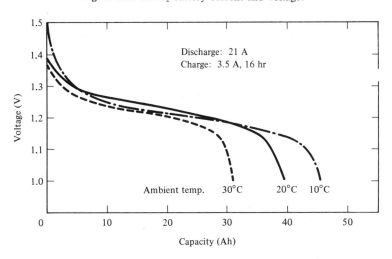

Figure 6.23 Ni-H₂ battery performance in discharge (Ref. 11).

Performance Degradation of Ni-H₂ Versus Ni-Cd Batteries

Some of the principal degradation modes in Ni-Cd batteries are eliminated in Ni-H₂ batteries. The known principal degradation modes in Ni-Cd batteries are hydrolysis and oxidation of the nylon separator material, cadmium migration, and electrolyte redistribution. The first two degradation modes are primarily time and temperature dependent. The electrode changes that cause electrolyte distribution problems appear to be primarily DOD dependent. Since Ni-H₂ cells have no cadmium electrodes and no nylon

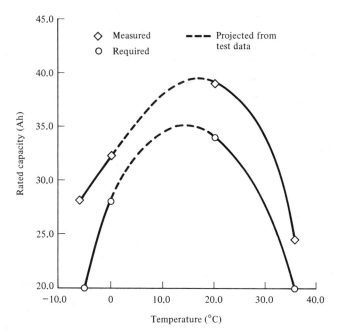

Figure 6.24 Effects of temperatures on Ni-H$_2$ cell capacity. (Courtesy of IN-TELSAT and FACC)

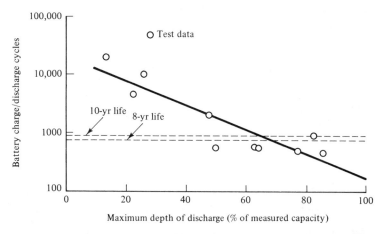

Figure 6.25 Ni-H$_2$ battery life test results.

separators, they suffer no apparent time- and temperature-dependent degradation. The platinum-catalyzed hydrogen electrodes undergo no significant changes with time or cycles. The asbestos separator material is much more stable in potassium hydroxide electrolyte than is nylon. Electrochemically impregnated nickel electrodes permit higher active material utilization than do conventional vacuum-impregnated types. The electrode stability, coupled with the high electrolyte affinity of the asbestos separator, strongly retards electrolyte redistribution problems. These factors are expected to make the operational life of Ni-H$_2$ batteries considerably longer than that of Ni-Cd batteries.

6.4 POWER CONTROL ELECTRONICS

The simplest configuration for an electric power subsystem would be to connect the battery, the solar array, and the load, all in parallel. The battery will discharge into the load during eclipse, and during the sunlight period the array will supply power to the load and to the battery for recharging. However, based on the solar array and the battery electrical characteristics, we can foresee some problems in this configuration.

The solar array output is described by a plot of current versus voltage (*I–V* curve). At each value of the voltage, there is only one current that the array can deliver. The *I–V* curve changes both due to seasonal variation in the array temperature and solar intensity, and due to radiation degradation of the solar cells, as shown in Fig. 6.26. The solar array voltage is maximum when the spacecraft comes out of eclipse due to the very low temperature of the solar cells. The solar arrays are normally designed to meet the load requirements at the end of life at summer solstice with the operating point at the maximum power point on the *I–V* curve. During the mission, however, the solar array provides excess power and the operating point is on the high-voltage portion of the curve. As shown in Fig. 6.26, if we keep the

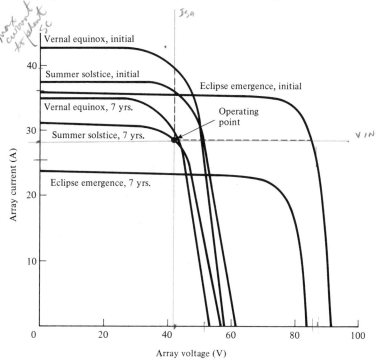

Figure 6.26 Solar array electrical characteristics.

load current constant, the voltage will vary from the desired 42 V to 85 V. Similarly, if we keep the operating voltage constant, the current will vary from the desired 29 A to 40 A. Hence we need a regulator which will find a suitable operating point on the array I–V curve and still provide the required load voltage and current.

Another problem is in the charging of the battery. The battery connected in parallel across the array will act as a shunt element and will force the system to operate at its charge voltage, accepting the difference between the load current and the available array current as charge current. Therefore, at the beginning of life, very high charge currents will go into the battery. These high charge currents will result in battery overheating and a reduction in the battery life. Hence we need a regulator to control the charging current to the batteries. The battery discharge voltage changes as a function of the discharge rate, temperature, and depth of discharge, as discussed in the preceding section. To keep the discharge voltage constant for a regulated bus, some type of battery discharge regulator will be required.

The commonly used regulators in a spacecraft power control system are as follows: (1) series dissipative regulator, (2) shunt dissipative regulator, and (3) pulse-width-modulated regulator.

Series Dissipative Regulators

A simplified diagram of a series regulator is shown in Fig. 6.27. The output voltage is sensed by a resistor divider. The signal is transmitted to an amplifier, which compares it to a reference voltage. The difference is amplified and inverted and transmitted to the pass transistor Q as a bias signal. The power dissipated in the pass Q transistor, as shown in Fig. 6.28, is

$$(V_{IN} - V_{REG})I_L = \text{power dissipated}$$

The characteristic curve of a series regulator is shown in Fig. 6.29.

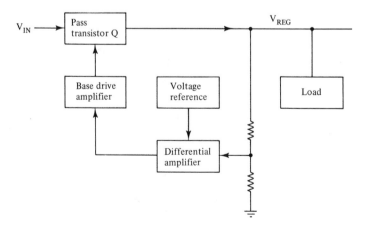

Figure 6.27 Series regulator.

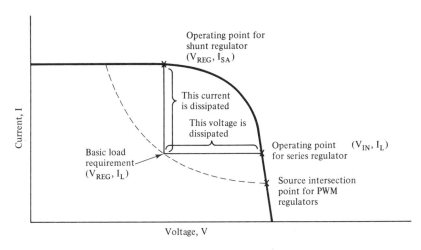

Figure 6.28 Solar array *I–V* curve with series and shunt regulator.

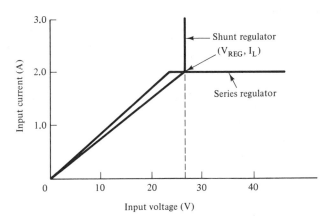

Figure 6.29 Characteristic curve of series and shunt regulator.

As shown in the figure, I_{IN} is increased to the point where regulator action takes place. I_{IN} is then a constant for all values of V_{IN}.

Shunt Dissipative Regulators

constant voltage

A simplified diagram of a full shunt regulator is shown in Fig. 6.30. The control elements are connected in parallel across the unregulated supply. The shunt element is a power transistor which receives its bias signal from the shunt driver amplifier in the mode controller. It acts as a current sink for a source current greater than load current demand at the system regulated voltage. The power dissipated in the full shunt regulator, as shown in Fig. 6.28, is

$$(I_{SA} - I_L)V_{REG} = I_{SC}V_{REG} = \text{power dissipated}$$

By not considering the battery function, the characteristic curve for a shunt

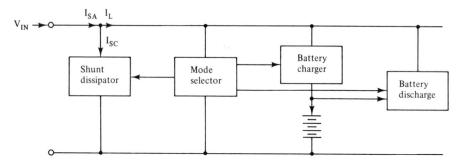

Figure 6.30 Shunt regulator.

regulator is shown in Fig. 6.29. It is a desirable approach where thermal design requires a "makeup" heat input for the condition when TWTs are turned off and no longer reject heat into the spacecraft system. In cases where thermal design requires a minimum dissipation from the bus voltage limiters, partial shunt limiters are preferred.

In a partial shunt regulator, the shunt element, which is a power transistor, is connected across a part of the solar array. Figure 6.31 shows the block diagram of a partial shunt regulator. The operation of the shunt regulator occurs when the solar array current (I_{SA}) exceeds the spacecraft load plus the battery current, causing the unregulated voltage, V_c, to rise. When the shunts are operating, they force the shunted portion of the solar array to operate in the current source region of its I–V curve. The output current from the shunted array exactly matches the spacecraft requirement. The expression for the current is

$$I_{SA} = I_l - I_{ps} = I_u$$

where I_{ps} is determined by the controller.

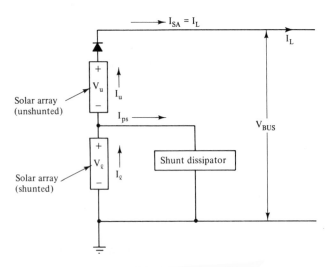

Figure 6.31 Partial shunt regulator.

used most frequently

Since the current available from the shunted portion of the solar array is less than I_l, the operating point of the unshunted array is in the voltage source region of its I–V curve, as shown in Fig. 6.32. The voltage across the shunted portion of the solar array assumes whatever value is necessary to maintain system regulated bus voltage. Thus the voltage is given by

$$V_l = V_{\text{BUS}} - V_u$$

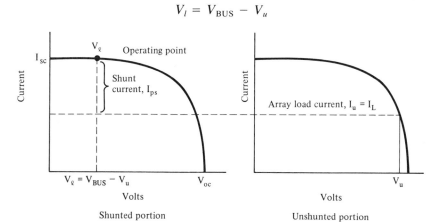

Figure 6.32 Partial shunt regulator I–V curve.

The partial shunt regulator results in low power dissipation, hence simpler electrical design and easier thermal control. It has the distinct advantage that it provides maximum system efficiency since it is open circuited at the end of mission when all the solar power is needed. The shunt regulator, however, introduces implementation complexities. A solar array normally consists of a number of solar cell strings with partial shunt regulators for each string. The solar cell layout must be designed to accommodate a "tap" point on each string, and these tap-point connections must be brought out into harnessing individually for grouping into one of several regulator units. These partial shunts can easily be sequenced so that all the shunts are either off or fully on, with only one unit operating in its linear region. This reduces the total power dissipation very significantly.

Pulse-Width-Modulated Regulators

Pulse-width-modulated regulators utilize power transistors in a switching mode rather than in an analog mode as in the series regulator. They generally offer higher efficiencies than do dissipative regulators, but at the cost in complexity, and a loss in frequency response, and output independence. It can be used to buck, boost, or buck-boost the input voltage.

A bucking regulator is used where the input voltage is always higher than the output voltage. The general configuration of this regulator is shown in Fig. 6.33. The output voltage is related to the input voltage by the ratio t_{ON}/t, where t_{ON} is the on-time of the switch and t is the total drive period. L_1 and C_1 form the input line filter. It should be noted that without the diode D, inductor L_2, and and capacitor C_2, the output voltage will be a

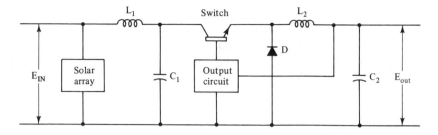

Figure 6.33 Pulse-width-modulated buck regulator.

square-wave form. These elements act as an integrator and smooth out the square-wave voltage.

A boost regulator is used where the input voltage is always less than the output voltage. The general configuration of this regulator is shown in Fig. 6.34. The operation of this regulator is the inverse of the bucking regulator. The ratio of output to input voltage is t/t_{off}, where t is the total drive period and t_{off} the off-time of the shunt switching element.

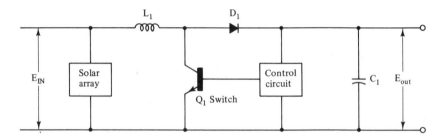

Figure 6.34 Pulse-width-modulated boost regulator.

Figure 6.35 illustrates the basic block diagram of a buck-boost regulator. The output voltage is related to the input voltage by the ratio $t_{\text{ON}}/t_{\text{off}}$, where t_{ON} and t_{off} are the on and off times of the switch, respectively. When a 50% duty cycle for the switch occurs, the input and output voltages are equal. When t_{ON} is greater than t_{off}, the circuit boosts the voltages, and it bucks the voltage when t_{ON} is less than t_{off}.

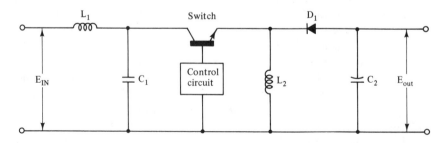

Figure 6.35 Pulse-width-modulated buck-boost regulator.

Partial Shunt Regulator Design

The power characteristics of a solar array for different mission conditions are shown in Fig. 6.26. The bus voltage is 42 V. The maximum voltage occurs at eclipse emergence when the solar array is coldest. Neglecting the voltage drop in transistors, the physical location of the tap point for the partial shunt is given by

$$\text{tap point (TP)} = 100 \left[1 - \frac{V_{\text{BUS}}}{V_{\text{OC}} \text{ (initial eclipse emergence)}} \right]$$

This represents a case in which the unshunted portion of the solar array voltage is equal to the desired bus voltage and the other portion is short circuited. For the solar array in this example, $V_{\text{BUS}} = 42$ V and $V_{\text{OC}} = 91$ V. Hence

$$\text{TP} = 100 \left(1 - \frac{42}{91} \right) \cong 54\%$$

6.5 SUBSYSTEM DESIGN

The electric power subsystem (EPS) is designed to meet spacecraft system requirements which are stated in terms of load profile, orbit, mission life, and system configuration. Initially, different design concepts are analyzed to achieve an optimum design of EPS for given system requirements. The trade-offs are performed among mass, cost, and reliability. There are several issues, such as bus voltage, single versus dual bus, unregulated versus regulated bus, solar cells and solar panel design, and fault protection, which are normally decided on the basis of past experiences of the manufacturer and the preference of the customer. This section provides an overview of these issues.

Load

The power requirements for each load element of the spacecraft system are evaluated for each major phase of the mission, beginning with switchover to internal battery power during lift-off through the on-orbit operation phases. The electric power system for a communications satellite is designed primarily for synchronous orbit load requirements. A typical breakdown of power requirements for a three-axis-stabilized communications satellite is shown in Table 6.4. The power requirements are given for three critical times during the year: autumnal equinox, summer solstice, and eclipse. In this case, the communications subsystem is fully operational during eclipse. The batteries that provide power during eclipses constitute approximately 40% of the electrical power subsystem mass. To reduce the EPS mass, some spacecraft are designed to have only a fraction of the communications subsystem on during eclipse. The variation in thermal control load requirements in Table 6.4 is mainly due to change in the power requirements for

TABLE 6.4 ELECTRIC POWER REQUIREMENTS IN SYNCHRONOUS ORBIT AT EOL (Watts)

Load	Autumnal Equinox	Summer Solstice	Eclipse
Communications	764	764	764
Telemetry, command and ranging	44	44	44
Attitude determination and control	38	52	38
Propulsion	1	1	1
Electric power	10	10	10
Thermal control	110	65	35
IR harness loss	10	10	9
Battery charging	98	29	0
Total	1075	975	901
Five percent load growth contingency	54	49	45
Power Source requirement	1129	1024	946
Solar array 10% design margin	113	102	95
Solar array output requirement with margins	1242	1126	

the battery heaters which are required to keep the battery temperature within a narrow range. The battery charging load also varies because of the requirements of a higher charge rate during equinox to fully charge the discharged battery after eclipse, and a lower charge rate during summer solstice for trickle charging. During the design phase, it is a common practice to use a 5% load contingency to take into account the uncertainty in the calculation of equipment load. For solar arrays, an extra 10% design margin is used for uncertainty in the radiation degradation and other power prediction factors. During autumnal equinox, the load requirements are higher than that in the summer solstice. The extra power requirements during equinox usually do not pose a problem because the available solar array power during equinox is higher than that at solstice.

Bus Voltage

The selection of bus voltage is frequently based on the desire to use existing equipment which has been proven on another satellite. The main power-switching semiconductors are limited to approximately 100 V by present technology, except for a few which operate up to 400 V. They normally set the upper limit for the bus voltage. The minimum voltage is determined by the distribution losses that can be tolerated. Low voltages, requiring high currents, will result in high losses. In the 1960s, bus voltages of 20 to 30 V were common. In the 1970s and early 1980s, bus voltages were of the order of 40 to 50 V. Bus voltages of 100 to 120 V are being designated for some spacecraft now, and will probably be quite common in the late 1980s and beyond.

The choice of the bus voltage level and range has a major effect on the power control electronics. In a regulated bus configuration, with separate solar array and battery regulators, the array and battery voltages may be chosen independently of each other. To simplify the battery charge and discharge regulators, the voltages should permit the use of either buck or boost regulators, rather than buck-boost types. For an unregulated bus, the designs of load regulators will be simplified if the voltage range of the battery discharge lies within that of the solar array. In general, load power conditioning will be eased if the solar array EOL voltage is made equal to the battery EOL discharge voltage.

For a lightweight solar array on a three-axis-stabilized spacecraft, the array voltage at eclipse exit may be more than double the normal array voltage and it decays in about 5 to 20 minutes as the array warms up. If a high level (e.g., 50 V) is chosen for the nominal array voltage, the eclipse exit voltage spike of 130 V can require careful attention to the load regulator design.

Single versus Dual Bus

In a single-bus system all the loads are connected to a single bus. In a dual-bus system, the steady-state loads are divided between two independent spacecraft power buses to maintain equal battery depth of discharge and to provide additional reliability, and protection against single point faults. In general, two complete sets of housekeeping equipment are installed with redundant units of the same types connected to opposite main buses. The failure of one unit will require use of the redundant unit, which is provided from the other power bus. In a dual-bus system, no single fault in the power subsystem or in the loads can affect more than half of the spacecraft system. A slight disadvantage of the dual-bus system is the constraint on the use of smaller redundant units to maintain power balance.

Unregulated versus Regulated Bus

In an unregulated bus like Intelsat IV, the voltage regulation is provided at each load separately. In Intelsat IV, the bus voltage varies between the upper voltage limit of 48 V, determined by the cold solar array emerging from an eclipse, and the lower limit of 23.8 V for the battery discharge control set point. When the discharge control turn-on set point is reached, the fully charged battery is switched directly to the bus. This places a step voltage increase of up to 10 V on the bus.

In a partially regulated bus with a direct energy transfer system, such as Intelsat V, the solar array voltage is regulated during the sunlight period. During the eclipse period, however, the battery discharge voltage is unregulated. For Intelsat V, the solar array voltage is regulated to 42 ± 0.5

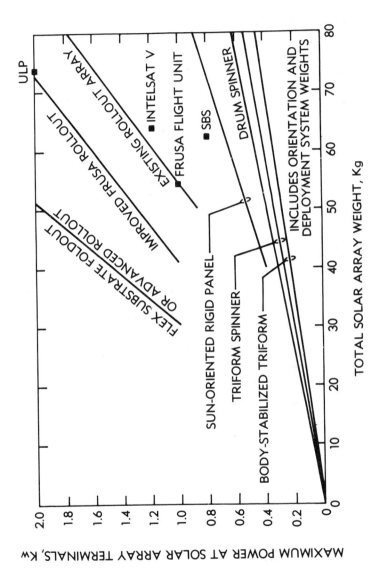

Figure 6.37 Solar array power at end of 7-year mission versus weight (summer solstice) (Ref. 15).

       Electric Power   Chap. 6

masses of Ni-Cd and Ni-H$_2$ batteries as a function of power capability during 1.2-hour eclipses are given in Figs. 6.38 and 6.39, respectively.

The mass of an electrical power subsystem will depend on the selected solar cells, solar panels, batteries, depth of discharge, and power control electronics. However, as a first approximation, Table 6.5 gives specific mass of an EPS for different power levels for a partially regulated and fully regulated bus. In a partially regulated bus, the solar array voltage during sunlight is regulated by a shunt regulator and the battery voltage is unregulated

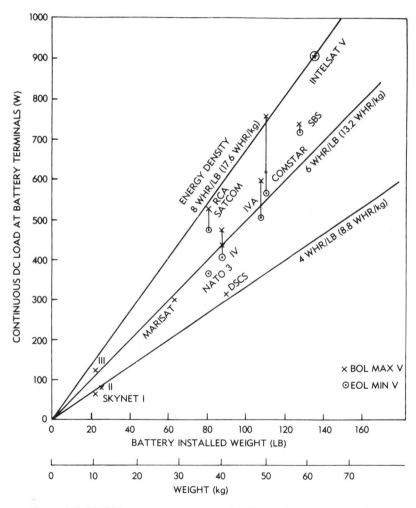

Figure 6.38 Ni-Cd battery power versus weight for synchronous spacecraft (based on 1.2-hour eclipse) (Ref. 15).

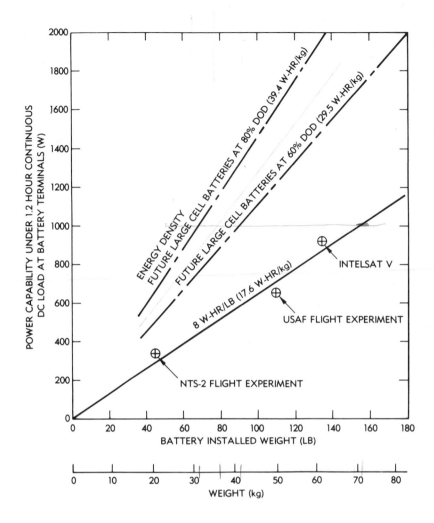

Figure 6.39 Ni-H$_2$ battery power density for synchronous spacecraft (Ref. 15).

during eclipse. In a fully regulated bus, the battery discharge voltage is also regulated. The solar array specific mass is based on regulated power at the end of 7 years (summer solstice). The battery specific mass is based on 70% depth of discharge for the Ni-H$_2$ batteries.

Fault Protection

Good electric power subsystems for spacecraft are designed in such a way that there is no single point failure. There are three types of failures:

TABLE 6.5 SPECIFIC MASS OF ELECTRIC POWER SUBSYSTEMS WITH PARTIALLY REGULATED DC BUS (7 YEARS LIFE AND 70% DEPTH OF DISCHARGE FOR Ni-H$_2$ BATTERIES) (gm/W)

Equipment	Three-Axis Stabilized				Drum Spinner	
	1.25 kW	2.5 kW	3.75 kW	5 kW	1.25 kW	2.5 kW
Solar array	50	42	33	25	125	70
Charge array	7.8	6.6	5.1	3.9	19.5	11
Shunt	7.5	7.5	7.5	7.5	7.5	7.5
Charge control	1.5	1.5	1.5	1.5	1.5	1.5
Battery	56.8	47.3	47.3	47.3	56.8	47.3
Discharge regulator	0.2	0.2	0.2	0.2	0.2	0.2
Total	123.8	105.1	94.6	85.4	210.5	137.5

SPECIFIC MASS OF ELECTRIC POWER SUBSYSTEMS WITH REGULATED DC BUS (7 YEARS LIFE AND 70% DEPTH OF DISCHARGE FOR Ni-H$_2$ BATTERIES) (gm/W)

Equipment	1.25 kW	2.5 kW	3.75 kW	5 kW	1.25 kW	2.5 kW
Solar array	50	42	33	25	125	70
Charge array	8.4	7.0	5.5	4.2	20.9	12
Shunt	7.5	7.5	7.5	7.5	7.5	7.5
Charge control	1.5	1.5	1.5	1.5	1.5	1.5
Battery	60.8	50.6	50.6	50.6	60.8	50.6
Discharge control	15.0	12.5	12.5	12.5	15.0	12.5
Total	143.2	121.1	110.6	101.3	230.7	154.1

(1) failures of power subsystem components, (2) failures in load components, and (3) harness failures.

Failures in power subsystem components.   The solar cells in a solar array can fail open, short, or short to ground. The design practice is to diode couple sections consisting of one or more solar cell strings into the main bus. Usually there are a large number of these sections. A single failure will be limited to the loss of one of these sections in the worst case.

The batteries have more complex failure modes. Because of the mass of batteries, cell- rather than battery-level redundancy must be used. A single cell in a battery can fail open, short, or short to ground. The case can rupture due to the evolvement of hydrogen gas. This failure mode can occur when the cell is overcharged and its voltage becomes too high or a cell's voltage is actually reversed. To protect against these failures, cells require individual monitoring. One control approach that has been used is the individual cell bypass circuit. The bypass circuit will bypass the cell if an indication is that hydrogen is evolving, or if the cell is an open circuit. In this simplest form, this circuit consists of four diodes. The cell voltage is the trigger for bypass circuit action.

Failures in load components.   The failure in load components can be short or open. If the load component has an open-circuit failure mode there is no danger to the power subsystem. However, a short-circuit failure could endanger the whole power system. In the simplest approach, each load is equipped with parallel redundant fuses. Alternatively, an undervoltage detector for each load equipment will preclude any danger to the equipment and is also used to sequentially unload the main bus in case of power deficit.

Harness failures.   Short-circuit failures in a wiring harness can be corrected by one of the following:

1. Dual or multiple bus
2. Diode isolation of sources
3. Double insulation system

In a dual-bus system, a fault in the power subsystem or in the loads can affect only half of the system. The main disadvantage of a split system is constraints on the use of small load elements to maintain power balance. However, main and auxiliary bus designs have also been used where there are only a few large loads.

Typical Spacecraft Power System Design

Let us consider a spacecraft with the following parameters:

Orbit: Synchronous orbit
Mission life: 7 years

Bus voltage: 42 V during sunlight and minimum 28
V during eclipse

Load: 1000 W, including load contingency, and
assume it to be constant

Array: Suntracking flat panels *← NORMAL INCIDENCE, NO COSINE LOSSES*

It is assumed that a system of two independent buses is required. Let us
first design the batteries.

Batteries. It is assumed that Ni-Cd batteries at a maximum depth
of discharge of 55% are used. The minimum discharge voltage of the battery
cell at the end of 7 years is assumed to be 1.1 V. The number of cells in
series is calculated on the basis of minimum battery discharge voltage V_{DB}
during eclipse. Let N be the number of cell in series. Assuming open-circuit
failure of one cell and bypass diode voltage drop V_{DD} to be 1.1 V, the
minimum discharge voltage V_{DB} equation becomes

$$V_{DB} = (N - 1) V_D - V_{DD}$$

Here $V_{DB} = 28$ V, $V_D = 1.1$ V, and $V_{DD} = 1.1$ V. Substituting these
quantities

$$28 = (N - 1) \times 1.1 - 1.1$$

or

$$N = 27.45 \cong 28$$

The minimum discharge voltage is

$$V_{DB} = 27 \times 1.1 - 1.1 = 28.6 \text{ V}$$

The power required by each bus will be half of the total power (i.e.,
500 W). Hence the battery for each has to supply 500 W of power for the
maximum of 1.2 hours. The required capacity of battery cell C is given by

$$C = \frac{Pt}{V_{DB} \cdot \text{DOD}} \quad ← FIGURE\ 6.20$$
$$Lifetime\ cycles$$

where C = cell capacity, Ah

     P = power, W

     t = maximum eclipse period, 1.2 hours for synchronous
orbit

  V_{DB} = battery minimum discharge voltage

 DOD = depth of discharge

Substituting the values of the parameters above gives us

$$C = \frac{500 \times 1.2}{28.6 \times .55} \approx 38 \text{ Ah}$$

During charging, the maximum allowable battery charge voltage is 1.5
V. It is assumed that an open-circuit failure of a battery cell during charge
is accommodated by three series-connected silicon diodes connected in

parallel with the cell. The voltage drop in each diode is 0.8 V. Hence the maximum charge voltage is

$$V_{BC} = 1.5 \times 27 + 3 \times 0.8$$
$$= 42.9 \text{ V}$$

The main bus voltage V_{BUS} is regulated to 42 ± 0.5 V. The lower limit of the bus voltage is 41.5 V. Let the battery charger voltage drop V_{CD} be 1.75 V. The boost voltage required by the charge array V_{CA} is given by

$$V_{CA} = V_{BC} - V_{BUS} + V_{CD}$$
$$= 42.9 - 41.5 + 1.75 = 3.15 \text{ V}$$

It is assumed that the charge current is applied sequentially to each battery on a 50% duty cycle. The charge rates, similar to Intelsat V, are assumed to be C/15 for equinox and C/45 for summer solstice. The C/15 corresponds to the current of 2.53 A (38/15) and C/45 corresponds to 0.844 A. Assuming the charging efficiency to be 90%, the total power required for charging each the batteries in high rate mode at equinox is

$$P_{charge} = \text{current} \cdot \text{voltage}$$
$$= 2.53 \times 42.9 \simeq 108 \text{ W}$$

The recharging time at high rate is based on returning the energy extracted from each battery during discharge. It can be calculated by

$$t_{recharge} = \frac{P_{discharge} \times t_{discharge}}{P_{charge} \times \eta}$$

$$t_{recharge} = \frac{500 \times 1.2}{108 \times 0.9} \simeq 6.2 \text{ hrs.}$$

Each battery usually remains on trickle charge during the remaining portion of the eclipse day.

The power required for battery charging at summer solstice is

$$P_{charge} = 0.844 \times 42.9 \cong 36 \text{ W}$$

Solar array design. The total load on the solar array will be the summation of the equipment load and the power required for charging the batteries. Generally, 10% solar array design margin is used to take into account the uncertainty in the radiation degradation and other design factors of the solar array.

The solar array design load at equinox is

$$1.1(1000 + 108) \cong 1{,}218 \text{ W}$$

The solar array design load at summer solstice is

$$1.1(1000 + 36) \cong 1140 \text{ W}$$

Since the power system is assumed to have two independent bus systems, the power provided by each bus is 609 W at equinox and 570 W at summer solstice. The solar arrays are designed in such a way that the load operating point at EOL is at the maximum power point of the solar array *I–V* curve.

Let us assume solar cells with the following characteristics at 25°C:

size = 2 × 4 cm   thickness = 0.025 cm

coverglass = cerium-doped microsheet of 0.015-cm thickness

$I_{mp} = 0.2966$ A $I_{sc} = 0.315$ A

$V_{mp} = 0.45$ V $V_{oc} = 0.548$ V

Solar array temperature:

Summer solstice: 39°C

Autumnal equinox: 49°C

Temperature coefficient at EOL:

$\alpha_I = 0.24$ mA/°C   *typical for Silicon*

$\alpha_V = -2.2$ mV/°C

The electrical characteristics mentioned earlier are based on 25°C temperature and standard solar intensity (135.3 mW/cm²) for a single-cell basis at BOL. The realistic solar array power is, however, calculated by considering several factors, such as assembly loss factors, environmental degradation factors, the seasonal variation of solar intensity, solar cell temperature, and random failures. The typical design factors for a synchronous satellite for a 7-year life are given in Table 6.6. It should be noted that the

TABLE 6.6 SOLAR ARRAY DESIGN FACTORS FOR A TYPICAL THREE-AXIS STABILIZED SPACECRAFT WITH 7 YEARS LIFE

Factors	Current		Voltage
Assembly Losses	0.96		
Module Assembly		0.99	1.0
Mismatch		0.99	1.0
Measurement Error		0.98	1.0
Panel Wiring Loss		1.0	0.005V/cell
Blocking Diode Dreop		1.0	0.9V
Array Wiring Harness and Slip Ring		1.0	0.9V
Environmental Degradation (7 years)	0.8154	0.935	
Micro-metroid		0.99	1.0
Ultra-violet		0.99	1.0
Adhesive and Cover Slide		0.98	1.0
Low Energy Protons		0.99	1.0
Radiation		0.875	0.935
Temperature Cycling and Random Failure		0.98	1.0
Solar Intensity Factor			
Summer Soltice		0.8885	1.0
Autumnal Equinox		0.9941	1.0
Effective Illumination factor			
Suntracking Flat Panel		1.0	1.0

effective illumination factor is unity for sun-tracking flat panels and $1/\pi$ for cylindrical panels in a spin-stabilized spacecraft.

Taking these design factors into account, the cell current at maximum power point EOL summer solstice I–V curve is given by

$$I = [I_{mp} + \alpha_I(T - 25)]K_A^i K_D^i K_s \ k_I$$

where I = solar cell current at maximum power point of EOL summer solstice

I_{mp} = solar cell current at BOL

α_I = temperature coefficient for current

K_A^i = design factor for assembly losses in current

K_D^i = design factor for environmental degradation in current

K_s = solar intensity factors, including incidence angle

K_I = Illumination factor

By substituting the summer solstice values of the parameters above, we get

$$I = [0.2966 + 0.24 \times 10^{-3} (39 - 25)] \times 0.96 \times 0.8154 \times 0.8885$$

$$= 0.2086 \text{ A}$$

The required current at summer solstice by each bus, each wing of the array, is

$$I_T = \frac{\text{power}}{\text{bus voltage}} = \frac{570}{42} = 13.572 \text{ A}$$

The number of cells in parallel for each wing is

$$N_p = \frac{I_T}{I} = \frac{13.572}{0.2086} \cong 66$$

The solar cell voltage is given by

$$V = [V_{mp} - \Delta V + \alpha_V(T - 25)]K_E^V$$

where V = solar cell voltage at EOL

V_{mp} = solar cell voltage at maximum power points, BOL

ΔV = panel wiring loss per cell

α_V = temperature coefficient for the voltage

T = operating temperature

K_E^V = radiation degradation factor for voltage

Substituting the values for the parameters above yields

$$V = [0.45 - 0.005 - 0.0022 (39 - 25)] \times 0.935$$

$$= 0.3873 \text{ V}$$

The bus voltage drops are assumed to be 0.9 V in the blocking diode and 0.9 V in the array wiring harness and slip ring. Hence the number of cells in series, N_S, is given by

$$N_S = \frac{\text{bus voltage} + \text{bus voltage drop}}{\text{cell voltage}}$$

$$= \frac{42 + 1.8}{0.3873} \cong 114 \text{ cells}$$

The solar cell current and voltage at EOL autumnal equinox are given by

$$I = [0.2966 + 0.24 \times 10^{-3} (49 - 25)] \times 0.96 \times 0.8154 \times 0.9941$$
$$= 0.235 \text{ A}$$
$$V = [0.45 - 0.005 - 0.0022 (49 - 25)] \times 0.935$$
$$= 0.3667 \text{ V}$$

The current per bus, or wing, is

$$I_T = 0.235 \times 66 = 15.51 \text{ A}$$
$$V_{\text{BUS}} = 114 \times 0.3667 - 1.8$$
$$= 40 \text{ V}$$

The total power output from the two buses $= 2 \times 15.51 \times 40 = 1240$ W. Since the design load at equinox is 1218 W, the power margin is 22 W.

Charge array. The voltage required by the charge array is 3.15 V. The maximum power voltage for a cell at EOL summer solstice, as calculated earlier, is 0.3873 V. Hence the number of cells in series for the charge array, N_C, is

$$N_C = \frac{3.15}{0.3873} \cong 9 \qquad \text{for other battery}$$

The charge current at summer solstice is 0.844 A (C/45) and it is 2.53 A at equinox (C/15) for each battery. At summer solstice EOL, the maximum power current per solar cell, as calculated earlier, is 0.2086 A. Hence the number of cells in parallel for charging, is

$$N_{\text{SC}} = \frac{0.844}{0.2086} \cong 5$$

At autumnal equinox EOL, the maximum power current per solar cell, as calculated earlier, is 0.235 A. Hence the number of cells in parallel for charging at equinox, charging rate 2.53 A is

$$N_{\text{SC}} = \frac{2.53}{0.235} \cong 12$$

Hence for charging the batteries, we need a charge array (5PX9S) for summer solstice and (12PX9S) for equinox. It will be convenient to have one string of 5PX9S and a second string 7PX9S. For summer solstice, we can use the 5PX9S string and for equinox we can put both the strings in parallel. Since there are two batteries, we can put a set of these strings on each wing.

Solar panel design. The number of the solar cells per wing is 66 parallel by 114 in series. The assumed layout of the solar cells is shown in Fig. 6.12. Allowing 10 cm for margin and 1 mm for intercell spacing, the width of the array (or length of the panel) = 114 × 2.1 + 10 = 250 cm.

Let us assume that we use ULP panels with six segments. Since we need 66 cells in parallel for the main array and a small area for the battery charge array, we divide the solar cells into two panels, in such a way that the first panel carries 36 strings and the second panel carries 30 strings for the main array and the charge array. Allowing 6 cm for margins and 1 mm for intercell spaces, the width of the panel = 36 × 4.1 + 6 = 153.6 cm. Each solar wing is assumed to consist of a yoke, one empty panel to clear the antenna shadow, and two panels for solar cells. The panel folding is arranged so that this outer panel containing the battery charge array is exposed to the sun during transfer orbit. The mass breakdown is shown in Table 6.7.

TABLE 6.7 SOLAR ARRAY MASS BREAKDOWN (kg)

Component	Mass	Number per Wing	Number per Array	Total Mass
Solar cells plus covers	4.0	2	4	16.0
Panels	3.18	2	4	12.72
Empty frame	1.7	1	2	3.4
Panel hinges	.15	6	12	1.8
Yoke	1.8	1	2	3.6
Yoke hinge	0.7	1	2	1.4
Hold-down mechanism	3.5	1	2	7.0
Harness	3.0	1	2	6.0
				51.92

PROBLEMS

6.1. A dual-spin-stabilized spacecraft needs 2 kW power. The design life is 10 years. The bus voltage is 29 V. The solar cell characteristics at 25°C are: I_{mp} = 0.39 A; V_{mp} = 0.454 V; size = 1.8 × 6.2 cm. The radiation degradation factors for 10 years are 0.9303 and 0.949 for I_{mp} and V_{mp}, respectively. The other design factors can be used from Table 6.6. The temperature of the solar panels are estimated to be 0°C, 6°C, and −63°C for summer solstice, autumn equinox, and post-eclipse, respectively. The temperature coefficients of the solar cell at EOL are α_I = 0.24 mA/°C and α_V = −2.2 mV/°C. Determine (a) the number of solar cells in series, (b) the number of solar cells in parallel, (c) the total solar panel area, (d) the mass of the solar panel, and (e) the power at EOL post-eclipse.

6.2. Design a solar array for a three-axis-stabilized spacecraft with the same power requirements and solar cell characteristics as in Problem 6.1. The temperature of the solar panel is estimated to be 37°C, 45°C, and −160°C for summer solstice, autumn equinox, and post-eclipse, respectively.

6.3. A spacecraft needs 2 kW power during a solar eclipse for a maximum period of 72 minutes. The minimum bus voltage is 28 V. Design an energy storage system consisting of two Ni-H$_2$ batteries, each with a number of cells in series. Assume that EOL minimum cell discharge voltage is 1.07 V and the maximum depth of discharge is 70%. Determine (a) the number of cells in series, (b) the ampere-hour capacity of each cell, (c) the ampere-hour capacity of each battery, and (d) the mass of the battery system.

6.4. Design a charge array for the battery system defined in Problem 6.3. The charge array is in series with the main solar array, which is defined in Problem 6.1. The maximum cell voltage at charge is 1.5 V. The charge rates are C/70 during trickle charging and C/7 for high-rate charging.

6.5. For a spacecraft, the electrical characterisics of the solar array are given in Fig. 6.26. The load requirement is 1.2 kW at 42 V bus voltage. Compare power dissipation in the series and full shunt regulators at vernal equinox, summer solstice, and eclipse emergence for BOL conditions.

6.6. Assume a shunt regulator is across a half solar array whose characteristics are given in Fig. 6.26. The load requirement is 1.2 kW at 42 V bus voltage. Determine the power dissipation in the shunt regulator at BOL vernal equinox.

6.7. A DBS spacecraft needs 2.5 kW power. Assume a regulated dc bus. Estimate the EPS subsystem mass, using Table 6.5, for the following cases: (a) three-axis-stabilized spacecraft with full-load (2.5 kW) requirement during eclipse, (b) three-axis-stabilized spacecraft with half-load (1.25 kW) requirement during eclipse, and (c) spin-stabilized spacecraft with half-load requirement during eclipse.

REFERENCES

1. *Basic Photovoltaic Principles and Methods,* SERI/SP-290-1448, Solar Information Module 6213, Feb. 1982.

2. *Solar Cell Array Design Handbook,* Vol. I, National Technical Information Service, N-77-14193, Oct. 1976.

3. C. E. Backus, "Solar Cells," The Institute of Electrical and Electronics Engineers, Inc., New York, 1976.

4. G. W. Sutton, *Direct Energy Conversion,* McGraw-Hill, New York, 1966.

5. F. S. Osugi, "Electrical Power System—The Intelsat IV Spacecraft," *COMSAT Technical Review,* Vol. 2, Fall 1972.

6. H. S. Rauschenbach, *Solar Cell Array Design Handbook,* Van Nostrand Reinhold. New York, 1980.

7. F. H. Esch, W. J. Billerbeck, and D. J. Curtin, "Electric Power Systems for Future Communication Satellites," IAF-76-237, XXVII Congress International Astronautical Federation, Anaheim, Calif., Oct. 1976.

8. J. A. Allison, R. A. Arndt, and A. Meulenberg, "A Comparison of the Comsat Violet and Non-reflective Solar Cells," *COMSAT Technical Review,* Vol. 5, No. 2, Fall 1975.

9. F. C. Treble, "Solar Array for the Next Generation of Communication Satellites," *Journal of the British Interplanetary Society,* Vol. 26, pp. 449–465, 1973.

10. J. T. Radecki, *Power,* Hughes Aircraft Company Geosynchronous Spacecraft Case Histories, Vol. I, Jan. 1981.

11. J. F. Stockel, and others, "NTS-2 Nickel Hydrogen Battery Performance," *Proceedings,* AIAA 7th Communication Satellite Systems Conference, Apr. 1978.

12. W. Scott and D. Rusta, NASA Reference Publication 1052, *Sealed Nickel–Cadmium Battery Applications Manual.* Publ. DRC, 1979.

13. W. Billerbeck and W. Baker, "The Design of Reliable Power Systems for Communications Satellites," AIAA/NASA Space Systems Technology Conference, Costa Mesa, Calif., June 1984, AIAA Paper No. 84-1134-CP.

14. W. Billerbeck, "Minimizing Spacecraft Power Loss Due to Single-Point Failures," *Proceedings,* 19th Intersociety Energy Conversion Engineering Conference, San Francisco, Aug. 19, 1984, p. 345.

15. W. Billerbeck, "Spacecraft Electric Power," presented as a part of AIAA National Capital Section Professional Study Seminar Series, at University of Maryland, Nov. 8, 1982.

16. W. Billerbeck, "Electric Powers for State-of-the-Art Communications Satellites," *Proceedings,* NTC 73, Nov. 1973, pp. 19D-1 to 19D-12.

7

COMMUNICATIONS

7.1 INTRODUCTION

This chapter presents an overview of the communications subsystem, commonly referred to as the payload of the spacecraft. To understand the communications subsystem of a spacecraft, one begins with a communication link. A communication link starts with the source signal. The source signal could be a human voice that has been converted into an electrical analog waveform by the microphone, or it could be the signal generated by a color TV camera or it could be a series of binary digits (ones and zeros) generated by teletype machines or data terminals. For transmission over long distances, it is inconvenient and expensive to keep the source signal in its original form. Usually, this electrical source signal, or as it is commonly called by communications engineers, the "baseband signal," travels only a short distance down the cable before it reaches a center where the signal is processed for further transmission. (The term "baseband" is also used by communications engineers in a wider sense than that described here. Generally, it refers to signals of relatively low frequencies.) As an example, a voice signal leaving the telephone handset reaches the local switching office as its first destination. If it is a local call, the signal is switched to the receiving party's telephone and there will be no further signal processing. However, if it is a long-distance call, the signal has to be processed and amplified along the way using the most economical means of transmission. If the signal is carried over the air via microwave or satellite transmission, it is more convenient to use radio frequencies for transmission. Radio frequencies are at much higher frequencies than the source signal. For example, a voice signal spreads from about 300 Hz to about 4000 Hz, and a microwave signal can be operating in the 4×10^9 Hz or 4 GHz band. Thus a means of carrying the source signal in the microwave frequency band is necessary. This process of converting the source signal into a signal suitable for transmission over air in a microwave frequency band is in fact modulation and up-conversion. This type of signal processing is common to both terrestrial microwave transmission and satellite microwave transmission. Figure 7.1 illustrates the elements of a simplified communication link: (a) via terrestrial microwave transmission and (b) via satellite microwave transmission.

It can be observed that a satellite link is really a microwave link with only two hops (i.e., an uplink and a downlink), whereas the terrestrial line-of-sight microwave link consists of a number of hops, depending on the distance covered. One line-of-sight microwave hop can span up to about 80 km before the distant microwave tower disappears below the horizon. On the other hand, a satellite link can link up two earth stations situated on nearly opposite sides of the earth. The cost of the satellite link is therefore insensitive to distance. This makes satellite link ideal for long-distance (e.g., across the ocean) communications.

Figure 7.1 also shows that the satellite does not use the same frequency for the uplink and the downlink. The microwave signal has to be amplified and processed on-board the satellite for retransmission. Signal processing (in conventional satellites) consists of a simple frequency translation from the uplink frequency to the downlink frequency. For efficient retransmission and amplification of the signals, the signals are separated into individual channels or transponders for final amplification on-board the satellite before retransmission back to earth. A simplified block diagram of an active satellite repeater is shown in Fig. 7.2. Frequency translation from the uplink to the downlink and signal amplification are shown in this figure.

This chapter begins with the basic definitions and units used by communications engineers. Then it discusses the elements of a typical communication link, starting with the source signal and tracing the signal through modulation and transmission via a communications satellite. Having given the introduction, the chapter then deals with the subject in more detail by first considering some aspects of the Radio Regulations as laid down by the International Telecommunication Union (ITU) regarding satellite communications. The evolution of the satellite communications subsystem is to some extent traceable to the provisions of the ITU Radio Regulations. Next, this chapter considers the fundamentals of electromagnetic radiation (i.e., radio waves and the effects that our earth atmosphere has on the propagation of radio waves). This is followed by a discussion of the concept of thermal noise and the concept of carrier-to-noise ratio, which are very basic and important in the design of any communication link. Having developed these basic and essential ideas, this chapter then discusses practical design aspects of a satellite communications subsystem. The satellite communications subsystem will be separated into two parts: the antenna subsystem and the repeater or transponder subsystem. The antenna subsystem is essentially a passive subsystem; all the active elements are found in the transponder subsystem. The design of these subsystems is illustrated by

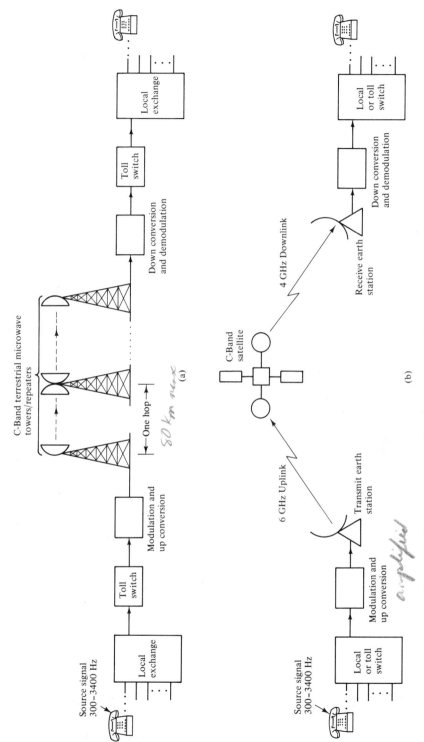

Figure 7.1 Elements of a simplified communication link: (a) terrestrial microwave communications system; (b) satellite microwave communications system.

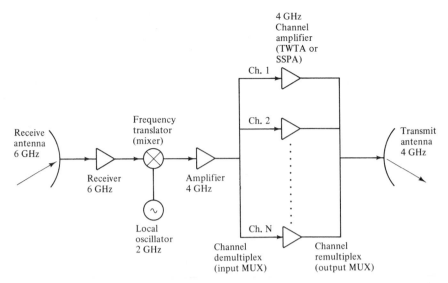

Figure 7.2 Simplified block diagram of a satellite communication subsystem (payload).

actual examples from existing spacecraft. After the discussion of the hardware aspects of the communication subsystem, the chapter concludes with a discussion on the communication aspects of the communication subsystem: access techniques, modulation techniques, and the communication capacity of the communications subsystem.

7.2 BASIC UNITS AND DEFINITIONS IN COMMUNICATIONS ENGINEERING

One of the most basic quantities in communications engineering is signal power. This is usually expressed in the well-known units of watts or milliwatts. Since the communications engineer has to multiply and divide quite often, it is more convenient to express quantities in a logarithmic form or in the form of the logarithm of a ratio. Thus, if the signal power S is expressed in the logarithmic form with respect to 1 watt, then

$$S(\text{dBW}) = 10 \log_{10} \frac{\text{signal power}}{1 \text{ watt}} \qquad (7.1)$$

Similarly, if the signal power S is expressed in the logarithmic form with respect to 1 milliwatt, then

$$S(\text{dBm}) = 10 \log_{10} \frac{\text{signal power}}{1 \text{ milliwatt}} \qquad (7.2)$$

Thus one can convert dBW to dBm by adding 30 dB to the figure, or conversely, one can convert dBm to dBW by subtracting 30 dB from the figure.

Another useful quantity that communications engineers use is power flux density (PFD). For a propagating plane radio wave, this is a measure of the amount of electromagnetic energy passing through a given aperture or "window." Usually, this is expressed in watts per square meter or more conveniently in dBW/m^2. Thus the intensity of the radio wave coming toward a receiving station can be expressed in dBW/m^2. Since the radio wave may not consist of one single frequency (a single-frequency radio wave is a continuous wave—commonly known as a CW wave), and rather, it may contain a multitude of frequencies, the intensity of the radio wave must be specified in a certain frequency bandwidth (e.g., in $dBW/m^2/kHz$ or in $dBW/m^2/4kHz$, etc.). This expression of radio-wave intensity basically implies that the radio signal is a continuous spectrum as opposed to a discrete line spectrum. This is indeed the case for most radio signals and it is rather analogous to a human voice, which spreads from about 300 Hz to about 3400 Hz without discrete components.

7.3 FREQUENCY ALLOCATIONS AND SOME ASPECTS OF THE RADIO REGULATIONS

Since all countries on earth share the same radio-frequency spectrum, the ITU Radio Regulations provide the requisite coordination and regulations to ensure that everyone's need is met on an equitable basis. Thus the radio-frequency spectrum is heavily segmented so that each segment or frequency band of the spectrum is allocated by mutual consent for a specific purpose on a coordinated basis. Most of the frequency bands are allocated for more than one type of service and precautions in coordinations are therefore necessary to avoid mutual interference. The ITU classifies usages of the radio-frequency spectrum into the following broad categories:

- Broadcasting
- Fixed
- Fixed-satellite
- Broadcasting-satellite
- Mobile
- Amateur
- Space research radio astronomy
- Aeronautic radio navigation
- Meteorological
- Radio location
- etc.

Thus, as far as satellite communications is concerned, the majority of those services discussed in this chapter will fall under three main categories: fixed-satellite, broadcast-satellite, and mobile-satellite. Mobile-satellites can be further subclassified into maritime-mobile, aeronautical-mobile, and land-mobile.

Since the satellite transmitters may share the same frequency bands with other terrestrial services, a maximum power flux density (PFD) limit at the earth's surface (e.g., -148 dBW/m^2/4 kHz) is usually imposed on the satellite transmitters so that the satellites do not interfere with existing terrestrial microwave links. Thus satellite designers often do not have complete freedom to increase the satellite capacity by increasing the power transmitted by the satellite. However, there are certain bands reserved primarily for broadcasting satellites and for these bands the allowable PFDs are usually somewhat higher than those of the fixed-satellite services. These bands are usually reserved for direct-to-home satellite broadcasting services. For further details on this subject, the reader is referred to the ITU Radio Regulations (Ref. 1).

For as much as the radio-frequency spectrum is a natural resource, so is the geosynchronous orbit. There is only one orbit that allows a satellite in that orbit to remain fixed with respect to an observer on earth. Efficient use of the geosynchronous orbit is in everyone's interest. Current technology permits minimum spacing of adjacent satellites to about 3 to 5° (longitude) in the orbital arc. Closer spacing may cause potential mutual interference. This point will become clearer when the radiation pattern of an earth station antenna is discussed. It is sufficient to point out that future earth station antennas may allow closer than 2° satellite spacings. Even at 2° spacing, the maximum number of geosynchronous satellites transmitting and receiving in the same band can be only 180. Due to this limitation, satellite designers have devised solutions whereby the allocated satellite frequency bands are reused several times from a single orbital slot. More will be said on this important topic in Section 7.6.

Communications engineers have used acronyms to designate various frequency bands. Following is a list of these.

Low frequency (LF)	30 kHz –300 kHz
Medium frequency (MF)	300 kHz – 3 MHz
High frequency (HF)	3 MHz– 30 MHz
Very high frequency (VHF)	30 MHz–300 MHz
Ultrahigh frequency (UHF)	300 MHz– 3 GHz
Super high frequency (SHF)	3 GHz – 30 GHz
Extremely high frequency (EHF)	30 GHz –300 GHz

$$1 \text{ Hz} = \text{one hertz} = \text{one cycle per second}$$
$$1 \text{ kHz} = 10^3 \text{ Hz} \quad (\text{k} = \text{kilo})$$
$$1 \text{ MHz} = 10^6 \text{ Hz} \quad (\text{M} = \text{Mega})$$
$$1 \text{ GHz} = 10^9 \text{ Hz} \quad (\text{G} = \text{Giga})$$

In addition to these band designations, communications engineers commonly use letters to designate bands in the microwave spectrum. These designations have no official international standing, and various engineers have used limits for the bands other than those listed below.

Band Destination	Frequency Band (Approx.)
L Band	400 MHz– 1.5 GHz
S Band	1.5 GHz– 3.9 GHz
C Band	3.9 GHz– 6.2 GHz
X Band	6.2 GHz–10.9 GHz
K Band	10.9 GHz–36.0 GHz
Q Band	36.0 GHz–46.0 GHz
V Band	46.0 GHz–56.0 GHz

Frequently, satellite engineers call the 4/6-GHz band for fixed satellite services the C band, the 11/14-GHz band the Ku band, and the 20/30-GHz band the Ka band.

The following are some popular bands allocated for satellite communications, extracted and simplified from ITU Radio Regulations (Ref. 1).

Service	Band	Usage
Mobile satellite	1.535–1.559 GHz	Downlink
	1.6265–1.6605 GHz	Uplink
Broadcast satellite	2.5–2.69 GHz	Downlink
Fixed satellite	3.4–4.2 GHz	Downlink
	4.5–4.8 GHz	Downlink
	5.85–7.075 GHz	Uplink
	7.25–7.75 GHz	Downlink
	7.9–8.025 GHz	Uplink
	8.025–8.4 GHz	Uplink + downlink
	10.7–11.7 GHz	Downlink
Fixed + broadcast satellite	11.7–12.75 GHz	Downlink
Fixed satellite	12.75–13.25 GHz	Uplink
	14.0–14.5 GHz	Uplink
Broadcast satellite	14.5–14.8 GHz	Uplink
Fixed satellite	17.3–17.7 GHz	Uplink
Fixed satellite	17.7–18.1 GHz	Uplink + downlink
	18.1–21.2 GHz	Downlink
Inter satellite	22.55–23.55 GHz	Intersatellite
Fixed satellite	27.5–31.0 GHz	Uplink
Inter satellite	32.0–33.0 GHz	Intersatellite
Fixed satellite	37.5–40.5 GHz	Downlink

To conclude this section on frequency bands and frequency allocations, it is necessary to point out that there is a free-space wavelength associated with every radio frequency. It is sometimes more convenient to refer to the wavelength rather than the frequency of the radio waves. In free space (or air), the relationship is as follows:

$$\lambda = \frac{c}{f} \tag{7.3}$$

where λ is the wavelength, f is the frequency in hertz, and c is the velocity of light and is approximately equal to 3×10^8 m/s. In the microwave region (about 1 to 30 GHz), engineers often describe sizes of the microwave components in terms of wavelengths of the chosen operating frequency. Since passive microwave components can be scaled readily from one frequency of operation to another, it makes sense to express component dimensions in wavelengths rather than in meters or centimeters. For example, the gain of a circular microwave dish antenna whose diameter is expressed in terms of wavelength is independent of the frequency of operation. This is explained in Section 7.5.

7.4 ELECTROMAGNETIC WAVES, FREQUENCY, AND POLARIZATION SELECTION FOR SATELLITE COMMUNICATIONS

Basic Concepts of Electromagnetic Waves

Electromagnetic waves are traveling waves of the electric and magnetic fields that vary sinusoidally with time and with distance in the direction of propagation. They are analogous to the acoustic traveling waves in air. This acoustic analogy does not extend to the vectorial (or directional) nature of the electric and magnetic fields; that is, the electric or magnetic fields have orientations in space due to the "lines of force" nature of these fields. These lines of force are quite evident when static electric or magnetic fields are examined. Thus, in general, electric or magnetic fields can be resolved into three orthogonal components in space, for example, in Cartesian coordinates, E_x, E_y, and E_z and H_x, H_y, and H_z, where E and H represent the amplitude electric and magnetic fields, respectively. In vectorial form,

$$\mathbf{E} = E_x\mathbf{i} + E_y\mathbf{j} + E_z\mathbf{k} \tag{7.4}$$

$$\mathbf{H} = H_x\mathbf{i} + H_y\mathbf{j} + H_z\mathbf{k} \tag{7.5}$$

Fundamental electromagnetic theory tells us that if the propagation of the plane wave is in the z direction, $E_z = H_z = 0$ (i.e., the fields are transverse to the direction of propagation). In free space, E and H are related by a simple constant,

$$E = Z_0 H$$

where $Z_0 = \sqrt{\mu/\varepsilon}$ is the characteristic impedance of free space and it is 377 Ω, and μ and ε are respectively the permeability and permittivity of the propagation medium. Thus, most of the time, communications engineers need only work with either the E or the H field in free space. Now choosing to express the electromagnetic waves in the E fields only, for a plane wave propagating in the z direction,

$$\mathbf{E} = E_x\mathbf{i} + E_y\mathbf{j} \tag{7.6}$$

and both E_x and E_y have sinusoidal variations with time. If $E_y = 0$, the wave is said to be horizontally polarized, and if $E_x = 0$, the wave is said to be vertically polarized. The communications engineer has the choice of sending the signal via either horizontally polarized or vertically polarized electromagnetic waves. E_x and E_y components are orthogonal to each other and they are mutually independent of each other. Figure 7.3 illustrates the transmission of the horizontally and vertically polarized waves. Both the vertically and the horizontally polarized waves are known as linearly polarized (LP) waves. This is only one of the many possible orthogonal sets of polarized electromagnetic waves. Another very popular orthogonal set consists of circularly polarized waves (Fig. 7.4). If

$$\mathbf{E} = E_0 \sin\left(\omega t + \frac{\pi}{2}\right)\mathbf{i} + E_0 \sin(\omega t)\mathbf{j} \qquad (7.7)$$

where t is time and $\omega = 2\pi f$, f being the frequency, and if one looks into the direction of propagation (z direction), the resultant E field (or vector)

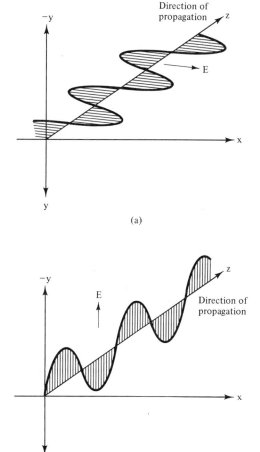

(a)

(b)

Figure 7.3 (a) Horizontally polarized wave; (b) vertically polarized wave.

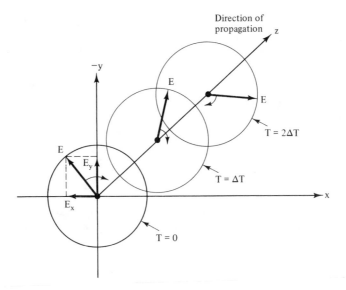

Figure 7.4 Right-hand circular polarization (RHCP).

has a constant magnitude and it rotates clockwise as it progresses down the z direction. This is known as a right-hand circularly polarized wave (RHCP). Equation (7.7) shows that an RHCP wave can be generated if both E_x and E_y are present and equal in magnitude but with E_x leading E_y by 90° in time phase. The left-hand circularly polarized wave (LHCP) is similarly generated by having E_y leading E_x by 90° in time phase, that is,

$$\mathbf{E} = E_0 \sin(\omega t)\mathbf{i} + E_0 \sin\left(\omega t + \frac{\pi}{2}\right)\mathbf{j} \qquad (7.8)$$

Thus LHCP is "orthogonal" to RHCP. In the most general situation, the E_x component is not equal to the E_y component, and these two components may have a difference in time phase other than 90°. In that case, the wave will be elliptically polarized. The orientation of the major axis of the polarization ellipse (as traced out by the tip of the rotating E vector) as well as the ellipticity of the ellipse are dependent on the relative time phase and the relative magnitude of the E_x and E_y components. The axial ratio of an elliptically polarized wave is defined by the power ratio between the major and minor axes of the polarization ellipse, usually expressed in decibels. It can now be seen that linear polarization is a special case of elliptical polarization where the ellipse becomes infinitely thin, and circular polarization is another special case where the ellipse becomes perfectly round.

Frequency and Polarization Selection

The communications engineer is usually faced with the problem of selecting the frequency and the polarization for the satellite communication link. Since the ITU has already segmented the radio spectrum into bands allocated for specific services, it appears, at first sight, that frequency band selection should be straightforward. However, it was shown in Section 7.3 that numerous bands have been allocated for satellite communications and most of these bands are shared with other services. Communications engineers usually take into account a large number of factors before deciding on the satellite operating frequencies and polarizations. Some of these factors will no doubt translate into mechanical and spacecraft considerations. The following are the more important factors that face the satellite designers and systems engineers at the definition phase of a satellite system.

Propagation Effects

Since the geosynchronous orbit is above the earth's atmosphere, line-of-sight satellite communication paths must traverse the entire atmosphere at various incident angles (or elevation angles for the earth stations). At low frequencies, the ionosphere literally prevents radio waves from penetrating it. At high frequencies, the atmosphere gives rise to loss mechanisms that result in propagation fade. This is especially true in the presence of precipitation (e.g., rain) in the communications path. Figure 7.5 shows the absorption loss of the clear sky atmosphere for radio waves.

Figure 7.6 shows the effects of rain and clouds on a satellite communication path. It can be seen that the most useful window of the spectrum for satellite communication is between 100 MHz and 30 GHz.

In addition to signal fades, precipitation in the atmosphere causes depolarization of an otherwise perfectly polarized electromagnetic wave. For example, a perfect linearly polarized wave may become elliptically polarized after traversing a rainstorm. Or a perfect RHCP wave may contain an LHCP component in addition to loss in signal strength after propagating through the atmosphere containing precipitation. In less likely circumstances, a sandstorm can also give rise to fade and depolarization. The ratio of the wanted polarization component to the unwanted orthogonal component is called the cross-polar discrimination ratio (XPD) and is usually expressed in decibels. It will become clear later why depolarization is important when spectrum reuse via orthogonal polarization is discussed.

At frequencies below the X band, Faraday rotation, caused by the ionosphere, is another propagation phenomenon of considerable importance. While the ionosphere does not appreciably attenuate the electromagnetic wave at frequencies above 100 MHz, the E field of a linearly polarized wave is twisted or rotated as it passes through the ionosphere without being depolarized to any large extent. The amount of rotation (Faraday rotation) is a function of the frequency of the wave. Table 7.1 gives some typical numbers for the rotation of the E field at selected frequencies. It is obvious that at L and S bands, LP radio waves for satellite communications is undesirable unless adaptive receiving systems are used to follow the rotation of the E field. Therefore, it is quite common that L-band satellite systems do not use LP but use CP radio waves for communication. For the C band, the choice is not clear between LP and CP, and both LP and CP systems are currently being used. For the X and K bands, there is little reason for

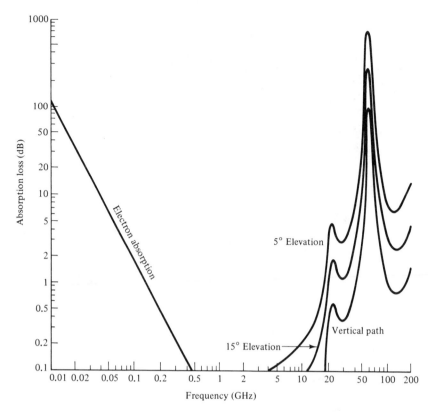

Figure 7.5 Absorption in the atmosphere caused by electrons, molecular oxygen, and uncondensed water vapor (Ref. 11).

adopting a CP system for fixed satellite services unless it is anticipated that either the transmitter or the receiver does not have a fixed orientation (e.g., a satellite designed to tumble in space). Engineers designing broadcast satellites may still choose CP to minimize the problem of earth receiving antenna feed alignment at the time of installation (i.e., avoidance of the necessity to rotate the microwave antenna feed to align with the incoming polarization of the electromagnetic wave).

Interference Effects

As mentioned before, most of the frequency bands allocated for satellite communications are also allocated for other services. Communications engineers designing a satellite system must consider the possibility of mutual interference in the process of selecting the frequency bands. In some cases, earth stations may have to be located well outside urban centers in a "quiet zone" to avoid interference. Microwave or cable back-haul will be needed to connect these earth stations to the urban switching centers. If, however,

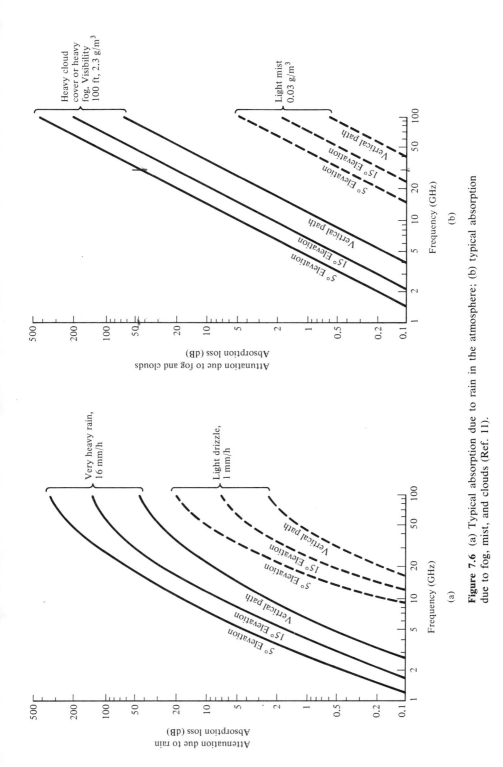

Figure 7.6 (a) Typical absorption due to rain in the atmosphere; (b) typical absorption due to fog, mist, and clouds (Ref. 11).

TABLE 7.1a ESTIMATED MAXIMUM IONOSPHERIC EFFECTS IN THE UNITED STATES FOR ELEVATION ANGLES OF ABOUT 30 DEGREES ONE-WAY TRAVERSAL

Effect	Frequency Dependence	100 MHz	300 MHz	1 GHz	3 GHz	10 GHz
Faraday Rotation	$1/f^2$	30 rot.	3.3 rot.	108°	12°	1.1°

TABLE 7.1b FARADAY ROTATION AT 1 GHz EXCEEDED FOR GIVEN PERCENTAGES OF TIME

Time percentage	99	90	50	10	1	0.1	0.01	Period
Rotation (degrees)	3	7	23	43	64	75	82	1979/80
	1	2	7	14	30	41	47	1977/78

the frequency bands are not already congested, rooftop-type earth station antennas can be placed directly on top of switching centers or "nodes" of the communications network to save the costs of back-haul connections. For most industrialized nations, the C bands are already congested to the point that many earth stations have to be located outside urban areas. However, the upper X bands or the low K bands may still allow rooftop-to-rooftop satellite link. A lot of satellite operators are currently exploiting these "newer" bands in this manner (e.g., the Canadian Anik-C system, the SBS system in the United States, and the Australian Aussat system).

Equipment Availability/Maturity

The choice of operating frequency bands may have profound effects on the hardware or equipment. For example, it was mentioned that microwave engineers design components in terms of their wavelengths of operation. It is therefore necessary to use either larger or else lossier components at lower frequencies. This is so because the ohmic losses of the passive microwave components increase as the sizes of the components decrease. Also, as mentioned earlier, the gain of a microwave dish antenna is proportional to the square of the diameter of the antenna when the diameter is expressed in wavelengths. Thus an antenna designed for a specific gain will have a smaller diameter if the frequency of operation is higher. This point is explained in more detail in Section 7.5.

Although ITU regulations do not specify the pairing of uplink and downlink frequency bands for a satellite, the uplink and downlink bands are usually paired up in a satellite for the ease of equipment design; that is, the uplink and downlink are usually not too far apart in frequency, so that the same antenna can be used for both transmit and receive purposes. On the other hand, if the uplink band is too close to the downlink band, there may be difficulties in achieving the required transmit/receive isolation, which is usually in the order of 110 dB or more for an antenna that transmits and receives at the same time. This is true for both the satellite antenna and for the earth station antenna.

The communications engineer will also be faced with the costs of the equipment used in the entire satellite communications system. Usually, microwave equipment at lower frequencies is more mature and hence less costly than that at higher frequencies. Design and planning engineers must trade equipment maturity and cost with duration of operation before obsolescence.

Policy and Regulations

Government policies or regulations may require satellite operators to choose one frequency band over another. This is often done for specific long-term goals or purposes. In such cases, satellite designers have little choice but to optimize the design for the given frequency bands. Techniques such as depolarization compensation, adaptive uplink power control, and encoding techniques can be used to overcome some of the inherent link problems. Some of these techniques are discussed later in this chapter.

7.5 LINK CONSIDERATION

Concepts of Antenna Radiation Pattern and Antenna Gain

A microwave radiator (or antenna) does not necessarily radiate energy with equal intensity in all directions, but if it does, it is called an isotropic radiator. The gain of an isotropic radiator is defined to be unity (i.e., 0 dB). If, however, an antenna concentrates its radiated energy in a particular direction within a small solid angle, the antenna will have gain over the isotropic antenna in that direction. Figure 7.7 illustrates the polar radiation diagram of an isotropic antenna and also that of an antenna with high gain in a certain direction. The peak gain of the antenna is defined as the ratio of maximum radiated power intensity of that antenna (in a certain direction) to the radiated power intensity of a hypothetical isotropic antenna fed with the same transmitting power, that is,

$$\text{antenna peak gain}|_{dB} = 10 \log_{10} \frac{I_a}{I_0} \qquad (7.9)$$

where I_a = maximum radiated power intensity
I_0 = power intensity of an isotropic antenna

This definition of the gain of an antenna can be generalized to include the gain $G(\theta, \phi)$ of the antenna in directions other than that of maximum radiation, that is,

$$G(\theta, \phi)|_{dB} = 10 \log_{10} \frac{I(\theta, \phi)}{I_0} \qquad (7.10)$$

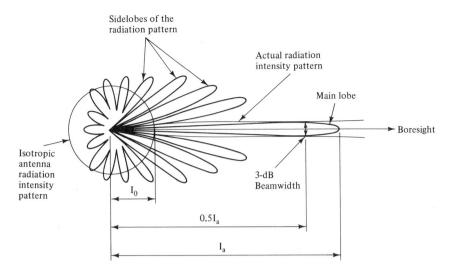

Figure 7.7 Polar radiation pattern of an antenna.

where θ and ϕ define the arbitrary direction in a polar (r, θ, ϕ) coordinate system and $I(\theta, \phi)$ is the radiated power intensity in the direction of (θ, ϕ). The peak gain, sometimes called the boresight gain of an antenna, is a measure of the antenna's ability to confine as much of the radiated power within as small a solid angle as possible (i.e., the higher the gain, the smaller is the solid angle). A good way to quantify this solid angle is the 3-dB beamwidth of the antenna. The 3-dB beamwidth of the antenna is defined by the two points on either side of the boresight in the polar radiation pattern where the gains are 3 dB (i.e., half power) below the maximum boresight gain of the antenna. This is illustrated in Fig. 7.7. Note that 3 dB below the maximum is equivalent to a radiation intensity where the E field is 0.707 times the maximum E field at boresight. The E field is expressed in volts per meters rather than in dBW/m^2, which is the unit used to describe radiated power intensity. Also, the 3-dB beamwidth (θ_1) in the θ direction may not necessarily be the same as the 3-dB beamwidth (ϕ_1) in the ϕ direction. A useful empirical formula that relates antenna gain to the two beamwidths is

$$\text{boresight gain} = \frac{30{,}000}{\theta_1 \times \phi_1} \qquad (7.11)$$

where gain here is expressed in real numbers (not in decibels) and θ_1, ϕ_1 in degrees. Equation (7.11) represents a fairly efficient antenna (i.e., $\eta \geq 70\%$; see the definition of η in the following section).

Figure 7.7 also illustrates the concepts of the "main beam" and "sidelobes" of an antenna. The main beam of an antenna is the main lobe of the radiation pattern centered around the boresight of the antenna. For example, the solid angle bounded by the 3-dB points can be considered the main beam of the antenna. In addition to the main peak of the radiation pattern formed by the main beam of the antenna, there are smaller peaks on either side of the main beam. These smaller peaks form the sidelobes of the radiation pattern. For an efficient antenna, the main beam peak gain should be as high as possible with respect to the isotropic gain, and the sidelobe peaks should be as low as possible. Practical antenna designs usually compromise between these two requirements. For a given antenna size, the ratio of the main beam peak gain to the sidelobe peak gain is one measure of the effectiveness of the antenna design. The importance of this ratio will become evident in later sections.

Fundamental antenna theory shows that the peak gain (or boresight gain) of a microwave antenna can be predicted from the aperture area A of the antenna. Here A is the projected area of the antenna reflector in a plane perpendicular to the direction of maximum radiation (i.e., the direction of the boresight). Figure 7.8 shows the projected aperture of an offset-fed reflector antenna. The term "offset" is used to describe an antenna geometry with the feed placed outside the main beam of the antenna. A discussion of the various basic geometries is given in Section 7.6.

An ideal antenna would have the following relationship:

$$G_{max} = \frac{4\pi A}{\lambda^2} \tag{7.12a}$$

where λ is the wavelength and G_{max} is the maximum boresight gain, and for a circular aperture

$$G_{max} = \left(\frac{\pi D}{\lambda}\right)^2 \tag{7.12b}$$

where D is the diameter of the antenna. G_{max} can, of course, be expressed in decibels above isotropic [if 10 times the log to the base 10 of the right-hand side of Eq. (7.12a) or (7.12b) is taken]. For a practical antenna, the

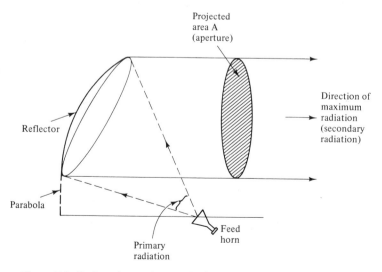

Figure 7.8 Projected area (aperture) of an offset-fed parabolic reflector.

boresight gain or the maximum gain will be less than that shown in Eqs. (7.12). The boresight gain G_0 can be expressed as

$$G_0 = \frac{4\pi A}{\lambda^2} \eta = \frac{4\pi A_{\text{eff}}}{\lambda^2} \qquad (7.13)$$

where η is the efficiency factor and A_{eff} is the effective area of the antenna. Usually, η is somewhere between 40 and 85% for a typical well-designed microwave antenna. Equation (7.13) is also useful to describe antennas that do not have an easily recognizable aperture, for example, the familiar Yagi TV antenna on top of many houses. In the above equation, A_{eff} can be considered as the "capture" area of a receiving antenna. Equation (7.13) applies to receive antennas as well as to transmit antennas. This concept of A_{eff} as a capture area for receiving antennas will be used to derive the basic link equations in the following section.

Power Transfer from a Transmitting Antenna to a Receiving Antenna

To assess the quality of the signal, the communications engineer must calculate the signal power that is being transferred from a transmitting antenna to a receiving antenna. Let G_T and G_R be the boresight gains of the transmitting and the receiving antennas, respectively. Assume that the transmitting and receiving antennas are properly aligned such that their boresights are collinear (i.e., they are looking at each other on-axis), as shown in Fig. 7.9.

Let P_T be the total power radiated by the transmit antenna. If the transmit antenna were isotropic, the power flux density ϕ at a distance r from the transmit antenna would be

$$\phi = \frac{P_T}{4\pi r^2} \qquad (7.14)$$

Since the transmit antenna has a gain of G_T over the isotropic antenna, the power flux density at the receiving antenna is actually

$$\phi = \frac{P_T G_T}{4\pi r^2}$$

From Eq. (7.13), the effective area A_R of the receiving antenna with an

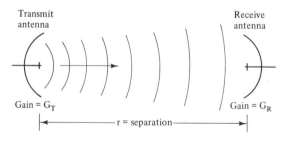

Figure 7.9 Power transfer from transmit antenna to receive antenna.

antenna gain of G_R is

$$A_R = \frac{\lambda^2 G_R}{4\pi}$$

Since A_R is the capture area of the receive antenna, the power P_R received by the receiving antenna is ϕA_R, that is,

$$P_R = \frac{P_T G_T G_R \lambda^2}{(4\pi r)^2} \tag{7.15}$$

Equation (7.15) is the fundamental equation governing power transfer between two antennas. The factor $(4\pi r/\lambda)^2$ is commonly called "path loss," P_L, by systems engineers. Although the well-known inverse-square law is frequency independent, this definition of path loss is frequency dependent, that is,

$$P_L = \left(\frac{4\pi r}{\lambda}\right)^2 \tag{7.16}$$

Thus, if Eq. (7.15) is written in the logarithmic form in decibels, then

$$\left.\frac{P_R}{P_T}\right|_{dB} = G_T|_{dB} + G_R|_{dB} - P_L|_{dB} \tag{7.17}$$

Example 7.1
The earth station antenna has an on-axis gain of 42.5 dB and the satellite antenna has a gain of 30 dB. The separation between the geosynchronous satellite and the earth station is 39,500 km. If the satellite delivers 10 W of signal power at 4 GHz, what is the received signal level at the earth station when atmospheric losses are neglected?

Solution At 4 GHz, the wavelength

$$\lambda = \frac{3 \times 10^8}{4 \times 10^9} \text{ m} = 0.075 \text{m}$$

Path loss P_L is

$$P_L|_{dB} = 20 \log_{10} \frac{4\pi \times 3.95 \times 10^7}{0.075} = 196.4 \text{ dB}$$

Transmitted power is 10 W or 10 dBW. Then according to Eq. (7.17), the received power P_R is

$$P_R = 10 + 42.5 + 30 - 196.4 \text{ dBW}$$

$$= -113.9 \text{ dBW or } -83.9 \text{ dBm}$$

The received signal power is approximately 4 pW. This example is fairly typical of a domestic satellite downlink for TV-receive-only earth stations using a 4.5-m receiving antenna.

Equivalent Isotropically Radiated Power

In the discussion above, it is observed that the gain of the transmit antenna in decibels is always added to the transmitted power expressed in decibels above 1 watt, in other words, in dBW. Systems engineers find it convenient to express the product of the transmit antenna gain and the total radiated power and call it equivalent isotropically radiated power

(EIRP), that is, power that appears to have radiated from an isotropic antenna. This is a very common term in satellite communications and is given by

$$\text{EIRP}|_{\text{dBW}} = \text{antenna gain}|_{\text{dB}} + P_T|_{\text{dBW}} \qquad (7.18)$$

Thus, in the numerical example above, the satellite EIRP is 10 + 30 = 40 dBW. Once the EIRP of a satellite transmitter is known, it is a simple matter to calculate the power flux density at the surface of the earth. It is left as an exercise for the reader to show that the power flux density (PFD) in dBW/m² at the earth's surface can be written as

$$\text{PFD} = \text{EIRP of satellite}|_{\text{dBW}} - 162\ \text{dB} \qquad (7.19)$$

at the subsatellite point, and

$$\text{PFD} = \text{EIRP of satellite}|_{\text{dBW}} - 163.5\ \text{dB} \qquad (7.20)$$

at an earth station farthest away from the satellite if atmospheric losses are neglected.

Concepts of Thermal Noise, G/T, and C/N

Life for the communications engineers would not have been as interesting if there had not been such a thing as noise. Noise could potentially corrupt the transmitted signal in a communication link. It must be emphasized at the outset that thermal noise is only one of the chief contributors to satellite link degradation. There are other contributors, such as distortion in the transmission equipment, nonlinearities in active devices, and interference from other transmissions. However, thermal noise has been shown to be one of the major contributors to link degradation in satellite communications. The name "thermal noise" suggests that the noise is of thermal origin. Indeed, the origin of thermal noise stems from random motions of electrons within the devices in the transmission path. A quantum mechanical model leads to the following equation for the thermal noise power density P_N in watts per hertz bandwidth (Fig. 7.10):

$$P_N(f) = \frac{hf/2}{e^{hf/(kT)} - 1} \qquad \text{W/Hz} \qquad (7.21)$$

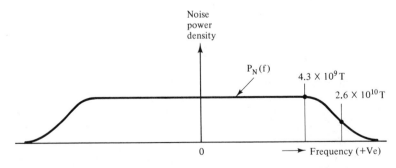

Figure 7.10 Thermal noise power density versus frequency.

where f is the frequency in hertz, h is Planck's constant $= 6.62 \times 10^{-34}$ J·s, k is the Boltzmann constant $= 1.38 \times 10^{-23}$ J/K, and T is the absolute temperature of the device or the noise source in Kelvin. Equation (7.21) gives the available noise power and this power is the actual noise contribution if the source and load are conjugate matched (i.e., maximum transfer of power from source to load). Also, at microwave frequencies and at moderate temperatures, $hf \ll kT$. Equation (7.21) reduces to

$$P_N(f) = \frac{kT}{2} \quad \text{W/Hz}$$

and if the communication system does not distinguish between negative and positive frequencies (as is the case),

$$P_N(f) = kT \quad \text{W/Hz}$$

For a finite filter bandwidth or receiver bandwidth of B hertz, the total noise power P_N received is

$$P_N = kTB \tag{7.22}$$

Equation (7.22) is the fundamental approximate noise equation used by communications engineers to calculate the noise power in a certain communication channel. It is emphasized here that Eq. (7.22) cannot be used to calculate noise power in optical links. Thermal noise is of secondary importance in optical devices, where quantum noise and shot noise dominate.

If thermal noise produced by different devices are uncorrelated, they can be summed on the basis of a simple power addition. For example, in a receiving system, the effective antenna thermal noise can be added directly to the effective receiver amplifier thermal noise. It is therefore convenient to develop the concept of effective thermal noise temperature T_e of microwave devices. This is illustrated in Fig. 7.11.

A noisy amplifier can be considered to have an effective input noise temperature followed by an ideal noiseless amplifier with gain as shown in Fig. 7.11(a). For a cascade of amplifier stages (or other microwave devices), where T_{ei} is the effective noise temperature of the ith amplifier, the total effective noise temperature T_s at the input to the cascade is [Fig. 7.11(b)]

$$T_S = T_{e1} + \frac{T_{e2}}{g_1} + \frac{T_{e3}}{g_1 g_2} + \cdots + \frac{T_{en}}{\displaystyle\prod_{i=1}^{n-1} g_i} \tag{7.23}$$

If this amplifier chain is connected to a receiving antenna that has an antenna noise temperature of T_a, the total effective noise temperature, T_T, is

$$T_T = T_a + T_s \tag{7.24}$$

Thus the effective input noise level P_N at the antenna–receiver interface [Fig. 7.11(c)] is

$$P_N = kT_T B \tag{7.25}$$

With the help of Eq. (7.25), the calculation of the carrier-to-noise ratio (C/N) can now be carried out if the carrier level is known. Using the numerical example 7.1, if it is assumed that the antenna effective noise

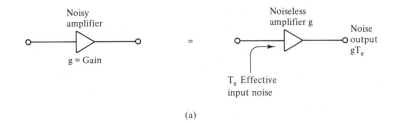

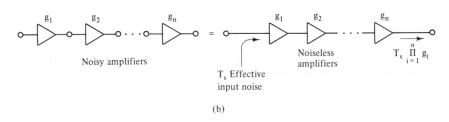

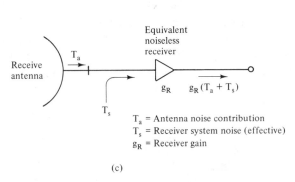

Figure 7.11 (a) Representation of thermal noise in a microwave amplifier; (b) representation of thermal noise in a cascade of microwave amplifier; (c) representation of thermal noise in a microwave receiving system.

temperature is 60 K and that T_s of the receiver is 100 K, the total noise power P_N in a bandwidth of 30 MHz is

$$P_N = 1.38 \times 10^{-23} \times (60 + 100) \times 30 \times 10^6 \text{ W}$$

i.e. $P_N = -228.6 + 22 + 74.8 \text{ dBW}$

or $N = -131.8 \text{ dBW}$ where $P_N = N$

The carrier level C was determined to be -113.9 dBW; then the carrier-to-noise ratio C/N is

$$\frac{C}{N} = -113.9 - (-131.8) = 17.9 \text{ dB}$$

This is a realistic C/N ratio for a TV signal delivered by a typical C-band satellite and received on the 4.5-m antenna and amplified by a good solid-state microwave receiver.

Before leaving this section on link considerations, two other quantities or figures of merit commonly used by communications engineers are now defined. Instead of quantifying the amplifier noise performance by the effective input noise temperature shown in Fig. 7.11(a), the noise figure N_F of an amplifier is defined by

$$N_F = 1 + \frac{T_e}{T_0} \qquad (7.26)$$

where T_0 is the standard noise temperature and is equal to 290 K. Equation (7.26) can be rewritten as

$$T_e = T_0(N_F - 1) \qquad (7.27)$$

Thus, given the noise figure, the effective noise temperature can be found, and vice versa. The noise figure N_F can also be expressed in decibels, that is,

$$N_F|_{dB} = 10 \log_{10} N_F \qquad (7.28)$$

Another figure of merit commonly used by communications engineers is the G/T of a receiving system. Referring to Fig. 7.11(c), the G/T of the receiving system is defined as follows:

$$\frac{G}{T} = \frac{G_A}{(T_A + T_S)} \qquad (7.29a)$$

$$\frac{G}{T}\bigg|_{dB} = G_A|_{dB} - 10 \log_{10} (T_A + T_S) \qquad (7.29b)$$

where G_A is the antenna gain and T_A and T_S are, respectively, the antenna and receiver noise temperature in Kelvin. Since G_A is related to A_{eff} of the receiving antenna, the G/T ratio of the receiving system at a given frequency can be used to predict the C/N of the link if the receive power flux density ϕ is known. For example, the carrier level C is

$$C = A_{eff}\phi$$

that is,

$$C = \frac{\lambda^2}{4\pi}G_A\phi$$

The noise power N is given by Eq. (7.25) as (note that $P_N = N$)

$$N = k(T_A + T_S)B$$

Therefore,

$$\frac{C}{N} = \frac{\lambda^2 G_A\phi}{4\pi kB(T_A + T_S)} = \frac{\lambda^2\phi}{4\pi kB}\frac{G}{T} \qquad (7.30)$$

For given ϕ, λ, and B, the C/N ratio is directly proportional to the G/T ratio of a receiving system. The G/T ratio of a receiving system is therefore a measure of the sensitivity of the receiving system.

Finally, readers should observe that all the discussions above on link considerations apply equally well for any pair of transmitting and receiving microwave stations, satellite–earth stations or terrestrial microwave repeater stations.

7.6 COMMUNICATIONS SUBSYSTEM OF A COMMUNICATIONS SATELLITE

Having discussed the fundamental concepts and relationships between the various quantities used to characterize a microwave link, this section is devoted to the design and design concepts of the communications system of a typical communication satellite. Figure 7.12 shows a communications subsystem or payload within a communications satellite. Figure 7.12(a) shows a communications payload using separate antennas for receive and transmit, while Fig. 7.12(b) shows a single antenna used for both transmit and receive. The configuration shown in Fig. 7.12(a) allows separate optimization of the two antennas for the two separate functions of receive and transmit. Electrically, the design of the antennas in Fig. 7.12(a) would be more straightforward than that in Fig. 7.12(b), but at the expense of the weight and complexity of an additional antenna on board the spacecraft. While the concept in Fig. 7.12(b) avoids the mechanical complications of an additional antenna, the antenna must now be optimized over two separate spacecraft frequency bands (i.e., the transmit band and the receive band) and additionally, an extra microwave device D must be used to "duplex" the transmit and receive signals to interface with the transponder. As shown

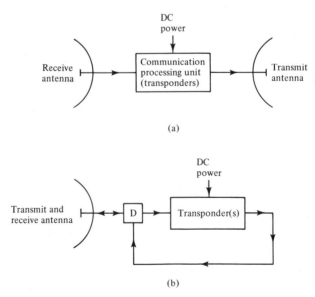

(a)

(b)

Figure 7.12 (a) Simplified block diagram of the payload of a communications satellite; (b) simplified block diagram of the payload of a communications satellite.

in Fig. 7.12, the term "transponder" is used to describe the communication processing channel unit within the payload. Thus the communications subsystem comprises the communications antenna(s) and the transponders.

A discussion of microwave satellite (transmit or receive) antennas will now precede the discussion on the transponders.

Spacecraft Antennas and the Reuse of the Allocated Spectrum (Bandwidth)

There are certain significant differences between the design of a satellite microwave antenna and the design of an earth station microwave antenna. The satellite antenna must be light and must survive in the space environment, where temperature extremes are usually encountered; whereas the weight of an earth station antenna is not a significant factor and the earth station survives in a completely different environment. The satellite antenna is usually designed to provide communications "coverage" over an area or a certain landmass, but the earth station antennas in most instances need to "cover" only a single point (i.e., the satellite in the sky). The satellite antenna designer is very often constrained by fairing envelopes of the launch vehicles and sometimes may have to resort to stowed and deployable designs to meet the various launch vehicle constraints. However, the earth station antenna designer is very much concerned with sidelobe levels (see Fig. 7.7) because of the fear of adjacent satellite interference. Until recently, satellite antenna designers have concentrated only on the antenna (gain) performance within the required main beam coverage area. Now, spectrum reuse techniques via spatial beam isolation have forced satellite antenna designers to come up with low-sidelobe designs as well. The technique of spectrum reuse via spatial beam isolation is illustrated in Fig. 7.13.

Figure 7.13(a) shows a satellite antenna designed to cover the entire visible surface of the earth all at once. This allows the allocated spectrum (by ITU) to be used only twice [i.e., via two orthogonal polarizations (see Section 7.4)]. If the required coverage can be split into subareas, a multibeam antenna can be designed so that the allocated spectrum is reused a number of times. In Fig. 7.13(b), the spectrum is reused eight times (four beams each with two orthogonal polarizations), assuming, of course, that sufficient mutual isolations between the four antenna beams can be maintained. This assumption is valid if the sidelobes of each beam are kept low enough such that effectively each of the main beams of the antenna covers each of the service areas and practically nowhere else.

Intelsat has pioneered the use of such multibeam antennas for increasing the capacity of the individual satellite. Recently, for example, Canada (Anik-C) and Australia (Aussat) have used orthogonal polarization to double the available bandwidth for domestic satellite systems. Figure 7.14 shows the typical Intelsat V spacecraft antenna coverages and Fig. 7.15 shows the Anik-C/D and Aussat coverages. Figure 7.16 shows the Intelsat VI antenna coverages. Sixfold frequency reuse will be achieved in this arrangement.

It should be noted from the figures that each of the main beams is of

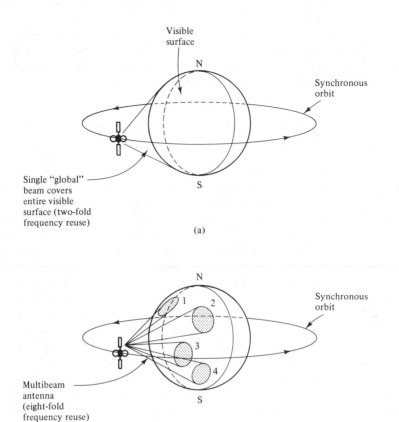

Figure 7.13 (a) Global beam coverage; (b) spot beam coverages.

a complicated shape designed to match the projected landmass to be covered. Thus the cross sections of the satellite antenna main beams are not necessarily circular or elliptical. In addition, the sidelobes of each of the main beams are carefully controlled so as not to interfere with the other main beams having the same sense of polarization. This is indeed an advancement from the simple "global" beam antennas used on early Intelsat spacecraft. It demonstrates the importance of the role of satellite antennas in spectrum-reuse techniques for current and future communications satellites. It can be seen that the number of times that the allocated spectrum (bandwidth) can be reused is predominantly antenna dependent (i.e., the number of times that the spectrum can be reused is ultimately limited by the aggregate of the sidelobe levels of all the other beams that have the same sense of polarization and also the aggregate of the cross-polarization levels of all the other beams that have the orthogonal sense of polarization).

To conclude this brief discussion on satellite antenna design concepts for spectrum reuse, it is worthwhile noting that antenna designers have

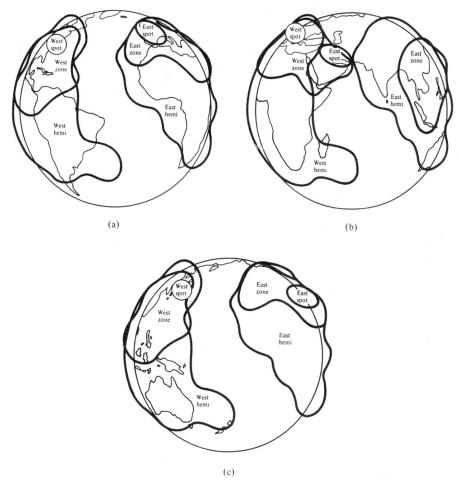

Figure 7.14 Intelsat V antenna coverages (hemi and zone C-band spots Ku band): (a) Atlantic Ocean; (b) Indian Ocean; (c) Pacific Ocean. (Courtesy of INTELSAT)

thus far managed to devise sophisticated multibeam antenna design techniques permitting the reuse of the allocated frequency bands a small number of times. It can be shown that, in theory, there is no limit to the number of times that the spectrum can be reused. In practice, that number is limited by the size, weight, and complexity of the satellite antenna (i.e., by achievable polarization purity and beam isolation). Certain administrations around the world are currently planning on advanced concepts whereby the spectrum can be reused 10 to 100 times or more. This will certainly ease the problem of orbital crowding that is currently happening. The following section discusses and explains some of the antenna configurations and design concepts presently being used in modern communications satellites.

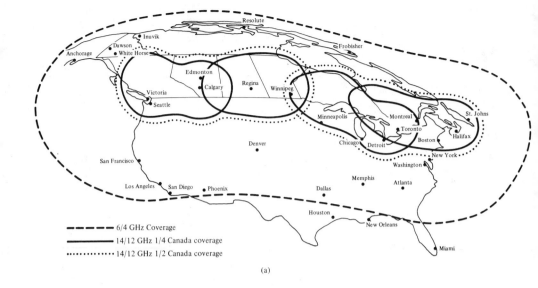

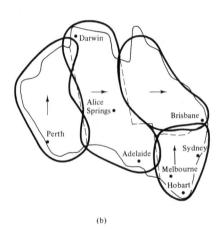

Figure 7.15 (a) Canadian Anik C and D antenna coverages (Courtesy of IN-
TELSAT); (b) Aussat transmit spot beams. (Courtesy of AUSSAT Pty. Ltd.)

Basic Types of Microwave Antennas for Satellites

As indicated earlier, the electromagnetic spectrum available for satellite
communication stretches from about 100 MHz to about 30 GHz. Such a
wide range of frequencies certainly offers the possibility of many different
design approaches for the antenna designers.

Below 1 GHz, the antennas could be of the dish (reflector) type or it
could be the wire type most often seen on rooftops of a lot of houses.
Because Faraday rotation of the linearly polarized E field is significant
below about 4 GHz, communications engineers normally use circularly
polarized waves rather than linearly polarized waves for these frequencies.

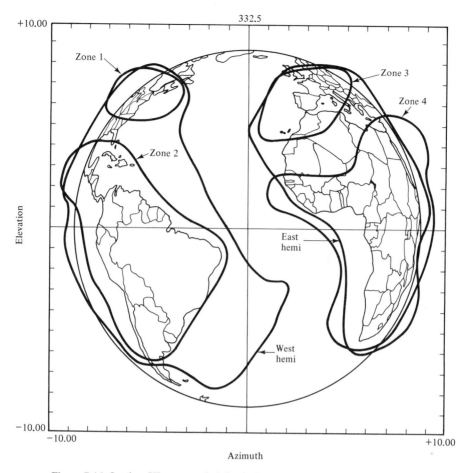

Figure 7.16 Intelsat VI spacecraft Atlantic Ocean region C-band coverages, six-fold frequency reuse, east and west hemi RCHP, zones LCHP. (Courtesy of INTELSAT)

A very popular antenna that generates circularly polarized waves is the helical antenna shown in Fig. 7.17.

Above 1 GHz, the most popular antenna designs are the parabolic-dish-type reflector antennas. These designs offer efficient and mechanically simple solutions to the antenna problem. Figure 7.18 shows that these designs are derived from the well known designs of optical telescopes. For the purpose of eliminating blockage by the antenna feed itself, offset-fed

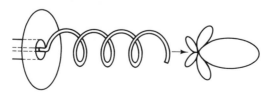

Figure 7.17 Helical antenna and its radiation pattern.

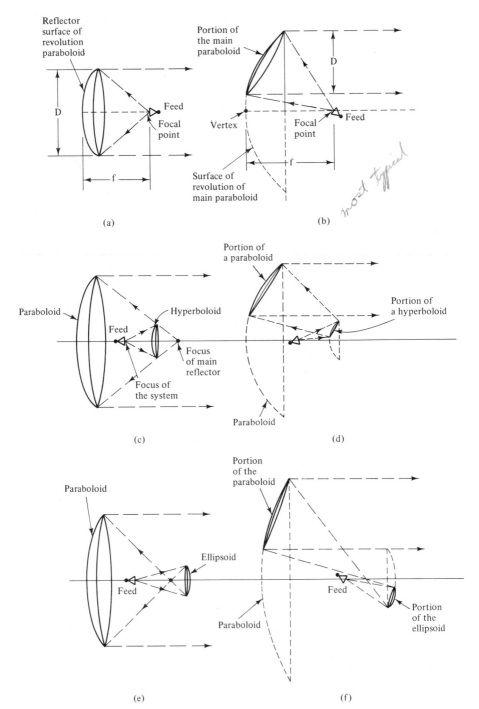

Figure 7.18 (a) Focal-fed symmetrical parabolic antenna; (b) offset-fed paraboloid (no feed blockage); (c) center-fed symmetrical Cassegrain antenna; (d) offset-fed Cassegrain antenna; (e) center-fed symmetrical Gregorian antenna; (f) offset-fed Gregorian antenna.

designs have become the mainstay of satellite antennas. In particular, the simple offset-fed paraboloid antenna has become the most popular design for microwave satellite antennas. This simple design avoids feed blockage and does not require the additional subreflector that is present in some other designs. In contrast, earth station antennas up until now have mostly adopted the symmetrical Cassegrain design or the symmetrical Gregorian design. This divergence in design approaches for satellite and earth station antennas is largely due to (1) the large difference in the antenna sizes (i.e., earth station antennas are typically many times bigger than the satellite antenna), and (2) the simplicity that is required in the satellite antenna in order to survive the launch, the deployment in orbit, and the space environment during the lifetime of the spacecraft.

It can be observed that in all the designs shown in Fig. 7.18, the reflector (or dish) is actually fed by a primary source antenna. The primary source antenna is in fact another antenna in itself. This leads to the discussion of some simple and popular horn antenna designs used for feeding dishes or sometimes used directly as the radiating device. Figure 7.19 shows four common types of horn antennas.

While the reflector antennas offer relatively high antenna gains efficiently, the horn antenna is a simple way of achieving low-gain radiation. Typically, horn antenna 3-dB beamwidths vary from 10° to about 120°, whereas reflector antennas are seldom used if the 3-dB beamwidth is greater than about 5 to 10°. Reflector antennas are used mostly when the 3-dB beamwidths are in the order of 5° or less. For example, horn antennas are used to provide 18° field-of-view earth coverage and reflector antennas are used for the zone and the hemispherical beams in the Intelsat V series of spacecraft. The helical antenna shown in Fig. 7.17 can be used as a primary radiator to feed a reflector or it can be used directly as a low-gain antenna (approximately 12 dB gain) for frequencies below 4 GHz.

Microwave energy above 1 GHz is very efficiently "piped" from one point to another via waveguides or hollow metallic tubes of either rectangular or circular cross section. These waveguides are the high-frequency equivalents of a twisted pair of wires or coaxial cables commonly used for lower frequencies. A flared section of the simple rectangular or circular waveguide forms a horn radiator. A horn antenna can be visualized as a transition from the guided mode of propagation of the electromagnetic energy within the "pipe" to the free-space unimpeded mode of propagation of the electromagnetic energy. Both simple rectangular and circular horn antennas suffer from high-sidelobe radiation. A more modern and efficient horn radiator uses a flared circular horn with corrugated interior wall. The corrugations are transverse to the propagation of the microwave energy from the throat to the open end of the horn. This corrugated circular horn antenna has the following advantages: (1) it has very low sidelobes; (2) the antenna beam is circularly symmetric around the boresight axis; and (3) it has very low cross-polarization levels. The only drawback is its size; it is somewhat larger than the simple conical horn for the same 3-dB beamwidth. Finally,

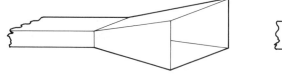

Rectangular pyramidal horn

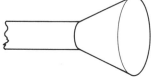

Circular conical horn

Corrugated conical horn (Courtesy of COMSAT)

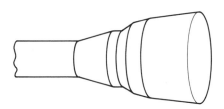

Multiflare conical horn "Potter horn"

Figure 7.19 Horn antennas.

a simple horn with steps and flare changes along its length, the Potter horn (Ref. 2), is a compromise between the simple conical horn and the corrugated conical horn. The Potter horn has improved sidelobe and cross-polarization levels over the simple conical horn, but its physical size is a little smaller than the corrugated horn. All four types of horn shown in Fig. 7.19 are commonly used to feed reflector antennas described in Fig. 7.18.

An important parameter of a reflector antenna is its f/D ratio, f being the focal length of the paraboloid that contains the reflector and D the diameter of the reflector, as illustrated in Fig. 7.18. There are electrical and mechanical reasons why certain f/D's are better than others. If the f/D of an antenna is made too small, say less than about 0.5, the reflector has poor scanning properties, that is, a beam cannot be displaced from its nominal boresight position simply by displacing the feed horn from the focal point without much loss of performance (e.g., degradation in gain, sidelobe, and cross-polarization). If, on the other hand, the f/D is made too large, say more than about 2.0, the scanning property improves at the expense of a much larger mechanical structure and a much larger primary feed horn. The primary feed size increases with f/D because the angle subtended by the reflector decreases with increasing f/D and the primary beam pattern of the feed must be reduced in beamwidth in order to illuminate the reflector efficiently without excess spillover losses. Thus most simple offset-fed satellite antennas have f/D ratios between 0.8 and 1.6. This range of f/D's is good for most offset designs for satellite applications.

For the dual offset reflector antenna (i.e., the offset Cassagrainians and the offset Gregorians), the design trade-off is not as straightforward. Since there is "magnification" involved in the dual offset optics, the size of the feed is dependent on the apparent f/D ratio. The overall size of the antenna is more or less dictated by the focal length and the diameter of the main reflector for most practical cases. A detailed discussion on the geometry of these "folded" optics is given in Ref. 3.

Techniques for Achieving Shaped Satellite Antenna Beams

In Section 7.5, the efficiency η of an antenna was defined as the ratio of the effective capture area (receiving antenna) of the antenna to the actual physical aperture of the antenna. The definition of efficiency η used in Eq. (7.13) was valid for simple "pencil"-beam antenna, antennas with either circular or elliptical or near-elliptical beam cross sections. For highly shaped beam antennas, such as the ones used in Intelsat V or VI spacecraft, aperture efficiency η is a less meaningful quantity. This is so because a relatively large aperture is needed to produce a beam cross section that is neither circular nor elliptical. The antenna is not designed to cover a single point but rather, a certain irregular coverage area. This is indeed the case for most satellite antennas that are designed to cover specific landmasses from the geostationary orbit. In these cases, the efficiency of the antenna is better defined in terms of the beam efficiency (η_B). Since an isotropic antenna covers 4π steradians or 41,253 square degreès, the maximum theoretical gain G_m achievable by an ideal antenna that covers a given Ω square degrees would be given by

$$G_m = \frac{41,253}{\Omega} \tag{7.31a}$$

$$G_m|_{dB} = 10 \log_{10} \frac{41,253}{\Omega} \tag{7.31b}$$

Equation (7.31) thus assumes that the antenna produces a "sectoral" beam, that is, a beam with constant gain across the required coverage area and with no sidelobe outside that required coverage area. It is obvious that such an antenna cannot be realized in practice. For example, if the area to be covered is a rectangle, the ideal antenna would produce a radiation pattern that looks like the one in Fig. 7.20(a). The practical antenna, on the other hand, would produce a radiation pattern that looks like the one shown in Fig. 7.20(b).

If, in this case, G_e is the minimum (edge) gain of the practical antenna that provides the rectangular coverage, the beam efficiency η_B is defined as

$$\eta_B = \frac{G_e}{G_m} \quad \text{— from 3d}\beta \text{ beamwidth} \tag{7.32}$$

Equation (7.32) is valid for antenna beams of arbitrary cross section. The beam efficiency η_B is now a measure of how close the practical antenna can come to the theoretical limit.

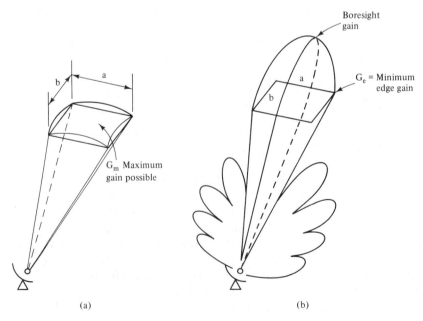

Figure 7.20 (a) Ideal pattern; (b) realistic pattern.

There are two common techniques for the design of antennas that produce arbitrary beam cross sections. The simpler approach (for reflector antennas) is to use the simple offset design in Fig. 7.18 and distort the reflector from the true paraboloidal shape. This design approach is simple but does not offer enough flexibility (i.e., degrees of freedom) for shaped beams of complicated shapes (e.g., those shown in Fig. 7.14 through Fig. 7.16). The other common technique is to use small circular component beams packed together to form the required beam shape. This is achieved by using an undistorted offset paraboloidal reflector fed by an array of feed horns. Figure 7.21 illustrates the basic idea of how a number of circular feed horns packed together radiating simultaneously can form a shaped beam of arbitrary cross section. Figure 7.22 shows the antenna feed array arranged in an inverted image of the beam coverage at the focal plane of the undistorted offset fed reflector. Each feed horn generates one circular component beam and the feed array is fed via a beam-forming network (BFN). The BFN takes the signal to be transmitted and divides it in such a way that each feed horn in the array is excited by an appropriate amount of the original signal. Since each of the feed horns in the feed array forms one single-component beam in the far field, there will be as many feed horns as there are component beams. Also, the reflector and the feed horn are sized to produce the component beam efficiently. Thus each individual feed horn illuminates the entire reflector and not a portion of it. The amount of excitation for each feed horn is adjusted in both amplitude and phase (an RF signal has both amplitude and time phase) to maximize the radiation within the coverage area while simultaneously minimizing the radiation (sidelobes) outside the coverage area.

The beam efficiency η_B can be improved by increasing the number of component beams within the given coverage area. However, a few things happen when the number of component beams is increased: (1) for a fixed

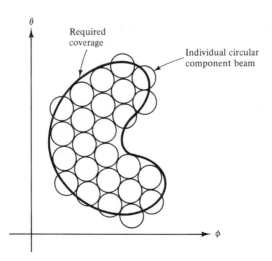

Figure 7.21 Shaped beam formed by the use of component beams.

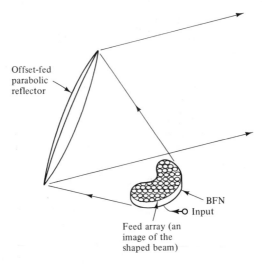

Offset-fed
parabolic
reflector

BFN
O Input

Feed array (an
image of the
shaped beam)

Figure 7.22 A shaped beam antenna.

coverage area, the size of each component beam decreases; (2) the reflector size must increase accordingly; (3) the number of feed horns increases; (4) the BFN becomes more complex; and (5) to maintain a certain f/D, the focal length of the reflector will increase accordingly. Thus the beam efficiency η_B can be increased (assuming lossless BFN) only at the expense of a bigger and more complicated antenna. Since the beam-edge gain roll-off of the shaped beam so derived is essentially the roll-off of the component beam, a larger number of component beams will also give the advantage of a faster beam-edge gain roll-off. In observing the statements above, it is clear that the number of component beams to be used in the design is dictated by (1) the beam-edge roll-off, for example, the roll-off required in a spectrum-reuse multibeam situation; (2) the desirable beam efficiency; (3) the tolerable complications, size, and weight in the feed array and BFN; and (4) the allowable losses in the BFN circuitry. The current satellite antenna technology allows a separation of 1.4 times the component beam size between any two shaped beams in a multibeam frequency reuse design. This results in a reasonable antenna realization and a minimum of 27-dB sidelobe isolation. Figure 7.23 illustrates this beam separation.

As a numerical example, suppose that a rectangular coverage area and a square coverage area of the dimensions and separation shown in Fig. 7.24 are required in a dual-beam spectrum-reuse situation at 4 GHz. It is necessary to define the approximate size of an offset-fed reflector antenna system to satisfy the requirements.

Since the minimum separation between the two coverage areas is 1.5°, the component beam must be $1.5/1.4 = 1.07°$. This will yield the approximate diameter D of the reflector as follows, assuming a circular reflector is used.

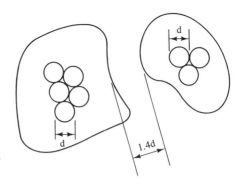

Figure 7.23 Minimum separation 1.4 times component beam diameter.

The gain G_0 of the component beam according to Eq. (7.11) is

$$G_0 = \frac{30,000}{1.07 \times 1.07} = 26,203$$

Then according to Eq. (7.13), and noting that $A = \pi D^2/4$,

$$26,203 = 0.7 \left(\frac{\pi D}{\lambda}\right)^2, \text{ by assuming } \eta = 0.7$$

At 4 GHz, $\lambda = 7.5$ cm; then solving for D, we obtain

$$D = \sqrt{\frac{26,203}{0.7}} \times \frac{7.5}{\pi} = 462 \text{ cm} \quad \text{or} \quad 4.62 \text{ m}$$

If f/D is chosen to be 1.0, $f \doteq 4.6$ m. Since the feed array is an inverted image of the beams, the vertical dimension y of the feed array will be approximately

$$y \doteq \frac{4.6 \times \tan 6°}{\text{BDF}}$$

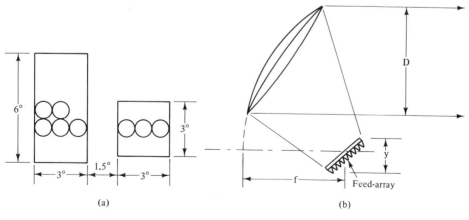

Figure 7.24 (a) Example of a coverage requirement; (b) example of a feed array.

Where the beam deviation factor (BDF) is assumed to be about 0.9,

$$y \doteq \frac{4.6 \times \tan 6°}{0.9} = 54 \text{ cm}$$

The BDF is the ratio between the angular displacement of the antenna beam boresight to the angular displacement of the antenna feed from the nominal antenna axis, and this is discussed in more detail in the following section. The horizontal dimensions can be estimated in a similar manner.

Example of a Satellite Antenna System

Reference 4 gives a summary of the overall design of the antenna subsystem for the Intelsat VI. Sixfold frequency reuse is achieved in this design in the C band and in addition, two steerable spots are available in the Ku band. Figure 7.25(a) and (b) show, respectively, the antenna subsystem

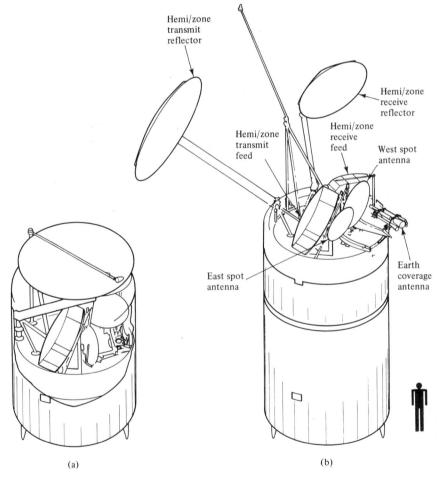

Figure 7.25 Intelsat VI antenna farm: (a) stowed; (b) deployed. (Courtesy of INTELSAT and HAC)

stowed for launch and deployed for on-orbit use. The C-band transmit reflector is 3.2 m in diameter and the C-band receive reflector is 2 m in diameter. An f/D of 1.3 was chosen by the manufacturer for this particular design to reduce scanned beam degradation. The transmit feed array measures approximately 2 m across and it consists of 145 Potter feed horns. On-orbit reconfiguration of some of the beams is possible through the use of switchable BFNs.

Effects of Mechanical Distortion in Reflector Antennas

Mechanical distortions of reflector antennas can be broken down into various components that are well understood or characterized. For the simple offset-fed reflector antenna, the reflector and the feed assembly are usually secured separately on the spacecraft body. Figure 7.26 illustrates the effects of an equivalent feed displacement due to either initial alignment error or relative motion between the reflector and the feed in the course of the lifetime of the spacecraft. If the feed displacement is lateral, i.e. perpendicular to the axis of the offset paraboloid [Fig. 7.26(a)] there is beam (boresight) displacement with only minor beam degradation. The beam deviation factor BDF is defined as

$$\text{BDF} = \frac{\theta_2}{\theta_1} \tag{7.33}$$

If the feed is displaced along the axis of the offset paraboloid, Fig. 7.26(b), the beam is defocused, resulting in gain loss, beam broadening, and increased sidelobe levels. The two types of feed displacement result in two quite different effects. In the design of a satellite antenna, all these factors must be taken into account to allow sufficient margins in the performance. For high-gain antennas, lateral feed displacement has significant effects on the pointing of the antenna beam, whereas defocusing is of minor importance if the f/D is 1.0 or larger.

Reflector distortions are usually treated by using a best-fit equivalent paraboloid with superimposed surface errors. Figure 7.27 illustrates this

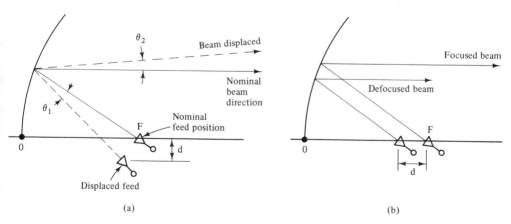

(a) (b)

Figure 7.26 (a) Effects of lateral feed displacement; (b) effects of de-focusing.

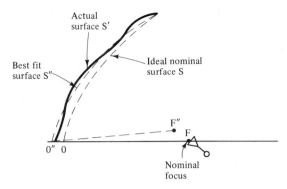

Figure 7.27 Concept of a best fit paraboloid.

technique. Having obtained a best-fit paraboloid S'' to the actual surface S', the vertex O'' and F'' can now be located. The effects of the equivalent feed displacement can be treated as suggested above and the surface errors can be broken down into systematic errors and random errors. The treatment of systematic errors is complex and it is beyond the scope of the discussion here, but the effects of random surface errors can be quantified approximately by the following relationship. The average gain degradation due to random surface errors has been derived by Ruze (Ref. 5) as

$$\frac{G}{G_0} = \exp\left[-\left(\frac{4\pi\sigma}{\lambda}\right)^2\right] \qquad (7.34)$$

where σ is the rms surface error in wavelengths, G_0 is the gain of a perfect reflector, and G is the gain of a reflector with surface error. Typically, an acceptable gain degradation would be 0.25 dB, corresponding to an rms error of 0.02λ. Figure 7.28 gives the BDF for various f/D ratios.

Satellite Transponder Design

Having discussed the basic design concepts for communication satellite antennas, it is now appropriate to turn to the microwave repeater or transponder that processes the signals received by the receive antenna and

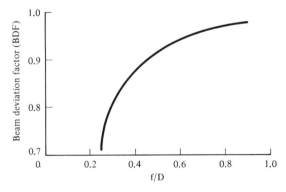

Figure 7.28 BDF versus f/D.

Communications   Chap. 7

transmit the processed signal via the transmit antenna. Figure 7.12 gave a simple diagram of the communication payload. The transponder portion of that block diagram is now expanded in Fig. 7.29 to give more details. There are several processing functions to be carried out by the transponder. It may be recalled from Sections 7.1 and 7.3 that the uplink frequencies are necessarily different from the downlink frequencies. One of the main functions of the satellite transponders is to transpose the uplink frequency to downlink frequency. This function is accomplished in the frequency translator or mixer, item 3 in Fig. 7.29. A conventional C-band communications satellite would have the uplink in the band 5.925 to 6.425 GHz and the downlink in the band 3.7 to 4.2 GHz. Thus a stable on-board local oscillator must provide the difference frequency of 2.225 GHz to down-convert the uplink frequencies to the downlink frequencies. The example in Fig. 7.29 shows a "single"-conversion system; that is, the conversion from uplink frequencies to downlink frequencies is achieved in one step. Other designs may employ a "double"-conversion system, where the uplink is first down-converted to a lower intermediate frequency (IF), say, 1 GHz, and then up-converted to the downlink frequency via an "up-converter." The arrangement in Fig. 7.29 is simpler and more common. Double-conversion systems are used only when there are other requirements or constraints that cannot be met by a single-conversion system.

Following the flow of the microwave signal through the communications subsystem, the bandpass filter (item 1) preceding the receiver (item 2) passes only the uplink signals received by the antenna and rejects out-of-band signals that may cause degradation in the receiver. The receiver amplifies the very weak signals to a certain convenient level so that down conversion can be effected without much degradation to the signal. Next, another amplifier (item 4) boosts the signal to a higher level before the various RF channels are demultiplexed into individual channels in the input demultiplexer, often referred to as the "input mux." More will be said later concerning the reasons behind this arrangement of having individual RF channels. Each individual channel is now amplified by a channel driver (item 6) and a high-powered transmitter (item 7). Sometimes an automatic level control circuit (ALC) is incorporated into the channel driver so that a constant signal level is presented to the input of the high-powered transmitter.

Two types of high-powered transmitters are commonly used, the traveling-wave-tube amplifier (TWTA) and the solid state power amplifier (SSPA) (Figs. 7.30 and 7.31). The use of SSPAs in space is a fairly recent development. Most C-band SSPAs are of the type that uses Galium Arsenide field-effect transistors (GaAs FETs). SSPAs promise good linearity, better reliability, and lower mass. Although the situation may change, TWTAs are still the most common high-powered transmitters used in present-day communications satellites. This is due mainly to the high RF power-to-dc power efficiency ($\approx 40\%$) offered by the TWTAs and also due to the accumulated knowledge of the behavior of space TWTAs that have been flown over the past 20 years.

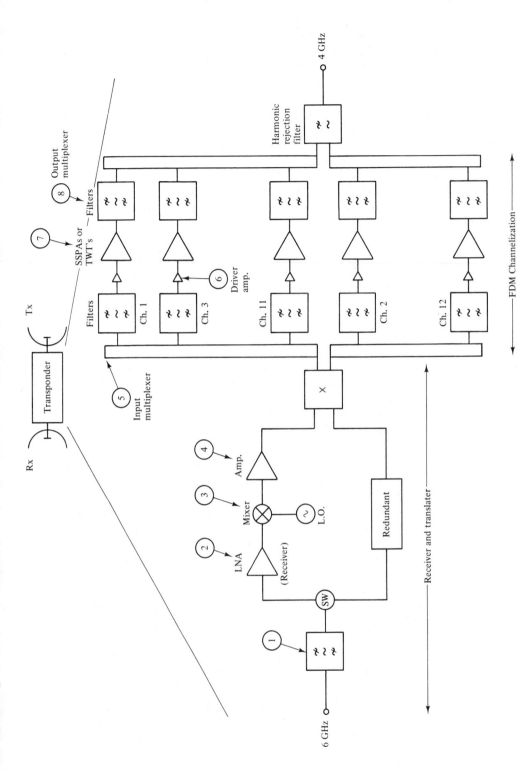

Figure 7.29 Typical satellite transponder.

Communications   Chap. 7

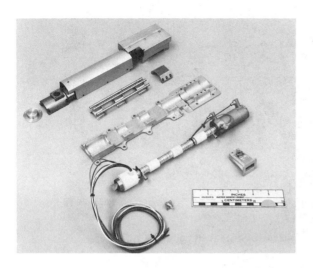

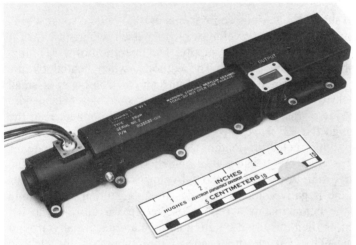

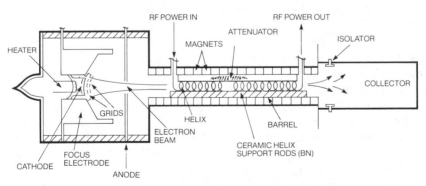

*VACUUM INSIDE TWT IS ABOUT 10^{-8} TORR NORMALLY

Figure 7.30 (a) Unpotted HAC 289HP TWT; (b) packaged HAC 289HP TWT; (c) description of operation of a TWT. (Courtesy of Hughes Aircraft Co., Electron Dynamics Division)

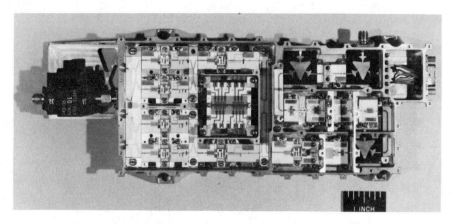

Figure 7.31 C-band IOW SSPA. (Courtesy of RCA)

Figure 7.30 shows the various components used to build a space TWT. In satellite jargon, a TWT is the traveling-wave tube itself, whereas a TWTA is the traveling-wave tube plus power supply. The operation of a traveling-wave tube is shown in Fig. 7.30(c). The cathode, when heated by the filament, emits electrons which are focused into an electron beam of small diameter. This electron beam travels through the center of a helical wire (the helix). The side of the helix closest to the cathode is connected to the RF input. Although the RF wave travels close to the speed of light in the helix (there is some dielectric loading), its forward velocity matches the velocity of the electrons because of its much longer path length in the helix. The positive side of the RF sinusoidal voltage waveform accelerates electrons while the negative side decelerates them, resulting in bunching of electrons in the beam that is a replica of the RF waveform. The RF signal in the helix is then absorbed by attenuating material, usually around the middle of the helix. On the other side of the attenuator, the electric field from the bunched electron beam recreates the initial waveform in the helix.

Since only a small amount of energy is required for bunching, the RF field generated in the helix by bunching is much greater than the input RF. Thus the RF induced on the output side of the helix is usually 40 to 60 dB greater than on the input side. The electron beam leaving the helix is collected by the "collector." This electron beam still has a considerable amount of energy after leaving the helix and for this reason "depressed" collectors (collectors at a negative voltage) are used so that less energy has to be used to bring these electrons up to the cathode voltage than would have to be used if the collectors were at ground potential. The use of a multicollector configuration further increases the TWT efficiency but at the expense of greater complexity and mass. Typical TWT dc-to-RF efficiencies range from 30 to 50%. When combined with a power supply to form a TWTA, the overall efficiency is usually between 25 and 45%.

Figure 7.31 shows one construction of an SSPA with the cover of the housing removed. To deliver 10 W of RF power at C band, the current technique relies on parallel operation of four microwave GaAs FETs, as shown. To match the gain of a TWT, which is on the order of 60 dB, multiple stages of GaAs FET amplifiers are required, and this is evident in Fig. 7.31. The achievable dc-to-RF efficiency of such an SSPA is about 27% which includes the efficiency of the power supply. This efficiency looks less attractive compared to the TWTAs. However, the SSPA offers better linearity and consequently for linear operation, less output back-off is required. Currently, the SSPA is practically as efficient as a TWTA for linear operation. The TWTA still has the advantage of being able to deliver high RF powers of hundreds of watts (e.g., 200 to 400 W) of RF for broadcast satellite applications.

The output signals at the outputs of all the high-powered transmitters are combined together in the output multiplexer (item 8, output mux) so that a single transmit antenna now transmits all the signals back to the earth.

7.7 SOME COMMON MODULATION AND ACCESS TECHNIQUES FOR SATELLITE COMMUNICATIONS

As shown in the following equation, a sinusoidal microwave signal $v(t)$ cannot transfer information unless some of its parameters are varied. A sinusoidal signal can be expressed as

$$v(t) = A \sin (\omega t + \phi) \tag{7.35}$$

where A is the amplitude of the signal, $\omega = 2\pi f$ is the angular frequency, f is the frequency in hertz, t is time, and ϕ is the time phase of the sinusoid with respect to some reference. There are three quantities associated with the sinusoidal wave: the amplitude A, the frequency f (or ω), and the phase ϕ. Any one of the three quantities can be varied or modulated by the source information so that the microwave sinusoidal "carrier" signal can now convey information.

If A is modulated by the source information, the modulation process is called amplitude modulation (AM). If the frequency f (or ω) is modulated, the modulation process is called frequency modulation (FM), and if the phase ϕ is modulated it is called phase modulation (PM). Pictorially, a rotating vector can be used to represent the sinusoidal microwave carrier, as shown in Fig. 7.32 in polar (r, θ) coordinates.

It can be seen that AM is orthogonal to either PM or FM; that is, AM modulates the vector in the radial r direction while PM or FM modulates the vector in the θ direction. Thus all modulations are resolvable into modulations of either r or θ component of the rotating vector, in other words, the amplitude and phase of the sinusoid. Also, there cannot be phase modulation without frequency modulation, and vice versa. This can be seen as follows. If Eq. (7.35) is written for PM as (ω_c = nominal steady-

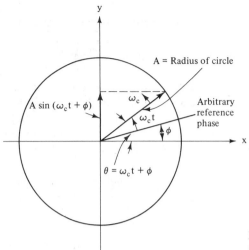

A = Radius of circle

A sin ($\omega_c t + \phi$)

ω_c

$\omega_c t$

Arbitrary reference phase

ϕ

$\theta = \omega_c t + \phi$

x

y

Figure 7.32 Representation of a sinusoid with the projection of a rotating vector on the y-axis.

state carrier frequency)

$$v(t) = A \sin [\omega_c t + \phi(t)] \qquad (7.36)$$

where the phase ϕ is a function of time t. The instantaneous phase of the sinusoid is

$$\theta = \omega_c t + \phi(t) \qquad (7.37)$$

Since the instantaneous frequency ω of the sinusoid is the derivative of the instantaneous phase θ, then

$$\omega = \frac{d\theta}{dt} = \omega_c + \phi'(t) \qquad (7.38)$$

where

$$\phi'(t) = \frac{d\phi}{dt}$$

Thus, as long as $\phi'(t)$ is not zero, the instantaneous frequency will be the sum of the steady-state carrier frequency ω_c and $\phi'(t)$.

For AM modulation, amplitude linearity of the transmission system becomes critical, whereas for FM or PM modulation, phase linearity is required. In the case of satellite transmission, maximum capacity per unit weight of the satellite is desirable. For this reason, the output transmitters are usually driven hard to produce the maximum available power to overcome thermal noise (i.e., to improve the C/N of the downlink). All transmitters experience amplitude nonlinearities at or close to saturation, as shown in Fig. 7.33. Thus, to maximize the available capacity of a spacecraft, it is necessary to choose a modulation technique that is more tolerant of amplitude nonlinearities. In addition, the modulation technique must provide some flexibility in trading off certain system parameters (e.g., trading bandwidth

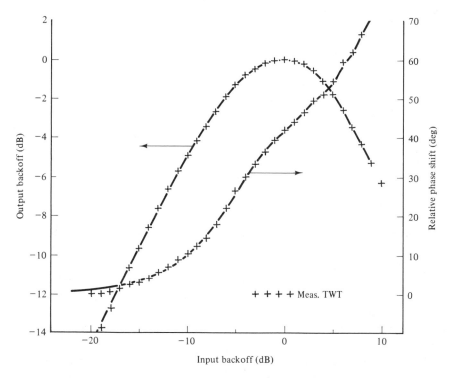

Figure 7.33 TWTA transfer characteristics measured at 3.8 GHz.

for RF power). With these thoughts in mind, let us examine AM and FM (or PM) modulation techniques in more detail.

AM modulation can be represented as follows by rewriting Eq. (7.35) as (A is scaling constant)

$$v(t) = A[1 + m(t)] \sin \omega_c t \qquad (7.39)$$

where $A[1 + m(t)]$ is the amplitude component of the microwave carrier and $m(t)$ is the modulating signal. Usually, $m(t)$ is a complex waveform derived from the source signal that has to be transmitted. For the present discussion, it is adequate to consider that the source signal is another sinusoid $\sin \omega_m t$, where ω_m is the frequency of the source sinusoid. Then Eq. (7.39) becomes

$$v(t) = A[1 + m \sin \omega_m t] \sin \omega_c t \qquad (7.40)$$

where $|m| \leq 1$ is the modulation index. Through trigonometric relations, Eq. (7.40) can be rewritten as

$$v(t) = A \sin \omega_c t + \frac{Am}{2}[\cos(\omega_c - \omega_m)t - \cos(\omega_c + \omega_m)t] \qquad (7.41)$$

It can be seen that an RF carrier or sinusoid, at ω_c, amplitude modulated by another sinusoid, at ω_m, generates two equal sinusoids ω_m away on either side of the RF carrier in the frequency domain. In the frequency domain one can represent the situation in Fig. 7.34.

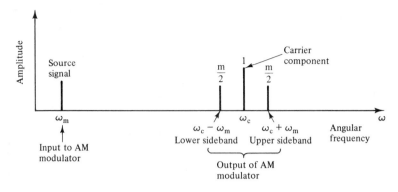

Figure 7.34 AM modulation: single sinsusoid in baseband.

Since the source information can always be broken down into a linear combination of sinusoids, it can be shown that AM modulation produces two sidebands that are exact replicas of the source information in the frequency domain (Fig. 7.35). The lower sideband is an inverted copy of the source signal or the upper sideband. The only flexibility in the AM process is the choice of the modulation index m in Eqs. (7.40) and (7.41). Since no information is carried by the carrier and because the upper sideband carries the same information as the lower sideband, this modulation technique is rather inflexible and wasteful of RF power. At best, $m = 1$, and for a source signal that is a single sinusoid, only one-sixth of the total RF power is present in each sideband. [The power of each component is the square of the voltage of the component shown in Eq. (7.41).]

Turning now to the FM technique, it was pointed out that the instantaneoous frequency ω of an RF carrier is the derivative of the phase θ of the RF carrier, that is,

$$\omega = \frac{d\theta}{dt} \tag{7.42}$$

If the modulating signal or source signal consists also of a sinusoidal waveform

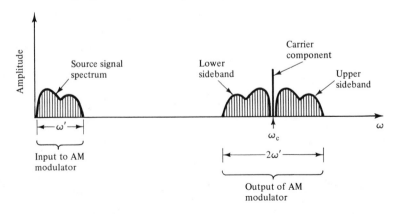

Figure 7.35 Spectrum of an AM-modulated signal.

sin $\omega_m t$ and if k is the peak angular frequency deviation in the FM process, then

$$\frac{d\theta}{dt} = \omega_c - k \sin \omega_m t \qquad (7.43)$$

where ω_c is the steady unmodulated carrier frequency of the RF. Integrating both sides with respect to t gives

$$\theta = \omega_c t + \frac{k}{\omega_m} \cos \omega_m t \qquad (7.44)$$

The use of $-k \sin \omega_m t$ rather than $k \cos \omega_m t$ to represent frequency modulation is a matter of convenience. Thus the resultant RF waveform $v(t)$ is

$$v(t) = A \cos \theta = A \cos \left(\omega_c t + \frac{k}{\omega_m} \cos \omega_m t \right) \qquad (7.45)$$

Again, the use of $\cos \theta$ rather than $\sin \theta$ for $v(t)$ is also a matter of convenience. It can be shown that functions of the type in Eq. (7.45) can be resolved into summation of sinusoids by the application of the following Bessel function identities:

$$\sin(\alpha + X \sin \beta) = \sum_{n=-\infty}^{\infty} J_n(X) \sin(\alpha + n\beta)$$

$$\cos(\alpha + X \sin \beta) = \sum_{n=-\infty}^{\infty} J_n(X) \cos(\alpha + n\beta)$$

$$\sin(\alpha + X \cos \beta) = \sum_{n=-\infty}^{\infty} J_n(X) \sin\left(\alpha + n\beta + \frac{n\pi}{2}\right)$$

$$\cos(\alpha + X \cos \beta) = \sum_{n=-\infty}^{\infty} J_n(X) \cos\left(\alpha + n\beta + \frac{n\pi}{2}\right)$$

Thus Eq. (7.45) can be rewritten as

$$v(t) = A \sum_{n=-\infty}^{\infty} J_n(X) \cos\left(\omega_c t + n\omega_m t + \frac{n\pi}{2}\right) \qquad (7.46)$$

where $X = k/\omega_m$. Using the identity that $J_{-n}(X) = (-1)^n J_n(X)$, $v(t)$ can be expressed as

$$v(t) = A \left[J_0(X) \cos \omega_c t \right.$$

$$+ J_1(X) \cos \left[(\omega_c + \omega_m)t + \frac{\pi}{2} \right]$$

$$+ J_1(X) \cos \left[(\omega_c - \omega_m)t + \frac{\pi}{2} \right] \qquad (7.47)$$

$$- J_2(X) \cos(\omega_c + 2\omega_m)t$$

$$\left. - J_2(X) \cos (\omega_c - 2\omega_m)t + \cdots \right]$$

It can be shown by further analysis that unlike the AM technique, if the source signal consists of more than one sinusoid, superposition of the sidebands created by individual source signal sinusoid is not applicable.

Thus the FM process is a "nonlinear" process, the analysis of which is very complex when the source signal is not a single simple sinusoid. Equation (7.47) gives the magnitudes of all the sidebands around the carriers at ω_c. With a suitable choice of k/ω_m, the carrier level can be zero for a particular case. Also, the sidebands, in theory, extend to infinity on either side of the carrier with decreasing amplitude. In Eq. (7.47), X is the modulation index.

For a practical communication link that uses the FM technique, the allowable bandwidth to carry the signal cannot be infinite. Thus significant sidebands are carried while the rest of the sidebands are truncated by filtering. It is therefore necessary to determine what this bandwidth should be. As it turns out, the value of n in Eq. (7.46), for which 98% of the power in the signal is included, always happens to be $n = X + 1$. Thus the necessary double sideband bandwidth B is

$$B = 2(X + 1)f_m$$

where $f_m = \omega_m/2\pi$. Recalling that $X = k/\omega_m$, then

$$B = 2(f_k + f_m) \tag{7.48}$$

where $k = 2\pi f_k$ and $2\pi f_m = \omega_m$, by definition. Equation (7.48) is the well-known Carson's rule bandwidth for FM signals, where f_k is the peak frequency deviation in hertz and f_m is the highest modulating source frequency, also in hertz. Since k (or f_k) is a free parameter, link performance can usually be improved by increasing k (or f_k) at the expense of using up bandwidth. Thus a trade exists between bandwidth and available RF power. This can be seen in Fig. 7.36, which shows the relationship between pre-detection signal C/N and post detection basebound signal S/N for a typical FM modulator.

It will be noticed that the trade-off between bandwidth and available RF carrier power exists only as long as the carrier to-noise ratio (C/N) at the input to the FM demodulator is above a certain threshold, as shown in Fig. 7.36. Once the input C/N drops below the threshold, the output signal to noise ratio S/N drops very rapidly. Most communication links attempt to maintain the input C/N several decibels above the threshold to ensure high-quality transmission.

To conclude the discussion on AM and FM techniques, it can be said that AM technique is usually unsuitable for satellite transmission. However, Carson bandwidth limited FM technique will result in an almost constant signal level (only the frequency is modulated and 98% of the power in the spectrum transmitted) and thus amplitude linearity is not critical. Furthermore, FM offers the advantage of a trade between bandwidth and carrier power. Until a few years ago, FM had been the principal modulation technique used in satellite communications.

Due to the widespread introduction of digital systems in terrestrial networks, digital modulation techniques have also become rather common. The rest of this section deals with one of the most common digital modulation techniques, the quaternary-phase phase-shift-keyed system (QPSK), also

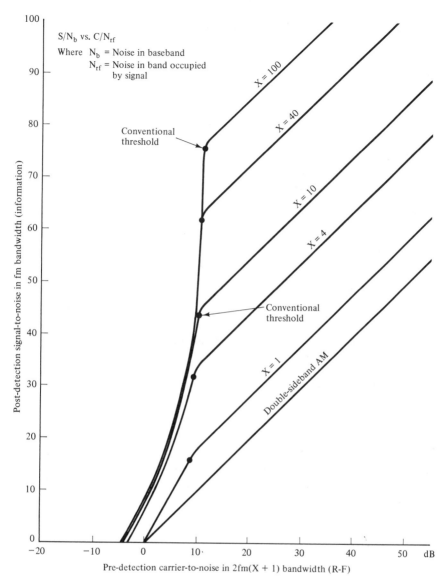

Figure 7.36 Signal-to-noise ratio in $2f_m(X + 1)$ bandwidth.

known as 4ϕ-PSK. Before QPSK modulation can be discussed, it is necessary to introduce the binary digital waveform, called the Non-Return to Zero (NRZ) waveform, and discuss how analog signals can be converted to this NRZ format.

Digital Encoding of Analog Signals

It is perhaps unfortunate that some of the digital encoding techniques are also called modulation techniques, which tends to confuse them with

RF carrier modulation techniques such as AM, FM, and QPSK modulation. For a source signal such as a voice or video signal, encoding means converting the analog signal to a digital binary waveform, commonly the NRZ waveform. The term "modulation" in this instance can be considered as the varying of a dc voltage to produce a binary digital waveform. For RF carrier modulation, the term "modulation" refers to the amplitude and/or phase modulation of the RF carrier. Since it is impractical to transmit the digital (NRZ) waveform over the air, the NRZ waveform is usually used to modulate an RF carrier at microwave frequencies for transmission.

The source analog signal thus undergoes two stages of processing (or modulation) at the sending end before final transmission over the air in a microwave link. At the receiving end, the source signal is recovered after two stages of processing (or demodulation) before delivery to the user.

Since the problem of digital encoding of an analog signal is a very broad subject, only a cursory treatment of the basic principles is possible. The reader is referred to Ref. 6, for example, for advanced and detailed treatment of the subject.

Pulse code modulation (PCM). This is a relatively well-established and common technique developed primarily for telephony. The idea is rather simple. If an analog voltage waveform generated by the microphone in the telephone handset could be sampled periodically, the voltages of these samples could be encoded in a digital binary word of a certain length. Figure 7.37 illustrates this technique.

In the particular example shown in Fig. 7.37, each voltage sample is encoded by a 4-bit binary word of ones and zeros. The 4-bit word would allow $2^4 - 1 = 15$ possible discrete voltage levels in the encoding process. The encoder must choose between one word or another if the sampled voltage is exactly the same as one of the selected voltage levels. All voltages in between two fixed levels will be encoded by the same digital word. Thus there will be quantization error in this process of encoding. At the receiving

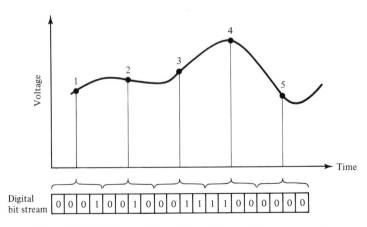

Figure 7.37 PCM encoding with a 4-bit word for each sample.

end, the electronic circuitry generates the voltage levels according to the information in the 4-bit words, and with the help of a bandpass filter, the original analog waveform is more-or-less reproduced to within the quantization errors introduced. In addition to the number of levels in the encoding process, the frequency of sampling plays an important role. Nyquist (Ref. 7) showed that for good reproduction of a waveform that contains frequencies up to f_{max}, the sampling rate must be at least twice f_{max}. For example, a telephone voice circuit contains frequencies up to about 3400 Hz; thus 8 kHz is a good sampling frequency according to Nyquist. Studies have also shown that an 8-bit word for encoding each voltage sample ensures good reproduction for the waveform (i.e., quite acceptable quality to the human ear). A conventional PCM encoding system would therefore require $8 \times 10^3 \times 8 = 64$ kbits/s for each voice circuit. This is indeed the bit rate adopted by most countries as the standard bit rate for one PCM voice circuit. Thus the PCM technique converts the analog waveform into a series of ones and zeros. If one is represented by $+1$ V and a zero is represented by -1 V, the resulting digital waveform is called a Non-Return to Zero (NRZ) waveform (Fig. 7.38).

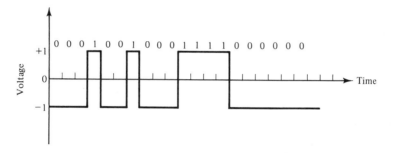

Figure 7.38 NRZ waveform for the digital bit stream in Fig. 7.37.

Delta modulation (Δ-mod). This is another encoding technique whereby an analog waveform is turned into a digital bit stream of ones and zeros. Instead of encoding the magnitude of each sampled voltage, sampling is done much more frequently and the encoder stores the previous sample and compares it with the current sample. The encoder sends out a one if the current sample exceeds or equals the previous sample and sends out a zero if the current sample is less than the previous sample. Figure 7.39 illustrates the basic principle in delta modulation. Thus the ones and zeros can again be represented by an NRZ waveform.

At the receiving end, the decoder steps up the voltage output by a fixed amount wherever a one is received and steps down by the same amount if a zero is received. Quantization noise can be reduced by filtering (i.e., smoothing out the stepped waveform). The problem associated with Δ-mod encoding/decoding lies with the finite step size introduced by the decoder. Slope overload can occur if the step size cannot follow the original waveform, resulting in waveform distortion.

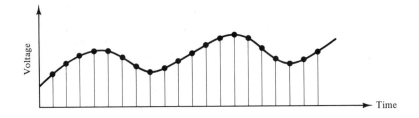

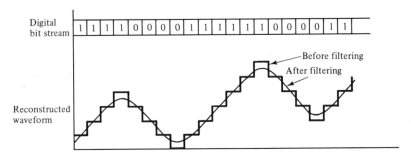

Figure 7.39 Delta modulation.

Modern Δ-mod techniques may overcome this problem with a fair degree of success depending on the encoding algorithm used. These techniques are often known as variable-slope Δ-mod or adaptive Δ-mod. Typically, a good encoding algorithm will allow a 32-kbit/s Δ-mod codec (encoder and decoder pair) to perform subjectively nearly as well as 64 kbit/s PCM codecs for a single voice circuit.

Differential PCM. To combine the virtues of the techniques of PCM and Δ-mod, there exists yet another encoding technique, called differential PCM (DPCM). In this case, the current sample is compared with the previous sample and only the difference between the two samples is encoded. The technique is illustrated in Fig. 7.40.

In the example shown, the difference between two adjacent samples is encoded in a 4-bit word. For a voice circuit, if the sampling rate is maintained at 8 kHz, the bit rate needed is $8 \times 10^3 \times 4 = 32$ kbits/s. If a 4-bit word is used to encode the difference between two consecutive samples, one of the four bits must be used to indicate the sign and only three bits ($2^3 - 1$ levels) are left for encoding the voltage difference. Again slope overload can happen if an advanced algorithm is not used. A popular advanced algorithm now being standardized by CCITT is called adaptive differential PCM (ADPCM). It has been shown that 32-kbit/s ADPCM performs just as well subjectively as 64-kbit/s conventional PCM. Thus the baseband capacity doubles with identical RF transmission requirements.

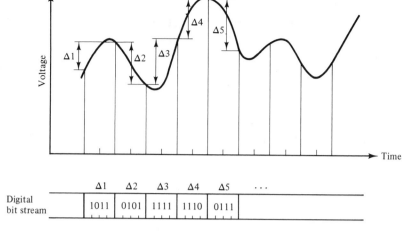

Figure 7.40 DPCM encoding.

QPSK modulation. Having discussed the encoding techniques of analog signals, we can now consider how the NRZ waveform is used to modulate an RF carrier using the QPSK modulation technique. A Fourier analysis of the time-domain NRZ waveform (Fig. 7.38) would yield a $(\sin x/x)^2$ spectrum in the frequency domain as shown in Fig. 7.41. The spectrum is

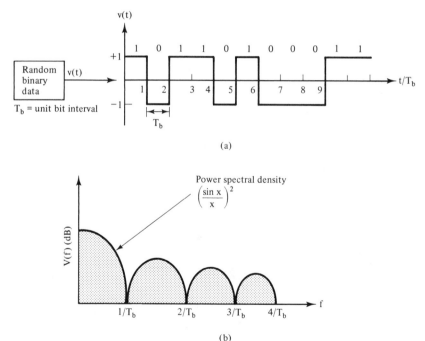

Figure 7.41 Power spectrum of an NRZ waveform: (a) time-domain representation; (b) frequency-domain representation.

continuous and without discrete components if the digital sequence is truly random (i.e., there are no correlated bits in the NRZ bit sequence). The NRZ spectrum is also infinite, although the sidebands would become insignificantly small as the frequency becomes infinite.

As the name implies, QPSK is in reality the superposition of two biphase PSK signals in time quadrature. Figure 7.42 shows the vector

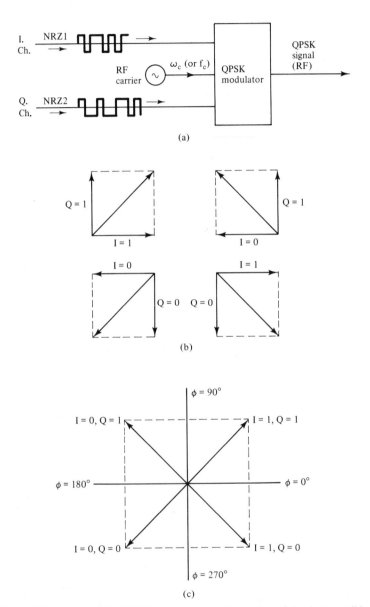

(a)

(b)

(c)

Figure 7.42 (a) Simplified QPSK modulator; (b) formation of the four possible phases of the RF carrier; (c) four phases of the QPSK modulated RF carrier, I & Q channels superimposed.

diagram of an RF carrier that is four-phase-PSK-modulated by two independent uncorrelated but synchronous NRZ inputs, commonly called the I and the Q channels (in-phase and quadrature channels). The I-channel NRZ biphase modulates the in-phase component of the RF carrier and the Q–channel NRZ independently biphase modulates the quadrature component of the RF carriers. These two signals are superimposed on each other to form the four-phase PSK-modulated RF carrier shown in Fig. 7.42(c). The phase of the RF carrier changes at T, $2T$, $3T$, ..., and can assume only one of the four phases shown in Fig. 7.42(c). To realize this situation, Fig. 7.42(a) is expanded to show the details. This is shown in Fig. 7.43. The local oscillator operating at a certain carrier frequency is supplied coherently to two mixers or signal multipliers. Each NRZ signal is multiplied with the local oscillator waveform within the mixer. A 90° phase shift in the path between the local oscillator and one mixer provides the phase quadrature needed in the quadrature channel. Since the NRZ signal can assume only values of $+1$ or -1, biphase modulation is automatically achieved in the mixer, where the NRZ signal is multiplied with the local oscillator's pure sinusoid. At the outputs of the two mixers, the two biphase modulated signals are now combined or superimposed on each other, giving rise to a (four-phase) QPSK modulated signal.

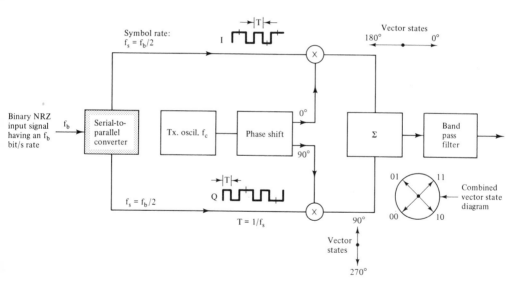

Figure 7.43 QPSK modulator block diagram.

It can be shown that the RF spectrum of a QPSK modulated signal retains the $(\sin x/x)^2$ shape around the carrier frequency but with no carrier component. This is shown in Fig. 7.44(a). As in the case of FM modulation, sideband truncation is needed in practice to limit the bandwidth of the signal to a finite value. This is shown in Fig. 7.44(b). Not too much degradation is incurred if the bandwidth B of the filter is greater than or equal to $1/T$

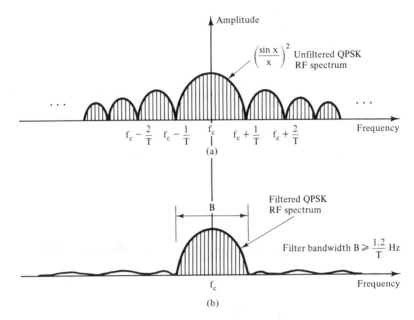

Figure 7.44 Unfiltered and filtered spectrum of a QPSK signal.

× 1.2, and contains most of the main lobe of the spectrum that lies between $f_c + 1/T$ and $f_c - 1/T$.

Thermal noise and other interference in the link will cause the carrier phase to jitter, and eventually at some instance there will be enough phase shift ($\geq \pi/2$) to cause an error. Theoretical bit error rate (BER) versus link C/N performance is shown in Fig. 7.45. It can be seen that at BER $\leq 10^{-4}$, the BER improves at the approximate rate of one decade per decibel in C/N improvement. A BER worse than 10^{-4} is considered an outage in most links. This is similar to the FM threshold in an FM link.

With the conclusion of this discussion of QPSK modulation, the two most common modulation techniques in satellite communications, FM and QPSK, have been described.

Example 7.2

The TV source signal from the camera has a highest component frequency of 4.5 MHz. If FM peak deviation of 4 MHz is used, what is the required RF bandwidth?

Solution   Using Eq. (7.48), the Carson bandwidth will be the required bandwidth and it is

$$B = 2(4.0 + 4.5) = 17 \text{ MHz}$$

Example 7.3

The symbol rate of a digital source signal is 60 Mbits/s (60×10^6 bits/s). If two such digital sources are to be combined to form a QPSK modulated RF carrier, what is the RF bandwidth required, and what is the required minimum C/N for a BER of 10^{-6}?

Solution   Figure 7.44(a) shows that the main lobe of the $(\sin x/x)^2$ spectrum resulting from such a modulation technique is $2/T$ hertz when T is the symbol rate.

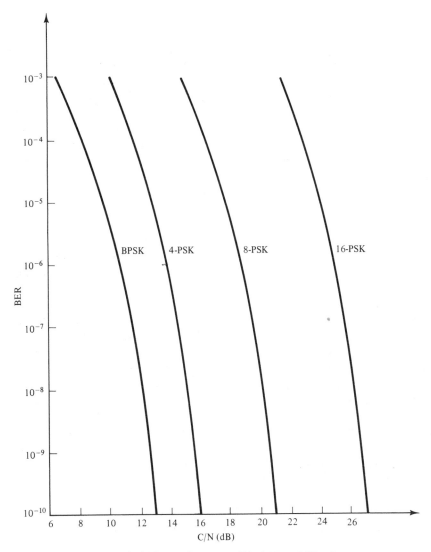

Figure 7.45 Theoretical BER performance of ideal *M*-ary PSK coherent systems. The rms C/N is specified in the double-sided minimum (Nyquist) bandwidth.

Here $T = (60 \times 10^6)^{-1}$ s. Therefore, $2/T$ hertz $= 60 \times 10^6 \times 2 = 120$ MHz. However, Fig. 7.44(b) shows that after appropriate filtering, the minimum required bandwidth in practice should be $1/T \times 1.2 = 72$ MHz. Thus a bit rate of $2 \times 60 = 120$ Mbits/s can be transmitted in an RF bandwidth of 72 MHz when QPSK is used as a modulation technique. This is indeed the case for Intelsat's TDMA system.

Figure 7.45 shows that a BER of 10^{-6} would require a minimum overall C/N of 13.5 dB where there is no further distortion introduced in the satellite. In practice, with a typical satellite transponder and a typical pair of modems (modulator and demodulator), the C/N required is approximately 16.0 dB, that is, an extra 2.5 dB is required to account for the non-linearities and distortions in the transmission path.

Access Techniques

This section deals with the problem of accessing the satellite from the ground and the problem of sharing the satellite's available capacity with other users of the same satellite. Section 7.3 pointed out that (ITU) allocated bandwidths for satellite communications were already predetermined. For example, one of the most common downlink bands in C band is the band 3.7 to 4.2 GHz, and a very popular C-band uplink is the band 5.925 to 6.425 GHz. Based on the examples in the preceding section, 500 MHz can carry either approximately 25 TV channels or about 800 Mbits/s of information. Very few individual users can, in fact, use up the entire 500 MHz of bandwidth on a full-time basis. Some sharing schemes are therefore necessary to allow access of the satellite by more than one user.

The block diagram in Fig. 7.29 already suggested one form of sharing arrangement (i.e., the allocated 500 MHz bandwidth is channelized to suit the users' needs). A common channelization scheme for C-band downlink is shown in Fig. 7.46. With a channel bandwidth of 36 MHz, depending on the modulation index chosen, either one or two TV signals can be transmitted in one channel or a 60-Mbit/s data stream can be transmitted in the channel, assuming, of course, that sufficient satellite power is used to overcome the thermal noise. Even 36 MHz may prove to be much more than required for small users. For example, 24 voice channels may require a bandwidth of less than 1.5 MHz. Thus any of the channels can be further subdivided on a frequency basis to allow small users to access the satellite directly. Figure 7.47 shows the details of how some of the RF channels might be used. This subdivision of the channel is done by the earth stations using small carriers; it is not done in the satellite. This process of sharing a satellite channel by subdividing the channel on the frequency basis is called frequency-division multiple access (FDMA).

Another way of sharing the channel is on the basis of time; each user uses the entire channel but for only a fraction of the time. To allow time sharing to happen, time is divided into consecutive frames and each user occupies the channel for only a fraction of the time in each of the frames. Figure 7.48 shows this arrangement for four users, A, B, C, and D, sharing channel 11 in Fig. 7.47. This process of sharing by multiple users on a time basis is called time-division multiple access (TDMA).

Each of the two sharing arrangements has problems. FDMA implies multicarrier operation within one satellite transmitter (the TWTA or SSPA). As shown in Fig. 7.33, satellite transmitters suffer from amplitude nonlinearities, which generate intermodulation products from the carriers present in the transmitters. To reduce the levels of the intermodulation products, the satellite transmitter must not be driven close to saturation (i.e., the transmitter must operate in the linear region). This reduces the total available RF power from the transmitter with an accompanying reduction in satellite capacity (for a certain C/N to be maintained). On the other hand, if TDMA is used, each user occupies the entire channel with a single carrier. No

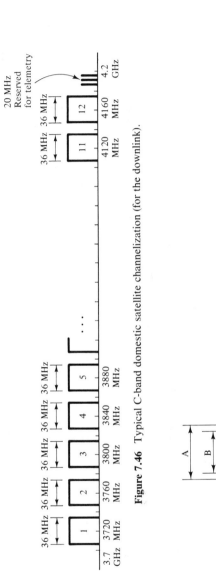

Figure 7.46 Typical C-band domestic satellite channelization (for the downlink).

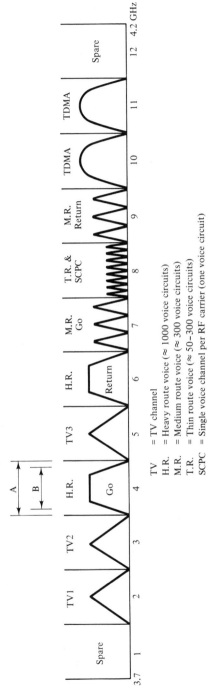

TV = TV channel
H.R. = Heavy route voice ($\approx$ 1000 voice circuits)
M.R. = Medium route voice ($\approx$ 300 voice circuits)
T.R. = Thin route voice ($\approx$ 50–300 voice circuits)
SCPC = Single voice channel per RF carrier (one voice circuit)
TDMA = Time-division multiple access
A = Channel separation = 40 MHz
B = Useable bandwidth = 36 MHz

Figure 7.47 Example of channel utilization (domestic satellites).

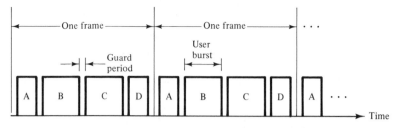

Figure 7.48 TDMA burst-time plan for four users, A, B, C, and D.

intermodulation products can be produced and hence the transmitter can be driven much closer to saturation. However, time sharing the channel implies accurate timing and synchronization of the "bursts" from the various TDMA user earth stations to prevent burst collision in the satellite. In view of the fact that the satellite does not stay absolutely fixed with respect to earth, this process of burst synchronization can present special problems.

FDMA results in higher space segment costs, and TDMA results in higher earth segment costs. Since the burst-time plan (BTP) used to coordinate user earth stations for TDMA is software controlled, TDMA is more flexible and adaptable to dynamic traffic requirements that change from hour to hour. Current status of the technology also dictates that TDMA be coupled with digital modulation. TDMA operates in the burst mode and since the source signal is usually continuous, some buffering arrangement must be used to convert the source signal from a continuous format to a burst format. This is quite easily accomplished by means of digital buffers if the source signal is in the digital format. The buffer reads in the data at a slow but constant rate and it is emptied at a much higher rate periodically, at the output.

Interference Considerations

Figure 7.7 shows a polar radiation pattern of a typical microwave antenna used in satellite communications. It is apparent from the figure that a practical antenna must have sidelobes of finite levels. An antenna with "close-in" sidelobes levels (sidelobes around the main beam) better than -35 dB down from the boresight peak gain is considered a high-performance antenna. The International Radio Consultative Committee (CCIR) (Ref. 8) to the ITU has generated reference earth station antenna sidelobe patterns for antennas with diameters above 50λ at θ degrees off boresight,

$$G(\theta) = 32 - 25 \log_{10} \theta \quad \text{dBi} \quad \text{(dB above isotropic gain)} \quad (7.49)$$

where $G(\theta)$ is the off-axis sidelobe gain of the antenna. Equation (7.49) is valid for $1° \leq \theta \leq 48°$. For $\theta > 48°$, $G(\theta) \leq -10$ dBi. In the near future, it is likely that Eq. (7.49) would be replaced by $G(\theta) = 29 - 25 \log_{10}\theta$. This would permit closer spacing of satellites.

Currently, Eq. (7.49) is interpreted by most administrations as the envelope of the peaks of the sidelobes. Figure 7.49 is a plot of Eq. (7.49).

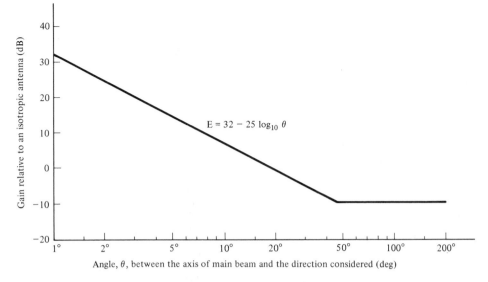

Figure 7.49 CCIR reference radiation diagram, $G(\theta) = 32 - 25 \log_{10} \theta$.

Using Eq. (7.49), the worst-case interference from satellites placed on either side of the "wanted" satellite can be estimated as shown in the following example.

Example 7.4

There are two "unwanted" or interfering satellites spaced at 3° intervals on each side of the "wanted" satellite. If all five satellites radiate with the same EIRPs toward the earth station, at the same frequency, what is the expected level of interference with respect to the wanted signal at 4 GHz if the earth station antenna is 10 m in diameter (Fig. 7.50)?

Solution First, it is to be noted that Eq. (7.49) is independent of the diameter of the earth station antenna as long as the diameter is over 50λ. At 4 GHz, λ = 7.5 cm, Eq. (7.49) is therefore valid for a 10-m antenna. The boresight gain of a 10-m antenna at 4 GHz can be estimated as follows using Eq. (7.13) and assuming $\eta = 0.7$:

$$G(0°) = 0.7 \left(\frac{\pi \times 10 \times 10^2}{7.5} \right)^2$$

$$= 122{,}822 = 50.9 \text{ dBi}$$

Then at $\theta = 3°$,

$$G(3°) = 32 - 25 \log_{10} 3 = 20.1 \text{ dBi}$$

and at $\theta = 6°$,

$$G(6°) = 32 - 25 \log_{10} 6 = 12.5 \text{ dBi}$$

Since there are two interfering satellites, one on each side equally spaced,

$$\frac{G(0°)}{2 \times G(3°)} = 50.9 - (20.1 + 3) = 27.8 \text{ dB}$$

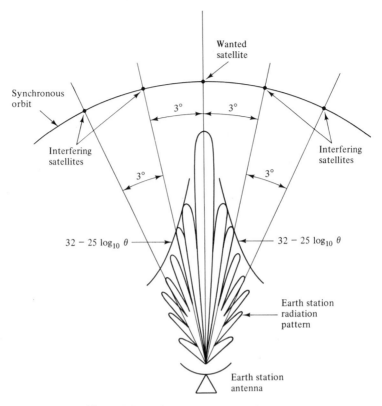

Figure 7.50 Adjacent satellite interference.

and

$$\frac{G(0°)}{2 \times G(6°)} = 50.9 - (12.5 + 3) = 35.4 \text{ dB}$$

If the 4-GHz received signal level is normalized to 1 W (0 dBW) for the wanted satellite, the interference will be -27.8 dBW and -35.4 dBW, respectively, or 1.66×10^{-3} W and 2.88×10^{-4} W, respectively. The sum of all the interfering signals gives a level $= 1.66 \times 10^{-3} + 2.88 \times 10^{-4} = 1.948 \times 10^{-3}$ W $= -27.1$ dBW. Thus the carrier-to-interference (C/I) ratio is 27.1 dB. In practice, a C/I ratio of 27 dB is considered acceptable.

This example has shown that the two immediately adjacent satellites contribute almost all of the interference if the satellite EIRPs are all the same. If the EIRPs of the adjacent satellites are higher than the EIRP of the wanted satellite, the C/I will deteriorate further by the amount equal to the difference between the wanted and unwanted EIRPs. Normally, without advanced processing techniques, a single-entry C/I of worse than 18 dB is considered undesirable, since the link is further degraded at the same time by thermal noise contributions. If the interference is noiselike, it is a simple matter to estimate the overall C/N of the link by the following:

$$\left(\frac{C}{N}\right)^{-1}_{\text{overall}} = \left(\frac{C}{N}\right)^{-1} + \left(\frac{C}{I}\right)^{-1} \qquad (7.50)$$

where the (C/N) term is due only to thermal noise and the (C/I) term is due only to noiselike interference.

It should be noted that the numerical example above assumes certain satellite EIRPs in the direction of the earth station. The situation used in the example is really a pessimistic situation where the satellite antenna coverages of the five satellites are all assumed to cover the same earth station. In practice, the adjacent interfering satellites may have antenna main beams that do not cover the earth station under consideration (i.e., there will be additional protection from the satellite antenna). In that case, for the same C/I, the EIRPs of the interfering satellites could be somewhat higher than the EIRP of the wanted satellite. In general, any interference situation must be examined with care. There is no general rule that can predict the interference level under all conditions.

7.8 SATELLITE CAPACITY AND THE SIZING OF SATELLITES

Capacity Consideration

In a thermal-noise-limited environment where the link degradation is due only to thermal noise that enters into the communication link, Shannon (Ref. 9) showed the following:

$$C = B \log_2 \left(1 + \frac{S}{N}\right) \qquad (7.51)$$

where C = maximum information capacity possible in the link
B = available bandwidth
S = signal power
N = thermal noise power

It can be seen that to achieve a certain communication capacity C, bandwidth B can be traded off with signal power S if thermal noise N is to remain constant for that particular link. As discussed in Section 7.7, AM techniques do not allow such a trade-off, and also, sufficient linear power must be available to establish a communication link. On the other hand, FM techniques with the modulation index as a free parameter do allow such a trade between B and S and furthermore, nonlinear power (i.e., amplitude nonlinearities) can be used if only one carrier is transmitted by the transmitter. Based on the discussion of QPSK modulation, it would appear at first sight that the trade-off between B and S is not available for QPSK modulation. However, in the digital world of communications, the source data can be further encoded up to a higher rate before transmission. This further encoding of the data at the source inserts redundant bits into the information bit stream so that some incurred errors at the receiving end can now be recognized, corrected and removed. Thus even though the actual predecoded BER is poor, the postdecoded BER could be much improved. Figure 7.44 shows that the bandwidth varies linearly with the symbol rate. Thus, using coding techniques in QPSK transmission, a trade also exists between B and S.

For a communications satellite, both B and S are scarce resources. In the early days of satellite communications, it was a good trade-off to use up the bandwidth to save power. This was due to the mass and size constraints of the early launch vehicles. As launch vehicles become cheaper and more powerful, the situation has changed somewhat. Bandwidth is now also becoming a scarce resource. Section 7.3 pointed out that most of the ITU-allocated bands for fixed satellite communications are bands shared with other services and this situation cannot be changed without the consent of other users. In addition, there are only 360° around the geosynchronous orbit, and due to earth station antenna resolving capabilities, it is unlikely in the near future that satellites can be spaced much closer than 2°. A 500-MHz allocated bandwidth can, in theory, be used only 180 (360/2) times unless other frequency-reuse techniques, such as those discussed in Section 7.6, are employed. As is already happening for some cases, thermal noise is not necessarily the major contributor to link degradation when advanced frequency-reuse techniques are used. It is almost certain that in the future, satellite links will be interference limited rather than thermal noise limited. While antenna designers are now trying to reduce antenna sidelobes and cross-polarization levels, communications engineers are trying to develop interference-tolerant modulation techniques to cope with the developing situation. Shannon's results in Eq. (7.51) can still be applied if interference is similar in nature to thermal noise. Even though signal power is now available in reasonable abundance, the eventual capacity available is very much dependent on the antenna frequency reuse technology. Thus far the K bands have not been fully exploited, but Figs. 7.5 and 7.6 suggest that the upper portions of the K band experience a somewhat higher percentage of outage due to atmospheric conditions and propagation fades. The future therefore poses an exciting challenge to engineers in satellite communications fields to exploit the natural resources of bandwidth and orbital arc to the fullest extent.

Sizing Communications Satellites

Two immediately identifiable major factors determine the size (i.e., mass and volume) of a communications satellite: design lifetime and the communications capacity. The design lifetime of a spacecraft determines the amount of fuel and extra redundant equipment to be carried by the spacecraft into the orbit. Communications capacity is the other main driver in sizing a spacecraft. Other significant factors that affect the size of a spacecraft are eclipse capability, which in turn determines the sizing of the on-board battery; and as spectrum-reuse techniques are being implemented, the size and mass of the main antenna has also become a driver. The last factor is to some extent linked to the required communications capacity of the satellite. Finally, but certainly not the least important element that sizes a spacecraft, is the earth segment capability. What is lacking in the spacecraft can, to some extent, be made up for by the earth stations. An example here is the difference between FDMA and TDMA access techniques. The lack of RF power in the spacecraft can be compensated for by using more advanced access techniques or larger earth station antennas. Thus, sizing a satellite requires a complete systems trade-off that includes the costs and capabilities of the earth segment.

If the earth segment already exists, the problem of sizing a satellite may converge to a solution relatively quickly. But if the earth segment does not exist, the sizing of a satellite may require a number of iterations that include the projected plan of future generations of satellites and earth stations. In general, each generation of earth stations will last between two and three generations of satellites, assuming that each generation of satellites is designed for a lifetime of 7 to 10 years. Thus it is probably logical to start with the planning of the earth segment if a completely new system is considered.

If a large number of earth stations are envisaged, low-cost earth stations coupled with a sophisticated satellite is probably the best trade-off. On the other hand, if the system is serving point-to-point trunking purposes, larger and more costly earth stations can be coupled with simpler satellites.

Given an earth segment configuration and capability, the next step would be to design a least-expensive space segment. Most of the satellite links are downlink limited (i.e., the noise contribution in the downlink dominates). It is a reasonable assumption that uplink power from earth stations is easier to obtain than downlink power from the satellite. Given the size and the G/T of the receiving earth stations, it is only a simple matter to estimate the required EIRP of the satellite. Knowing the coverage solid angle subtended by the landmass at the satellite, the satellite RF power can be calculated. Using the efficiency factors and the allowance of lifetime degradation, the beginning of a life (BOL) dc power requirement can then be established. At this point, the batteries for eclipse protection can be sized if the amount of eclipse protection is known. Finally, based on the mass of the communications payload, the mass of the batteries and the solar array plus power subsystem, and the mass of the attitude control subsystem, the size of the fuel tanks and the fuel loading can be estimated. The total mass of the satellite, not including the apogee motor, can then be estimated once the structural mass is determined.

Example 7.5

A domestic satellite system is required for TV program delivery. The C band has been chosen, and at least 24 TV channels must be available at an EIRP of 36 dBW at EOL. The size of the projected landmass is approximately 12 (2 × 6) square degrees. If 100% eclipse protection is required, what is the approximate size of the satellite?

Solution The first step is to evaluate the total dc power requirement and the size of the main antenna of the satellite. Since 500 MHz is available in the C band (3.7 to 4.2 GHz), it is convenient to use orthogonal polarizations to double the 500 MHz to a 1-GHz bandwidth. A 12-RF-channel-per-500-MHz channelization scheme (see Fig. 7.46) would be suitable. This will allow a total of 24 RF channels, each having a 36-MHz bandwidth.

Equation (7.11) is for an elliptical beam $\theta_1° \times \theta_2°$. We have a rectangular area of size $6° \times 2°$. In addition, if a pointing error of $\pm 0.1°$ is assumed, the peak gain of a shaped beam antenna is approximately

$$G_{max} = \frac{30,000}{6.2 \times 2.2}\left(\frac{\pi \times 3.1 \times 1.1}{6.2 \times 2.2}\right)$$

$$= 1727 \approx 32.4 \text{ dBi}$$

correcting for the difference in the coverages between an elliptical beam and a rectangular beam, and allowing for pointing errors by enlarging the beam dimensions by 0.2°. Note that the area of an ellipse is πAB where A and B are respectively the semi-major and semi-minor axes.

The edge gain would be $32.4 - 3 = 29.4$ dB. Assume that the loss between the antenna input and the transmitter output is 1.5 dB and allow 1 dB for the aging of the transmitter. The BOL RF power per channel is

$$P_1 = 36 + 1.0 + 1.5 - 29.4 = 9.1 \text{ dBW} = 8.1 \text{ W}$$

Assume a dc-to-RF efficiency of 35% for the transmitter. The total dc power required for all the transmitters (24 channels) would be $= 24 \times 8.1 \times 1/0.35 = 555$ W. Allow 150 W for housekeeping and the other electronic and microwave active devices. The total dc power requirement is then approximately 700 W.

The eclipse can last 72 minutes, so the batteries must be sized to support 700 W through at least 72 minutes with a depth of discharge compatible with the reliability requirements. Similarly, the solar array should be sized to support the total power requirement plus the power necessary for battery charging during eclipse seasons at the end-of-life.

The antenna size can be estimated as follows. The $6° \times 2°$ rectangular coverage is best approximated by three contiguous 2° component beams. The size of the reflector necessary for the component beam can be estimated from Eqs. (7.11) and (7.13):

$$\eta\left(\frac{\pi D}{\lambda}\right)^2 = \frac{30,000}{2 \times 2} = 7500$$

Solving for D and assuming that $\eta = 0.7$ and $\lambda = 7.5$ cm, we have

$$D = \sqrt{\frac{7500}{0.7}} \times \frac{7.5}{\pi} \approx 250 \text{ cm} \quad \text{or} \quad 2.5 \text{ m}$$

The exact size of the reflector in an optimum configuration needs sophisticated computer simulation before it can be determined. Therefore, determination of the optimum reflector focal length and diameter is beyond the scope of this section.

The example above illustrates the approximate size of a typical domestic satellite which is capable of delivering either 24 simultaneous TV programs or about 20,000 simultaneous telephone one-way circuits (10,000 duplex circuits).

Capacity Estimate Based on Current Encoding Techniques

As the technology advances, the capacity of a satellite tends to increase. In the early 1970s, it was considered normal to assume that one TV channel requires about 36 MHz of RF bandwidth and that one 36-MHz RF channel can support up to about 960 one-way voice circuits, or 480 duplex voice circuits. By the mid-1970s it was shown that with some compensation, one 36-MHz RF channel could really support two TV signals. Using TDMA and digital encoding techniques, the number of voice circuits carried in 36 MHz can be doubled to 2000 circuits quite easily. With the introduction of 16-kbit/s ADPCM, the capacity can be redoubled. Capacity estimates of satellite systems are therefore very dependent on the type of technology being applied and is also a function of the complexity on board a satellite (i.e., the number of frequency reuses) and the signal-processing technology at the earth stations. There are other factors and considerations associated with the determination of the capacity of a given satellite design. For example, the quality of the circuit must be taken into consideration. A high-quality signal transmission would require relatively more resources. Cost could be another factor when capital investment in a satellite system is evaluated. Since the satellite is shared by many users and the RF links are subjected to propagation fades, availability of the service is another factor that will determine the capactiy of a satellite.

If one were to leave aside the questions of satellite EIRPs, earth segment costs and performance, the available capacity of a satellite might be estimated very roughly in the following way by just noting the available bandwidth and by assuming the use of conventional QPSK modulation techniques. For example, a domestic satellite uses orthogonal polarizations to double the 500-MHz allocated bandwidth to 1000 MHz. It was pointed out in Example 7.3 that QPSK modulation can accommodate approximately 1.7 bits of information per hertz of RF bandwidth. Thus 1000 MHz would in theory support up to 1.7 Gbits/s of information. If the given 500 MHz is channelized, only 80% of the available bandwidth is usable, and the realistic information capacity would be $1.7 \times .8 = 1.36$ Gbits/s. The current encoding technique (ADPCM) requires 32 kbits/s for each voice circuit. Therefore, the maximum number of voice circuits that can be carried by the domestic satellite in this example is $1.36 \times 10^9/32 \times 10^3 = 42,500$ circuits, or approximately 20,000 duplex voice circuits simultaneously. If in the future, 16-kbit/s voice encoding proves feasible, the capacity of this satellite could double to 40,000 simultaneous duplex voice circuits.

For TV transmission, the current projection indicates that a good-quality TV picture could be transmitted using a 6.0-Mbit/s bit stream in the future. Our example of the domestic satellite can therefore potentially handle $1.36 \times 10^9/6 \times 10^6 = 226$ or approximately 200 TV pictures simultaneously. However, current cost-effective encoding techniques require about 25 to 40 Mbits/s to transmit a studio-quality TV picture; thus the domestic satellite in our example can handle only about 30 to 50 simultaneous high-quality TV pictures.

The reader is reminded that this method of estimating the available satellite capacity is rather approximate, since no consideration has been given to satellite EIRP, earth segment costs and performance, availability, and the exact quality of the transmission.

PROBLEMS

7.1. Based on the discussion in Section 7.4, show that the polarization is elliptical if $|E_x| \neq |E_y|$ and E_x and E_y assume arbitrary time phases. Find the orientation of the ellipse in space for any given $E_x < \phi_x$ and $E_y < \phi_y$, where ϕ_x and ϕ_y are the respective time phases of E_x and E_y.

7.2. Show that Eqs. (7.19) and (7.20) are correct for geostationary satellites, regardless of operating frequency.

7.3. Equation (7.30) shows the following relationship between C/N and G/T:

$$\frac{C}{N} = \frac{\lambda^2 \phi}{4\pi k B} \frac{G}{T}$$

Show that this equation can be reduced to

$$\left. \frac{C}{N} \right|_{dB} = \text{EIRP} + \frac{G}{T} + 228.6 - P_L - 10 \log_{10} B$$

where P_L is path loss in decibels and B is the bandwidth in hertz; all other quantities are in decibels.

7.4. For a satellite link, there will be a C/N associated with the uplink (i.e., C/N_U), a C/N associated with the downlink (C/N_D), and also an interference term (C/I). Assume that the interference can be treated as thermal noise, show that the overall link $C/(N + I)$ is given by

$$\left. \frac{C}{(N + I)} \right|_{overall} = \left(\frac{C}{N} \right)_U^{-1} + \left(\frac{C}{N} \right)_D^{-1} + \left(\frac{C}{I} \right)^{-1}$$

7.5. Estimate the minimum diameter of a single reflector needed to produce the spot beam coverages shown in the following figure. Assume that the frequency involved is 4 GHz and it is re-used between the circular and the elliptical coverages.

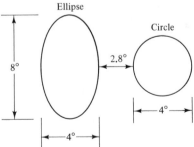

7.6. Calculate the satellite EIRP required to produce a BER of 10^{-6} at 120 Mbits/s if the earth station G/T is 40.7 dB/K. Assume that the downlink frequency is 4 GHz and that the QPSK modems need a 2.5-dB margin over the ideal theoretical performance.

7.7. Using Eqs. (7.11) and (7.13), show that -3-dB beamwidth θ_3 is related to the wavelength λ and the aperture diameter D by

$$\theta_3 = 1.15\frac{\lambda}{D}$$

if η is assumed to be 0.7 and θ_3 is expressed in radians rather than in degrees.

REFERENCES

1. *ITU Radio Regulations,* ITU, Geneva, 1982.

2. P. D. Potter, "A New Horn Antenna with Supressed Sidelobes and Equal Beamwidths," *Microwave Journal,* pp. 71–76, June 1963.

3. P. W. Hannan, "Microwave Antennas Derived from Cassegrain Telescope," *IRE Transactions on Antennas and Propagation,* pp. 140–153, Mar. 1961.

4. S. O. Lane, M. F. Caulfield, and F. A. Taormina, "INTELSAT VI Antenna System Overview," AIAA Conference on Communications Satellites, Orlando, Fl., Mar. 1984.

5. J. Ruze, "Antenna Tolerance Theory," *Proc. IEE,* Vol. 54, pp. 633–640, 1966.

6. N. S. Jayant, *Waveform Quantization and Coding,* IEEE Press, IEEE, New York, 1976.

7. H. Nyquist, "Certain Factors Affecting Telegraph Speed," *Bell System Technical Journal,* Vol. 3, pp. 324–326, Apr. 1924.

8. *CCIR Recommendation 465-1* and *CCIR Report 391-4,* Vol. IV, Part 1: *Fixed Satellite Service,* Geneva, 1982.

9. C. E. Shannon, "A Mathematical Theory of Communications," *Bell System Technical Journal,* Vol. 27, pp. 379–423 and pp. 623–651, 1948.

10. J. M. Wozencraft and I. M. Jacobs, *Principles of Communication Engineering,* Wiley, New York, 1965.

11. J. Martin, *Communications Satellite Systems,* Prentice-Hall, Inc., Englewood Cliffs, N.J. 1978.

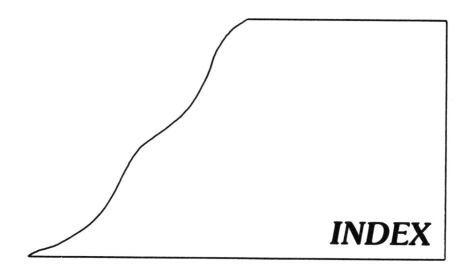

INDEX

Structural design verification tests, 250
  acoustic, 259
  notched test, 254
  random, 258
  shocks, 260
  sinusoidal test, 253
  static test, 252
Sun angle, 96
Sun sensor, 155

T

Thermal analysis, 280
Thermal control techniques, 292
  heat sinks, 297
  passive, 292
  thermal coating, 295
  thermal insulation, 295
Thermal noise, 402
Thermal radiation of the earth, 280
Thermal testing, 314
  infra-red simulation, 317
  solar simulation, 315
  thermal balance, 314
  thermal vacuum test, 319
Three-axis reaction wheel system, 149

Three-axis-stabilization, 130
Three-dimensional stress analysis, 195
Time division multiple access (TDMA), 442
Torsional stress, 193
Transfer orbit sensors, 155
Transponder design, 422
Travelling wave tube amplifier, 426

U

Unregulated bus, 367
Uplink, 384

V

Vacuum test, 319
Vibro-acoustic excitation, 257
View-factors, 274

W

Wavelength, 390
Whecon control system, 142